The thermomechanics of plasticity and fracture

The thermomechanics of plasticity and fracture

GERARD A. MAUGIN

Modélisation en Mécanique

Université Pierre et Marie Curie (Paris VI)

The right of the University of Cambridge to print and sell all manner of books was granted by Henry VIII in 1534. The University has printed and published continuously since 1584.

CAMBRIDGE UNIVERSITY PRESS

Cambridge

New York Port Chester

Melbourne Sydney

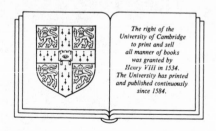

Published by the Press Syndicate of the University of Cambridge
The Pitt Building, Trumpington Street, Cambridge CB2 1RP
40 West 20th Street, New York, NY 10011-4211, USA
10 Stamford Road, Oakleigh, Victoria 3166, Australia

First published 1992

Printed in Great Britain at the University Press, Cambridge

British Library cataloguing in publication data
Maugin, Gérard A. *1944–*
The thermomechanics of plasticity and fracture.
1. Plasticity, Mathematics
I. Title
531.38

Library of Congress cataloguing in publication data available

ISBN 0 521 39476 7 hardback
ISBN 0 521 39780 4 paperback

AO

Contents

Preface

The present book is an outgrowth of my lecture notes for a graduate course on 'Plasticity and fracture' delivered for the past five years to students in Theoretical Mechanics and Applied Mathematics at the Pierre-et-Marie Curie University in Paris. It also corresponds to notes prepared for an intensive course in modern plasticity to be included in a European graduate curriculum in Mechanics. It bears the imprint of a theoretician, but it should be of equal interest to practitioners willing to make an effort on the mathematical side. The prerequisites are standard and include classical (undergraduate) courses in applied analysis and Cartesian tensors, a basic course in continuum mechanics (elasticity and fluid mechanics), and some knowledge of the strength of materials (for exercises with a practical touch), of numerical methods, and of elementary thermodynamics. More sophisticated thermodynamics and elements of convex analysis, needed for a good understanding of the contents of the book, are recalled in Appendices.

The book deals specifically with what has become known as the *mathematical* theory of plasticity and fracture as (unduly) opposed to the *physical* theory of these fields. The first expression is reserved for qualifying the macroscopic, phenomenological approach which proposes equations abstracted from generally accepted experimental facts, studies the adequacy of the consequences drawn from these equations to those facts, cares for the mathematical soundness of these equations (do they have nice properties?), and then, with some confidence, provides useful tools to designers and engineers. The second expression refers to the 'physical' approach which consists in justifying laws governing macroscopic phenomena on the basis of microscopic descriptions at a finer scale (that of the discrete world), and thus is intimately related to solid-state physics, metallurgy and, more generally, what is now called materials science. For example, with such a frame of mind, the slip deformation in crystal plasticity is viewed as resulting from the movement and generation of dislocations, and is generally irreversible due to the atomic potential barriers. Although my own research would incline me to consider this stimulating aspect of plastic phenomena, the present book deals practically exclusively with the macroscopic, smoothed-out scale, physics being used in the guise of irreversible thermodynamics in order to control the macroscopic evolution along the 'Arrow of time'. For the physical theory of plasticity and fracture I recommend Asaro (1983), Cottrell (1953), Honeycombe (1968), Mura (1982), Seeger (1958), Teodosiu (1975, 1982), Zener (1948), and the book of Polukhin *et al.* (1983).

Numerous problems are proposed by way of illustration or to provoke the thoughts of the reader. But the composition of the book is quite different from that of some 'classics' of plasticity theory. Just to quote two examples, while Hill's book of 1950 devotes some seventy pages among its first chapters to a general presentation and general theorems (not many at that time) and then proceeds for several hundred pages to solving problems analytically, including by the slip-line theory (in rigid–plastic bodies), Kachanov's book of 1974, another classic, presents many applications and then concludes with general theorems. Since then the emphasis has shifted, this being faithfully reflected in the theoretical and numerical developments reported in the present book: with the implementation of powerful numerical methods (finite elements, fast integration of systems of evolution equations), analytical solutions have often become outdated while general theorems have gained in importance, perforce. Therefore, most of the present book is rather general in approach with theoretical statements spread all over while some analytical solutions are given because of their everlasting pedagogical value. More on this evolution can be found in Drucker (1988).

Another development reflected in the title is the *thermomechanical* framework which, to the posthumous satisfaction of the great P. Duhem, has been fully integrated into the curriculum in continuum mechanics. This also fully agrees with the late H. Ziegler's point of view (Ziegler, 1963; Ziegler and Wehrli, 1987).

Some inevitable choices had to be made as, originally, the course was delivered in ten sessions of two hours. As a consequence, I think it fair to announce at the outset what the reader will *not* find in the book: non-associated flow rules, non-convex yield surfaces, slip-line theory, stability and wave problems in elastoplasticity, and applications to soil mechanics (except for a very few exceptions). In contrast, the reader *will* find hereinafter elastoplasticity with hardening, an introduction to elastoplasticity in finite strains, homogenization of elastic–plastic composites, coupling between plasticity and damage, path-independent integrals in fracture, numerical treatments of elastoplasticity problems, coupling with thermal fields and singularities, and applications to metals and crystals. That is, all through these pages the plastic potential coincides with the yield surface, which is always supposed to be convex: the normality rule applies and Drucker's 'postulate' holds true (see counterexamples in Houlsby (1981) and Richmond and Spitzig (1980); Mroz (1963), and Ziegler and Wehrli (1987)). Strain-rate dependence is not considered in general, but for comparison purposes. Damage and fracture enter the *same* thermodynamical framework as elastoplasticity. This owes much to the works of Nguyen Quoc Son (in elastoplasticity and fracture) and Pierre Suquet (for composites) from whom we have borrowed much material (especially, Nguyen Quoc Son (1981a) and Suquet (1987)). My admiration for those who have contributed much to this *nonlinear* science is expressed in the list of names given in the following 'Historical perspective', as well as in the Bibliography, obviously far from exhaustive but, in my opinion, doing justice in a balanced manner to North American, West European and East European contributions.

The reader may naturally wander through the pages and chapters (a practice I very much enjoy myself), but if any advice can be given at all the following order of reading may be an astute, if not efficient, one: Chapter 1, then Appendix 1, followed by Chapter 2, completed by Appendix 2, then Chapters 3 and 4 and Appendix 3, and the rest in order of appearance. Chapters 8, 9 and 10 can be skipped in a first reading. In my own practice the contents of the appendices, together with many of the problems, were left to the students as personal reading and homework or term papers.

It is hoped that the book will find its place both on the student's shelf and in the professional scientist's office. I would be most pleased if designers and engineers would consider it as a reference.

My thanks belong to those who have helped me to capture the flavour and peculiarities of elastoplasticity, for me a rather young science. I found efficiency, kindness, encouragement and advice in the CUP team on (Applied) Mathematics and Mechanics. Moreover, the 'home typist', Eleni, the only person who 'took' this course on elastoplasticity without taking first 'Continuum mechanics', has done an excellent job in typing the manuscript. Can I ask for more than that? I feel spoiled.

G.A. Maugin

Historical perspective[†]

A Landmarks in plasticity theory from 1870 to 1980:

Tresca (1872), Barré de Saint Venant (1871): Flow criterion; torsion problem

Bauschinger (1886): The 'Bauschinger' effect

Mohr (1900): Mohr's circle to express criterion of failure (cited by Timoshenko (1953, p. 286))

Prandtl (1903): Membrane analogy, Prandtl–Reuss relations (Prandtl, 1924; Reuss, 1939)

Huber (1904), Mises R. von (1913): Flow criterion

Hencky (1923): Slip-line theory

Hencky (1924): Hencky's laws

Odqvist (1933): Hardening parameter (Odqvist's parameter)

Melan (1938): Plastic potential

Melan (1938): Shakedown theory (static approach)

Bridgman (1940–1952): High-pressure deformation and flow

Ilyushin (1943): Method of successive elastic solutions

Zener (1948): Rheological models

Hill (1948): Maximal dissipation principle

Hodge, Prager (1949): Minimum principle

Ilyushin (1948): Stability postulate

Greenberg (1949): Minimum principle

Drucker (1951): Drucker's inequality

Symonds (1951): Limit analysis, collapse mechanisms

Koiter (1960): Shakedown theory (kinematic approach)

Ziegler (1963): Plasticity and nonequilibrium thermodynamics, nonlinear generalization of Onsager's relations

Lee (1969): Multiplicative decomposition of finite deformation gradient

Halphen and Nguyen Quoc Son (1975): Notion of generalized standard material

[†] Full references in Bibliography

Strang, Suquet, Temam (1979, 1980): Functions of bounded variations in elastoplasticity

B Landmarks in fracture theory from 1920 to 1970

Griffith (1920): Surface energy theory

Westergaard (1939): Singularity at crack tip

Irwin and Kries (1951): Elasticity theory of fracture

Eshelby (1951): Path-independent integral (also Cherepanov, 1967, Rice, 1968)

Barenblatt (1960): Elastic theory of cohesive forces at cracks

C Landmarks in damage theory (1858—1963)

Woehler (1858–1870): Cyclic loading, Woehler's curves

Norton (1929): Creep, Norton's law

Kachanov (1958): Notion of damage

Rabotnov (1963): Damage parameter

D Landmarks in related areas (useful for the subject matter of the present book)

Rayleigh, Lord (1873): Dissipation function

Goursat (1898): Biharmonic and analytic functions

Carathéodory (1909, 1925): Axiomatics of thermostatics

Kolossov (1909), Muskhelishvili (1953): Complex-function technique in plane problems

Duhem (1911): The Clausius–Duhem inequality

Sobolev (1936): function spaces of the 'Sobolev' type

Onsager, Casimir, Meixner, de Donder, etc. (1940, 1960): Nonequilibrium thermodynamics (see de Groot and Mazur (1962))

Bridgman (1943): The notion of internal variable

Fenchel (1949): Conjugacy and Legendre–Fenchel transformation (see Rockafellar (1970))

Coleman, Noll, Truesdell (1960–1970): 'Rational thermodynamics' (see Truesdell (1969))

Argyris and others (1965): Finite-element method

Notation

Vectors and Cartesian tensors

Real line: $\mathbb{R}$
Three-dimensional Euclidean space: $\mathbb{E}^3$
Open set of $\mathbb{R}^3$: Ω
Boundary of Ω: $\partial\Omega$
Closure of Ω: $\bar{\Omega}$
Orthonormal basis in $\mathbb{E}^3 : \mathbf{e}_i \cdot \mathbf{e}_j = \delta_{ij}$: $\{\mathbf{e}_i\} i = 1, 2, 3$
Kronecker delta ($=1$ if $i = j$, $=0$ otherwise): δ_{ij}
Vector: $\mathbf{V} = \sum_{i=1}^{3} V_i \mathbf{e}_i$
Einstein summation convention (on dummy indices): $A_j B_j = \sum_{j=1}^{3} A_j B_j = A_1 B_1 + A_2 B_2 + A_3 B_3$
Second-order tensor $\boldsymbol{\sigma} = \sum_{i,j=1}^{3} \sigma_{ij} \mathbf{e}_i \otimes \mathbf{e}_j = \sigma_{ij} \mathbf{e}_i \otimes \mathbf{e}_j$
Tensor product $\otimes$: $(\mathbf{A} \otimes \mathbf{B})_{ij} = A_i B_j$
Transposition $\boldsymbol{\sigma}^{\mathrm{T}} = \{\sigma_{ji}\}$ if $\boldsymbol{\sigma} = \{\sigma_{ij}\}$
Symmetric second-order tensor: $\boldsymbol{\sigma} = \boldsymbol{\sigma}^{\mathrm{T}}$ or $\sigma_{ij} = \sigma_{ji}$ (six independent components at most)
Symmetric part: $\boldsymbol{\sigma}_{\mathrm{S}} \equiv \{\sigma_{(ij)} \equiv \frac{1}{2}(\sigma_{ij} + \sigma_{ji})\}$
Anti-(or skew-) symmetric part: $\boldsymbol{\sigma}_{\mathrm{A}} \equiv \{\sigma_{[ij]} = \frac{1}{2}(\sigma_{ij} - \sigma_{ji})\}$
Scalar product of two vectors: $\mathbf{A} \cdot \mathbf{B} = \sum_{j=1}^{3} A_j B_j = A_j B_j$
Vector product of two vectors: $(\mathbf{A} \times \mathbf{B})_i = \sum_{j,k=1}^{3} \varepsilon_{ijk} A_j B_k = \varepsilon_{ijk} A_j B_k$

$$\text{Permutation symbol } \varepsilon_{ijk}: \begin{bmatrix} = & 1 & \text{if } i, j, k \text{ is an even permutation of } 1, 2, 3 \\ = & -1 & \text{if } i, j, k \text{ is an odd permutation of } 1, 2, 3 \\ = & 0 & \text{otherwise} \end{bmatrix}$$

Trace of a tensor $\boldsymbol{\sigma}$: $\operatorname{tr} \boldsymbol{\sigma} = \sum_{i=1}^{3} \sigma_{ii} = \sigma_{jj} = \sigma_{11} + \sigma_{22} + \sigma_{33}$
Deviator of a tensor: $\boldsymbol{\sigma}^{\mathrm{d}} = \boldsymbol{\sigma} - \frac{1}{3}(\operatorname{tr} \boldsymbol{\sigma})\, \mathbf{1} = \{\sigma_{ij} - \frac{1}{3}\sigma_{kk}\delta_{ij}\}$
'Inner' product of two tensors: $\boldsymbol{\sigma} : \boldsymbol{\varepsilon} = \sum_{i,j=1}^{3} \sigma_{ij}\varepsilon_{ji} = \sigma_{ij}\varepsilon_{ji}$.

Differential operators:

Nabla ('del' operator) $\nabla := \{\partial/\partial x_i; i = 1, 2, 3\}$, e.g., $\nabla\phi = \phi_{,i}\mathbf{e}_i$.
Gradient tensor of a vector field: $\nabla\mathbf{A} = \{A_{i,j}\}$

Divergence of a vector $\nabla \cdot \mathbf{A} = \mathrm{tr}\,(\nabla \mathbf{A}) = A_{i,i} = \sum_{i=1}^{3} A_{i,i}$
Curl of a vector: $(\mathrm{curl}\,\mathbf{A})_i = (\nabla \times \mathbf{A})_i = \varepsilon_{ijk} A_{k,j}$
Normal derivative: $\partial\phi/\partial n = (\mathbf{n} \cdot \nabla)\phi = \phi_{,i} n_i$
Laplacian of a scalar: $\nabla \cdot \nabla\phi = \nabla^2\phi = \phi_{,ii}$

Kinematics, strains:

Position (placement) in $\mathbb{E}^3$: $\mathbf{x}, \mathbf{X}$
Displacement vector: $\mathbf{u} = \{u_i\}$
Linear strain tensor: $\boldsymbol{\varepsilon} = (\nabla \mathbf{u})_S = \{\varepsilon_{ij} = \frac{1}{3}(u_{i,j} + u_{j,i})\}$
Deviator of $\boldsymbol{\varepsilon}$: $\boldsymbol{\varepsilon}^{\mathrm{d}}$
Elastic strain: $\boldsymbol{\varepsilon}^{\mathrm{e}}$
Plastic strain: $\boldsymbol{\varepsilon}^{\mathrm{p}}$
Viscous strain: $\boldsymbol{\varepsilon}^{\mathrm{v}}$
Equivalent strain: $\varepsilon_{\mathrm{p}} = (\frac{2}{3}\varepsilon^{\mathrm{p}}_{ij}\varepsilon^{\mathrm{p}}_{ij})^{1/2}$
Cumulated plastic strain: $\overline{\varepsilon^{\mathrm{p}}} = \int_0^t (\frac{2}{3}\dot{\varepsilon}^{\mathrm{p}}_{ij}\dot{\varepsilon}^{\mathrm{p}}_{ij})^{1/2}\,\mathrm{d}t$
Velocity vector: $\mathbf{v} = \{v_i = \dot{u}_i\}$
Partial time derivative: $\dot{A} = \partial A/\partial t$
Acceleration vector: $\boldsymbol{\gamma} = \dot{\mathbf{v}} = \ddot{\mathbf{u}}$
Strain-rate tensor: $\mathbf{D} = \{D_{ij}\} = (\nabla \mathbf{v})_S$
Rotation-rate tensor: $\boldsymbol{\Omega} = \{\Omega_{ij}\} = (\nabla \mathbf{v})_A$
Vorticity vector: $\boldsymbol{\omega} = \frac{1}{2}\nabla \times \mathbf{v}$
Finite deformation: $\mathbf{F}$
n-vector of internal variables: α

Forces, stresses:

Stress tensor: $\boldsymbol{\sigma}$
Deviator of stress: $\mathbf{s} = \boldsymbol{\sigma}^{\mathrm{d}}$
Invariants of stresses: $(\sigma_{\mathrm{I}}, \sigma_{\mathrm{II}}, \sigma_{\mathrm{III}})$ or $I_\alpha(\boldsymbol{\sigma})$, $\alpha = 1, 2, 3$
Mean stress: $\sigma_{\mathrm{m}} = \frac{1}{3}\,\mathrm{tr}\,\boldsymbol{\sigma}$
Normal stress: σ_{n}
Tangential stress: τ
Pressure: p
Elastic limit: σ_0 (yield stress)
Mises equivalent stress: $\sigma_{\mathrm{Y}} = (\frac{3}{2}\sigma^{\mathrm{d}}_{ij}\sigma^{\mathrm{d}}_{ij})^{1/2}$
Convex of plasticity: C
Plastic multiplier: $\dot{\lambda}$
Viscosity: η_{v}
Loading function: f or F
Body force: $\mathbf{f}$ or $\mathbf{g}$

Surface traction: $\mathbf{T}$
Hardening parameters: α, β
Hardening modulus: h
Force associated to $\boldsymbol{\alpha}$: $\mathbf{A}$
Tensor of elasticity coefficients (rigidities): $\mathbf{E}$
Tensor of elastic compliances: $\mathbf{S}$
Lamé coefficients: λ and μ
Hooke's modulus: E
Poisson's ratio: ν
Damage parameter: D

Thermodynamics:

Thermodynamic temperature: θ
Matter density: ρ
Internal energy per unit mass: e
Entropy per unit mass: η
Free energy per unit mass: ψ
Total internal energy: E
Total kinetic energy: K
Total entropy: $\mathscr{N}$
Total power: $\mathscr{P}$
Total dissipation: Φ
Internal energy per unit volume: $\mathscr{E}$
Entropy per unit volume: S
Free energy per unit volume: W
Source heat per unit mass: h
Heat flux vector $= \{q_i\}$: $\mathbf{q}$
Dissipation per unit volume: ϕ
Viscous dissipation: ϕ_{V}
Plastic dissipation: ϕ_{p}
Thermal dissipation: ϕ_{q}
Intrinsic dissipation: ϕ_{intr}
Dissipation function/potential: $\mathscr{D}$

1

Introduction to plasticity: experimental facts

1.1 Elastic and plastic behaviours

Although the underlying microscopic mechanisms are relatively complex, viewed on a macroscopic scale – as will be the case in this book – the plastic phenomenon can be quite simple. *Plasticity*, in particular, is characterized by the existence of a stress *threshold*, or plastic threshold, and the behaviour of the medium differs, depending upon whether the stress state is on the *inside* or *right on* this threshold. On the inside of the threshold, the medium is supposed to have a linear or nonlinear *elastic* behaviour. Typically, this elastic behaviour is characterized by a stress–strain response curve of the type sketched in Fig. 1.1 for a one-dimensional model. Loading from a natural stress-free state causes a reversible increase in the measure ε of strain. The unloading path in this diagram reproduces the loading path precisely in reverse, returning to the origin as the applied stress goes back to zero. In a more vivid way, it can be said that the material possesses the 'memory' of only *one* state, the natural free state. We remind the reader that elasticity derives from an energy density $W(\varepsilon)$ per unit volume. Here we consider only small strains $\boldsymbol{\varepsilon}$. Cauchy's stress tensor is obtained by

$$\boldsymbol{\sigma} = \frac{\partial W}{\partial \boldsymbol{\varepsilon}}, \qquad W = W(\boldsymbol{\varepsilon}). \tag{1.1}$$

In the *linear anisotropic* case W is a general homogeneous function of degree 2 of the components ε_{ij} of $\boldsymbol{\varepsilon}$ in a Cartesian reference frame,

$$W = \tfrac{1}{2}\boldsymbol{\varepsilon} : \mathbf{E} : \boldsymbol{\varepsilon} = \tfrac{1}{2}\varepsilon_{ij} E_{ijkl} \varepsilon_{kl}, \tag{1.2}$$

and thus

$$\boldsymbol{\sigma} = \mathbf{E} : \boldsymbol{\varepsilon} \qquad \text{or} \qquad \sigma_{ij} = E_{ijkl}\varepsilon_{kl}, \tag{1.3}$$

where E_{ijkl} is a fourth-order tensor of elasticity coefficients. In the *isotropic* case, which is very often considered in engineering, there are only *two* independent elasticity coefficients, and eqn (1.3) takes on the special form

$$\sigma_{ij} = \lambda(\varepsilon_{kk})\delta_{ij} + 2\mu\varepsilon_{ij}, \tag{1.4}$$

where λ and μ are Lamé's elasticity coefficients.

Nonlinear elasticity (curved part of the response in Fig. 1.1) may correspond to an energy W of order higher than 2 in the components of $\boldsymbol{\varepsilon}$. In general, however, the situation is more complex than that and the notion of Piola–Kirchhoff stress tensor must be introduced in the nonlinear, finite-displacement case (see Chapter 8). In the greater part of this book the elastic behaviour is described with sufficient accuracy by the simple equations (1.1)–(1.3).

Right on the plastic threshold, the mechanical behaviour of elastic–plastic materials is quite different from the elastic one. We must formulate new laws. For *perfectly plastic* media, the threshold is invariable; it is defined once and for all by the material's data and is therefore independent of the 'history' of the material. For a *not* perfectly plastic material, a material with so-called *hardening*, this threshold may evolve with the loading. This will bring some complications in the modelling.

More accurately, we should say that *we call 'plastic' the behaviour of a solid body acquiring permanent strains without cracking*, that is, without loss of the material's cohesion along certain surfaces. These permanent strains are produced starting from the plasticity threshold, or *elasticity limit*. This threshold is considered as a schematization.

An example of mechanical behaviour or response in simple traction for

Fig. 1.1. Reminder: elastic behaviour in a simple traction test

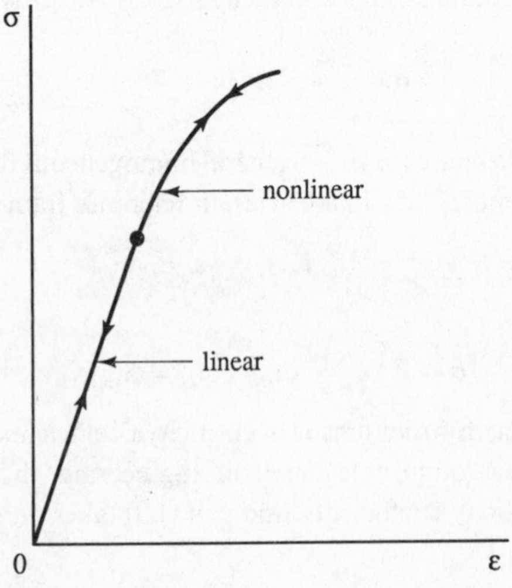

soft steel (with a low – below 0.2% – carbon content) is given in Fig. 1.2. The case of copper is also given for comparison purposes. In this figure, the plateau BC is called *the apparent yield stress limit* or *threshold*. It has been experimentally shown (experiments of P.W. Bridgman (1952) on the triaxial compression of nonporous solids and liquids) that the alteration in volume, and, consequently, the alteration of the mass per unit volume of a body, correspond to an elastic (reversible) strain, defined by an average pressure. So, in general, we leave out of consideration the insignificant alterations of the density caused by plastic strain while the *form* alteration is only due to the slip (or shear) strain. This means that, in a plastic regime, the only thing that enters into consideration is the *deviatoric* part of the strain (and, because of duality, also of the stress). We note (tr = trace)

$$\mathbf{e} = \text{deviator of } \boldsymbol{\varepsilon}, \quad e_{ij} = \varepsilon_{ij} - \tfrac{1}{3}(\operatorname{tr}\boldsymbol{\varepsilon})\delta_{ij}, \quad \operatorname{tr}\mathbf{e} \equiv 0, \tag{1.5}$$

and if τ denotes a tangential stress we consider the schematization of Fig. 1.3. Clearly, an experimentalist would describe this schematic behaviour by a succession of three 'regimes':

for $\tau < \tau_s$: $\tau = G\gamma$: this is Hooke's law in shear;

for $\gamma_e < \gamma < \gamma_s$: $\tau = \tau_s$: flow phase; the strain increases at fixed stress; it is said that the material *flows* plastically;

for $\gamma > \gamma_s$: $\tau = g(\gamma)\gamma$: hardening occurs; the *variable* quantity $g(\gamma)$ is called the *plasticity modulus*, while G was the *constant* elastic modulus; it is an experimental fact that

$$0 \leqslant g(\gamma) \leqslant G \tag{1.6}$$

Materials that exhibit a plastic regime before failure are said to be *ductile*,

Fig. 1.2. Real elastoplastic behaviour, soft steel behaviour in simple traction

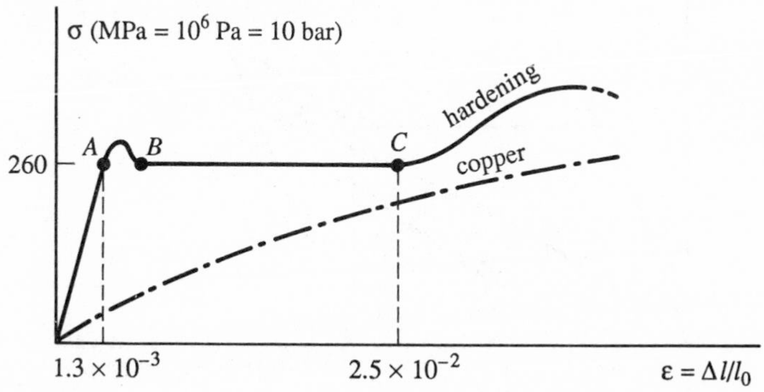

while materials that exhibit fracture while still in the elastic regime are called *brittle* materials. Obviously, the part AC in Fig. 1.3 corresponds to a *nonlinear* behaviour. In some cases the plateau AB is practically non-existent. In some other cases the straight linear-elastic part is very short or not even apparent. This is illustrated by the broken-line curve OD in Fig. 1.3 and even more vividly in Fig. 1.4 which demonstrates the typical variation in shape of a tension-test curve of an artificially prepared two-phase iron–silver material in terms of the phase concentration. Pure iron exhibits a response of the type of soft steel in Fig. 1.2, while pure silver is typically of the hardening type, like copper. But this is not all, as plasticity is essentially characterized by unloading paths that differ from the loading one.

Hardening When we unload *metallic* test samples, the return curve ABC in Fig. 1.5 is practically rectilinear. If we load again, the load curve CDE differs from the curve ABC. So, after an initial extension, the metal behaves as if it had acquired better elastic properties and a higher elastic limit, while at the same time it had lost, it is true, a great part of the plastic strain. This is the phenomenon of *work hardening.* We may also say that the point at which we stop loading defines instantaneously an elastic limit. Further loading after unloading will define a new instantaneous elastic limit and so forth.

The Bauschinger effect Work hardening is, as a rule, *oriented* in such a way that, generally speaking, the material, following a plastic strain, acquires a

Fig. 1.3. Schematic elastic–plastic behaviour

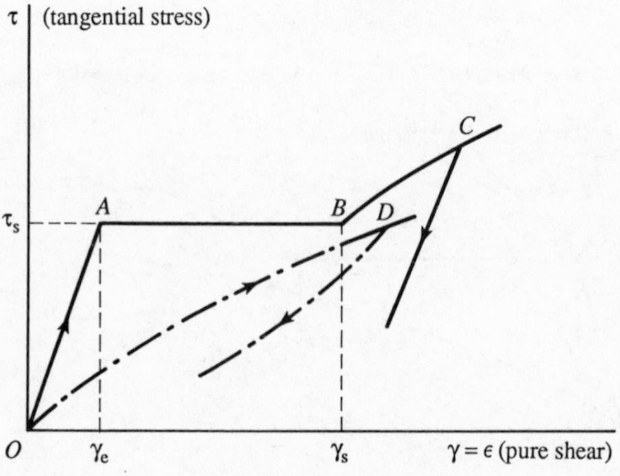

strain anisotropy; one of the manifestations of this phenomenon is the *Bauschinger effect*: a previous plastic strain with a certain sign diminishes the material's resistance with respect to the next plastic strain with the opposite sign. The plastic traction of a rod leads to a remarkable decrease of the yield limit of this rod when it is subsequently compressed again (Fig. 1.6). Here we have

$$|S_2'| < |S_1'|$$

with, in *all* cases,

Fig. 1.4. Tension-test curve of two-phase iron–silver for various phase concentrations (solid lines). Broken lines correspond to a three-phase model (after Bretheau *et al.*, 1991)

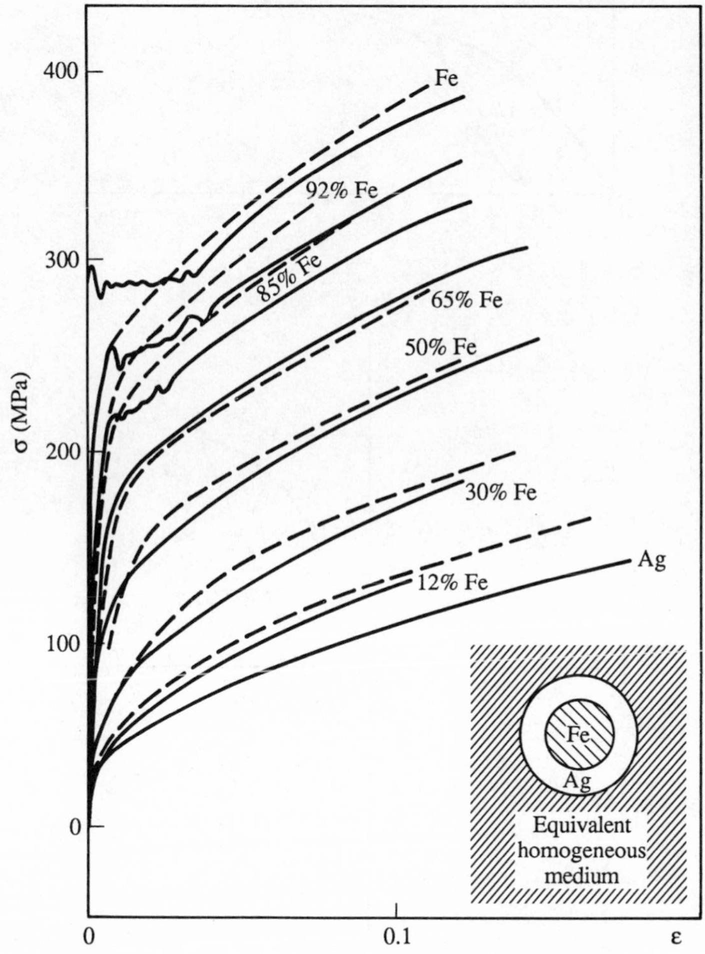

$$|S_2'| < |S_1'| + S_2 - S_1 .$$

In the case of metals, $S_2' = -S_1'$ and $S_2' > -S_2$. The discovery of the effect goes back to Bauschinger (1886).

Rest and annealing With the passage of time, we can observe the partial disappearance of the hardening; this phenomenon is called the material's *rest* and it becomes increasingly apparent as the temperature goes up. In

Fig. 1.5. Hardening

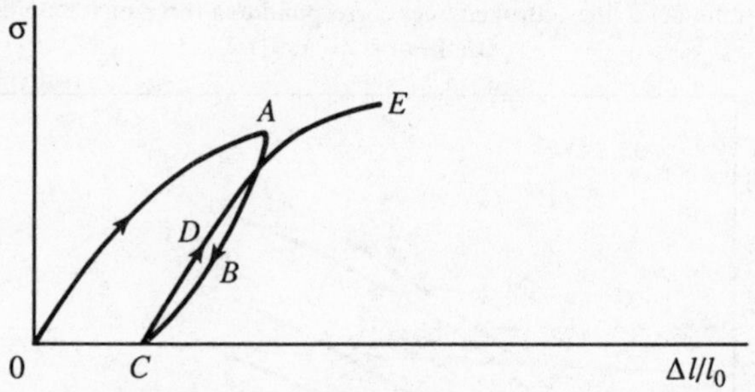

Fig. 1.6. Bauschinger effect

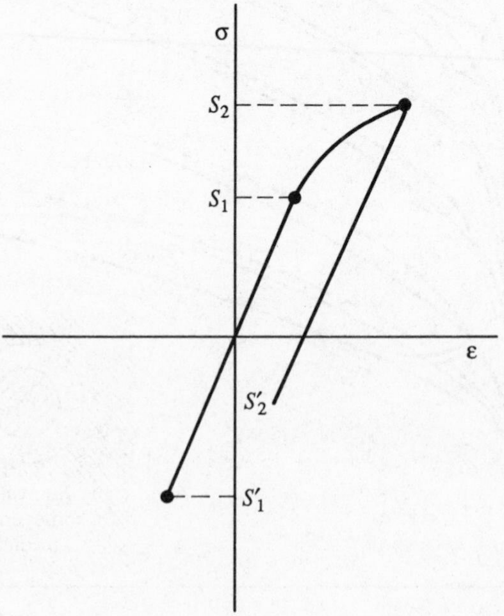

fact, the acquired hardening disappears under the effect of sufficiently high temperatures (we say that there is *annealing* of the material).

1.2 Influence of the strain rate

If the mechanical tests occur in ordinary time intervals and at room temperature, the mechanical properties of steel and of *brittle* materials in general (for example, heat-resistant materials) depend hardly at all upon the strain rate.

Fig. 1.7 illustrates the case for iron. This was well shown also in experiments by Manjoine (1944) on mild steel. Still, the velocity of the test is very important in the case of *ductile* materials (for example, lead, tin), in the case of lengthy tests on steel, copper and other metals under high-temperature conditions, and, finally, in the case of high strain rates. The effect of velocity depends to a great extent upon temperature, and, at rather low temperatures, it practically disappears. To conclude, in ordinary conditions, *the plastic strain of brittle materials is practically independent of the thermal motion of atoms and of the strain rate.* But, if velocity does play a part, then the behaviour is *viscoplastic.*

When there is neither viscosity, nor hardening, nor nonlinear elasticity, the scheme in Fig. 1.3 is reduced to *perfect elastoplasticity* (Fig. 1.8) whose rheological model is provided by a dry friction element. A word must be

Fig. 1.7. Influence of strain rate (iron)

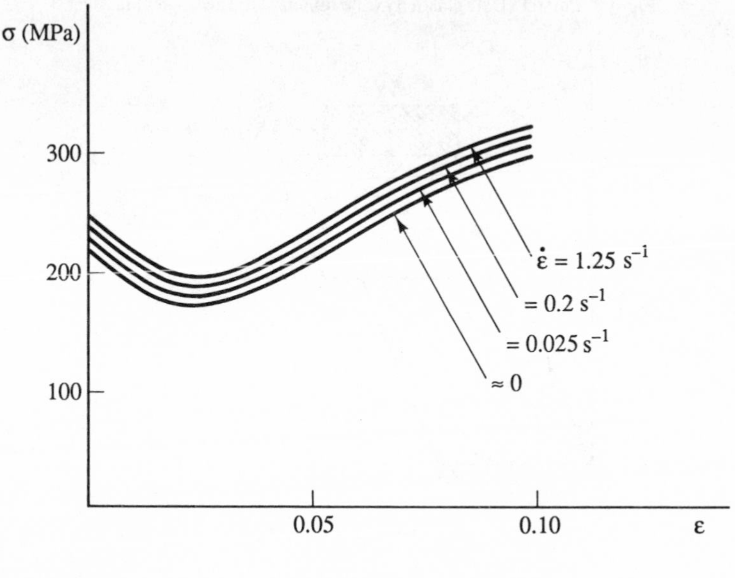

said about these rheological models. Very often we shall have recourse to them as they exhibit strong pedagogical and heuristic values. Their use in *rheology* (the science of what flows) has become general since Zener (1948). The combination in series or parallel of a few standard elements (spring, dashpot, dry friction, and others) provides a picturesque illustration of complex behaviours and the direct addition of stretches or forces, depending on the case, rapidly yields simple constitutive equations. In the present case, a spring (Hooke = H) and a dry-friction element are set in series. The *same* force is transmitted through the two elements. We call it σ. The total stretch is the sum of those in each element, i.e., $\varepsilon = \varepsilon^e + \varepsilon^p$, where $\varepsilon^e = \sigma/E$ if E is the spring constant. The dry-friction element slides only if a sufficiently intense force σ is transmitted to it so as to exceed, or indeed equal, the threshold of the dry-friction element σ_0. That is,

$$\left.\begin{aligned} &|\sigma| \leqslant \sigma_0, \qquad \exists \sigma = E \cdot \varepsilon^e, \\ &\sigma = \sigma_0, \qquad \varepsilon = \varepsilon^e + \varepsilon^p = \frac{\sigma_0}{E} + \varepsilon^p. \end{aligned}\right\} \tag{1.7}$$

Dry-friction elements are also called *Saint Venant* ideal (perfect) *plastic* elements (for short, SV elements). An SV element is such that

$$\varepsilon_{(SV)} = 0 \quad \text{if } \sigma < \sigma_0, \qquad \varepsilon_{(SV)} \neq 0 \text{ possibly} \quad \text{if } \sigma = \sigma_0. \tag{1.8}$$

Fig. 1.8. Perfect elastoplasticity (one dimension) rheological model

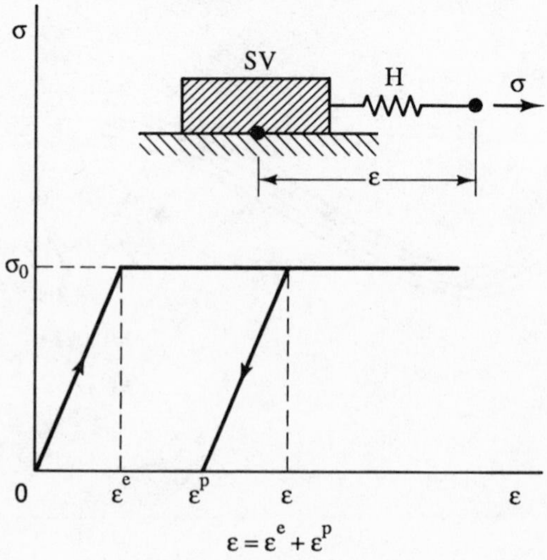

Note on creep At high temperatures we observe that plastic strain, under the effect of a relatively small stress, grows with time. This phenomenon, which is called *creep*, is expressed in certain cases by strains that increase with time, while the load remains constant, and, in some other cases, by the continuous decrease of stress, while the strain remains constant (*relaxation*). Creep is what determines the resistance and life duration of mechanical elements submitted to high temperatures. A creep curve is typically as shown in Fig. 1.9, exhibiting the regimes called primary, secondary, and tertiary creeps. In secondary creep $\dot{\varepsilon}$, the rate of strain, is nearly constant. Creep is hardly studied in this book. For this subject we refer the reader to Norton (1929) and Odqvist (1966), two pioneers in the field. Rabotnov (1969) and Cadek (1988) are also recommended references. Chapter 10 deals briefly with some aspects of creep in relation to damage (see below).

1.3 Other effects

The term *hysteresis* means that the loading and unloading curves do not coincide in spite of having similar upper and lower points in the load-versus-strain diagram. This may be due to plasticity (a phenomenon that, as we saw, is independent of velocity) or to viscoelasticity (where one or more characteristic times may interfere). We often consider a *cyclic* load

Fig. 1.9. The three regimes of creep in uniaxial constant load

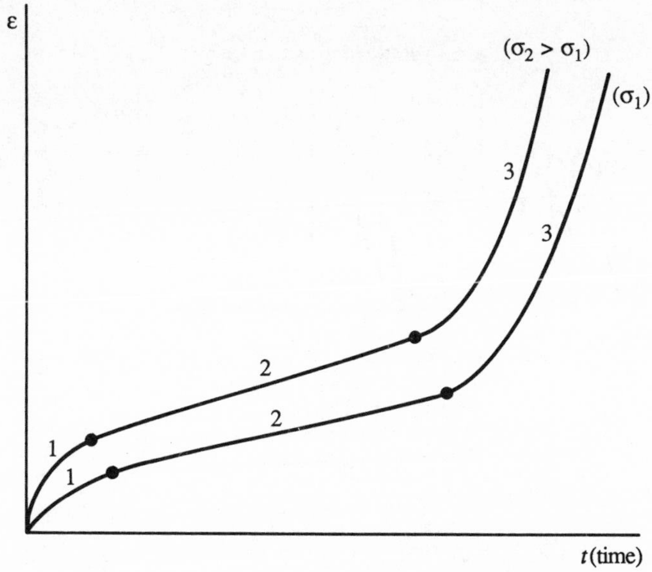

(i.e., a *time-periodic loading*). In that case, either the response curve does not close up and the strain increases at each period with the so-called *ratcheting* effect, until the point of fracture at point R (Fig. 1.10(a)), or else the curve closes up after a certain number of periods in a closed cycle, called the *hysteresis cycle*. In this case we say that there is *plastic shakedown* (Fig. 1.10(b)). The presence of angular points on the cycle (Fig. 1.10(c)) hints at the existence of plastic strains that are not affected by viscosity. If there are plastic strains, they are of opposite signs on the two halves of the cycle and, after a small number of cycles, fracture occurs as a result. When there is no plastic strain, the strains during the cycle, as long as there is no progressive cracking, are viscoelastic. If, moreover, the viscosity is negligible, then we have an elastic cycle, flattened, that is (Fig. 1.10(d)); then, after a certain number of periods, the material begins to react as an elastic material: in this case we say that there is *elastic shakedown*. If the strain is imposed cyclically, there is always *plastic shakedown*.

Damage This is an alteration of the elastic properties due to the fact that, in the course of loading, the *effective resisting* area diminishes as a result of the expansion of the voids and microcracks. This phenomenon, which results in the decrease in Hooke's modulus E, may be coupled up with plasticity (see Chapter 10). Fig. 1.11 provides an example pertaining to an aluminium alloy with a slight decrease in E. This decrease is much more drastic in materials like concrete that are not our concern.

Fig. 1.10. Ratcheting and shakedown

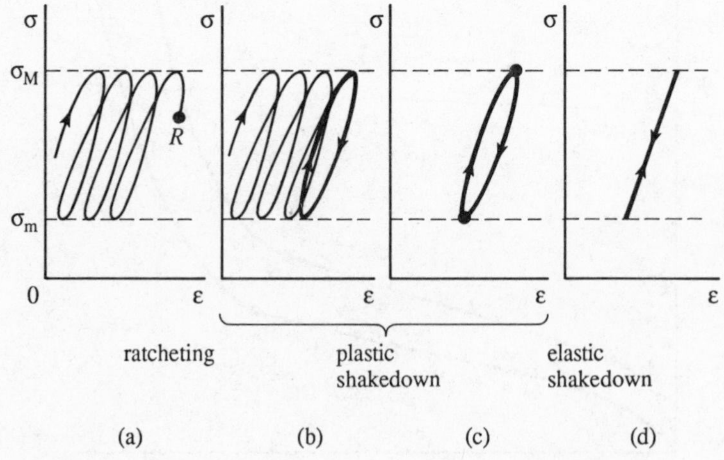

1.4 The plastic-hardening threshold: experimental data

The example of threshold or yield limit given in Fig. 1.8 for a one-dimensional model is elementary. Actually, most of the time, a state of *complex* stresses is preferably considered in the framework of a *three-dimensional* theory. Experimentally, the objects of most tests are thin-walled tubes. And the reason is that, by applying a combined traction, torsion and internal pressure, we can generate an arbitrary plane (or 'almost plane') stress state in the tube's walls (Fig. 1.12). This permits us to judge the law of plastic

Fig. 1.11. Hardening curve with unloadings (light aged annealed alloy AU4G); from Lemaître and Chaboche, 1985, p. 167 (reproduced from unknown source)

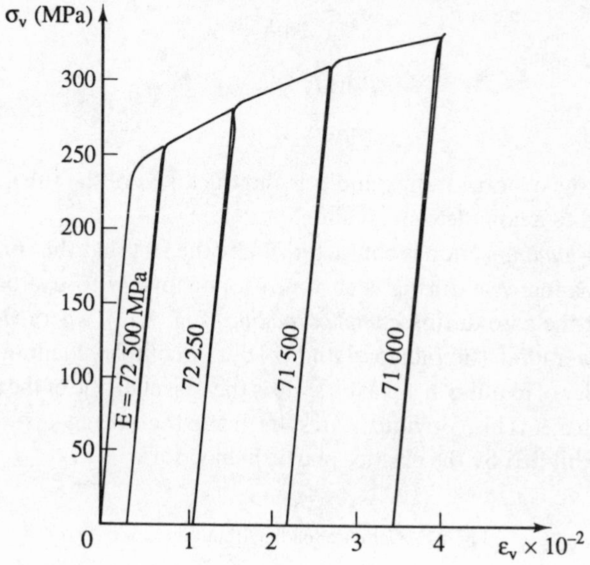

Fig. 1.12. Thin tube test

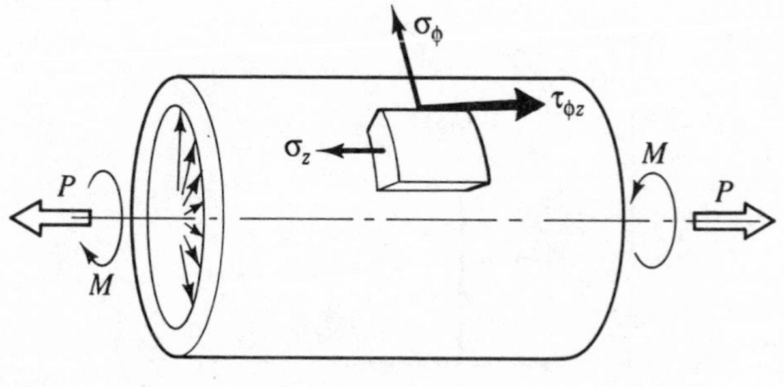

strain by measuring the strains on the tube (change in diameter, in length, in its torsion angle) and by comparing them to the known strain states. With the supplementary action of an external pressure we can follow the behaviour of the material during a state of triaxial stress.

For instance, using simple reasoning, in Fig. 1.12 we see that the combined action of the *axial traction P* and the *torsion moment M* yields a stress state

$$\sigma_\varphi \approx 0, \qquad \sigma_z = \frac{P}{2\pi a h}, \qquad \tau_{\varphi z} = \frac{M}{2\pi a^2 h}, \tag{1.9}$$

while the combined action of the *axial traction P* and the internal pressure p yields

$$\left.\begin{aligned} &\sigma_\varphi \approx p\frac{a}{h}, \qquad \sigma_z = \frac{P}{2\pi a h}, \qquad \tau_{\varphi z} = 0,\\ &\sigma_r = \mathrm{O}(p) \ll \sigma_\varphi, \sigma_z,\\ &h/a \ll 1, \end{aligned}\right\} \tag{1.10}$$

where a is the average radius and h is the thickness of the tube. Thus we have access to a complete stress state.

A *simple loading* action is characterized by the fact that the components of the stress increase during each test, proportionally to *one* parameter. This is not the case during *complex loading* (Fig. 1.13), where OO_1 corresponds to a 'radial' (simple) load and OAB to a complex loading (M, then P). The order of loading in a plastic state is the object of one of the problems in Appendix 3. This, obviously, results from the strong *essential* nonlinearity exhibited by the elastic–plastic behaviour.

Fig. 1.13. Complex loading (thin tube case)

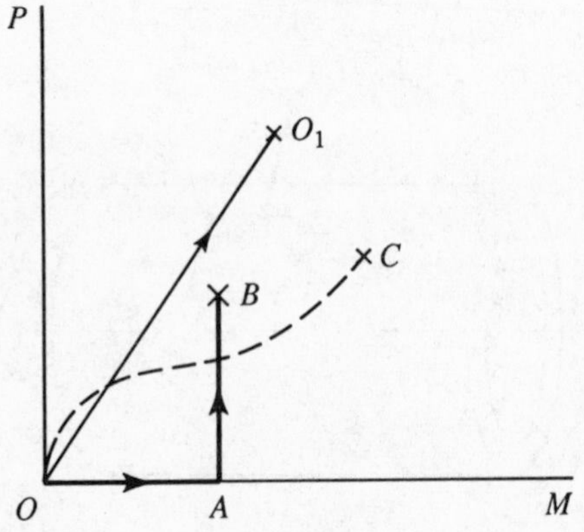

Returning now to the object of this section, the problem is to measure the *elastic limit* or *plastic-hardening* surface for complex stress states. The equation of this limit surface is written

$$f(\boldsymbol{\sigma}) = 0, \tag{1.11}$$

where $\boldsymbol{\sigma}$ is the stress tensor, whose Cartesian components, with respect to well determined axes, are σ_{ij}. Thus, in general, $f = 0$ is a surface of dimension 5 embedded in a six-dimensional space. The difficulty of a visual representation of such surfaces is obvious. Fortunately, in many cases the dimensionality can be tremendously decreased. For instance, we can already remark that only the deviatoric part of $\boldsymbol{\sigma}$ is of interest in the plastic regime so that only five out of six of the components of $\boldsymbol{\sigma}$ are relevant. In addition, the symmetry of the material will greatly reduce the complexity of the representation. For particular *structural members* it may be more astute to use a space of generalized stresses that are more suited to the particular type of structure than the usual stress components. The surface $f = 0$ is then called the *interaction surface* (see Sawczuk, 1989, and Problems in Chapters 1 and 2 herein). The case of isotropic bodies is particularly enlightening.

(i) For an *isotropic body*, the function f is form-invariant for arbitrary changes of the above-mentioned axes. *Changes of axes* are described by

$$x_i = \alpha_{ij}\hat{x}_j, \tag{1.12}$$

where the $\boldsymbol{\alpha} = \{\alpha_{ij}\}$ are orthogonal, i.e.,

$$\boldsymbol{\alpha}^{\mathrm{T}} = \boldsymbol{\alpha}^{-1}, \qquad \det\boldsymbol{\alpha} = +1, \tag{1.13}$$

where the upper T denotes the operation of transposition. The α_{ij} are real. We must verify then that

$$f(\sigma_{ij}) = f(\alpha_{ip}\alpha_{jp}\hat{\sigma}_{pq}) = \hat{f}(\hat{\sigma}_{pq}) = f(\hat{\sigma}_{pq}) \tag{1.14}$$

in such a manner that f must be a scalar-valued *isotropic* function of the tensor $\boldsymbol{\sigma}$, which, therefore, according to the Cauchy representation theorem (see, e.g. Eringen, 1967) can only depend upon $\boldsymbol{\sigma}$ by the intermediary of its principal invariants I_1, I_2 and I_3, i.e.

$$f(I_1(\boldsymbol{\sigma}), I_2(\boldsymbol{\sigma}), I_3(\boldsymbol{\sigma})) = 0, \tag{1.15}$$

or then, in a system of principal stresses,

$$F(\sigma_1, \sigma_2, \sigma_3) = 0. \tag{1.16}$$

The $(\sigma_1, \sigma_2, \sigma_3)$ coordinate system represents a stress space called the

Haigh–Westergaard stress space (Haigh, 1920, Westergaard, 1920), which obviously is more convenient than the six-dimensional space envisaged before.

(ii) For an *anisotropic body*, the function f of (1.1) depends upon the axes used and, in general, $\hat{f}(\hat{\boldsymbol{\sigma}}) \neq f(\boldsymbol{\sigma})$. For the plasticity of anisotropic media, and more particularly *composites* and fibre-reinforced media, we refer the reader to Spencer (1972), Rogers (1988, 1990) and the proceedings edited by Boehler (1985). An example of an analytical problem for anisotropic elastic–plastic bodies is given in Appendix 3. In the remainder of this section we consider exclusively *isotropic* bodies, or bodies that can reasonably be considered as such (e.g., polycrystals, alloys). Thus we assume that (1.15) or (1.16) holds true.

The domain $F < 0$ is called the *resistance domain* (or the domain of restrained elastic strains) whereas $f = 0$ or $F = 0$ is called the *yield limit-surface*. In most cases the influence of the average pressure on the change of form is so insignificant as to be negligible. It follows that the dependence of f on I_1 may be neglected and, instead of $\boldsymbol{\sigma}$, we may introduce the deviator $\boldsymbol{\sigma}^{\mathrm{d}}$, sometimes denoted by $\mathbf{s}$, of $\boldsymbol{\sigma}$.

$$\mathbf{s} = \boldsymbol{\sigma}^{\mathrm{d}} = \boldsymbol{\sigma} - \tfrac{1}{3}(\operatorname{tr}\boldsymbol{\sigma})\mathbf{I}, \qquad \text{i.e., } s_{ij} = \sigma_{ij}^{\mathrm{d}} = \sigma_{ij} - \tfrac{1}{3}\sigma_{kk}\delta_{ij}. \tag{1.17}$$

is such a way that (1.15) will be replaced by

$$\left.\begin{aligned} f(I_2(\boldsymbol{\sigma}^{\mathrm{d}}), I_3(\boldsymbol{\sigma}^{\mathrm{d}})) = 0, \\ I_2 = \operatorname{tr}(\boldsymbol{\sigma}^{\mathrm{d}})^2, \qquad I_3 = \operatorname{tr}(\boldsymbol{\sigma}^{\mathrm{d}})^3. \end{aligned}\right\} \tag{1.18}$$

According to the usual interpretation of I_2 and I_3, we see that f in its form (1.16) depends only upon the *differences of the main stresses*. Moreover, as long as the order of the main stresses is not fixed, the function F must be *symmetrical* in σ_1, σ_2 and σ_3 (we obtain the same function by any permutation of σ_1, σ_2 and σ_3). So the straight line Δ, where $\sigma_1 = \sigma_2 = \sigma_3$, must be an *axis* of ternary symmetry of the surface $\Sigma(F = 0)$ and the plane $\sigma_1 = \sigma_2$ – just like the planes $\sigma_2 = \sigma_3$ and $\sigma_3 = \sigma_1$ – must be a *symmetry plane* (see Fig. 1.14). The section of Σ then, by a plane P_Δ perpendicular to Δ, possesses the *trefoil* symmetries (without concave parts, see below), as in Fig. 1.15(a). So all we need to do in order to know thoroughly the surface Σ is to determine it in a dihedral angle equal to $\pi/3$. If, moreover, the properties of the material remain the same at compression and traction (no Bauschinger effect), we have

$$F(\sigma_1, \sigma_2, \sigma_3) = F(-\sigma_1, -\sigma_2, -\sigma_3), \tag{1.19}$$

and the flow surface must be *symmetrical* with respect to the straight lines, perpendicular to the axes σ_1, σ_2 and σ_3. Given the trefoil symmetry, the yield surface has now a symmetry of order 6 (hexagonal symmetry) and it is enough to determine experimentally a sixth of this 'curve' (Fig. 1.15(b)). Besides, experiment shows that this curve is always *convex* (see Appendix 2); it lies entirely on one side of the tangent, on every point, or of the basis line, if this line presents rectilinear sections; this does not mean that it cannot eventually present angular points (Tresca, Coulomb criteria).

Fig. 1.14. Haigh–Westergaard stress space (P_Δ plane and Δ-line; the case illustrated is Tresca's)

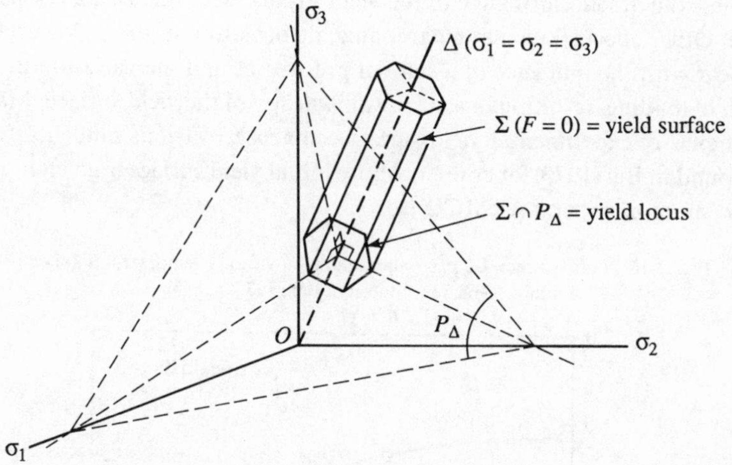

Fig. 1.15. Flow surface symmetries (isotropic body): (a) trefoil symmetry; (b) order-six symmetry.

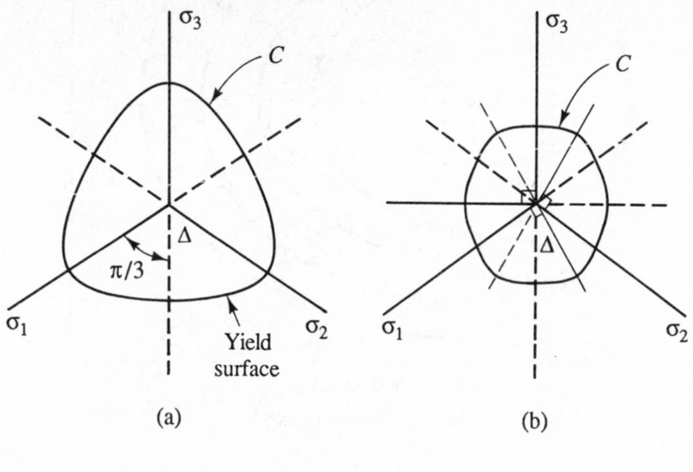

We also verify experimentally that the origin O ($\boldsymbol{\sigma}^d = \mathbf{O}$) is on the inside of this convex, because the state of plastic flow occurs for nonzero tangential stresses. In order to formulate the mathematical expression of the surface or the flow *criterion* all the above experimental facts should be taken into account. However, difficulties met in the experimental construction of yield surfaces should not be overlooked. For the sake of illustration we reproduce in Figs. 1.16 and 1.17 experimental yield surfaces at various temperatures and depending on the state of prestresses and load path for pure aluminium in a shear-stress-versus-tensile-stress plane. In Fig. 1.16 (which will be seen to be not too different from a von Mises criterion) we witness the drastic shrinkage of the yield surface with increasing temperature. Other effects (kinematic hardening, deformation of the yield surface) appear with the influence of a state of prestresses and the variation in the path of loading, resulting in a strong anisotropy of the yield surface. Many examples of experimental yield surfaces subject to various influences can be found in Bui (1970) for copper (whose initial yield surface is given in Fig. 1.19), aluminium, and ARMCO iron.

Fig. 1.16. Yield surface of pure aluminium at various temperatures (after Phillips, 1968; reproduced in Bell, 1973, p. 683).

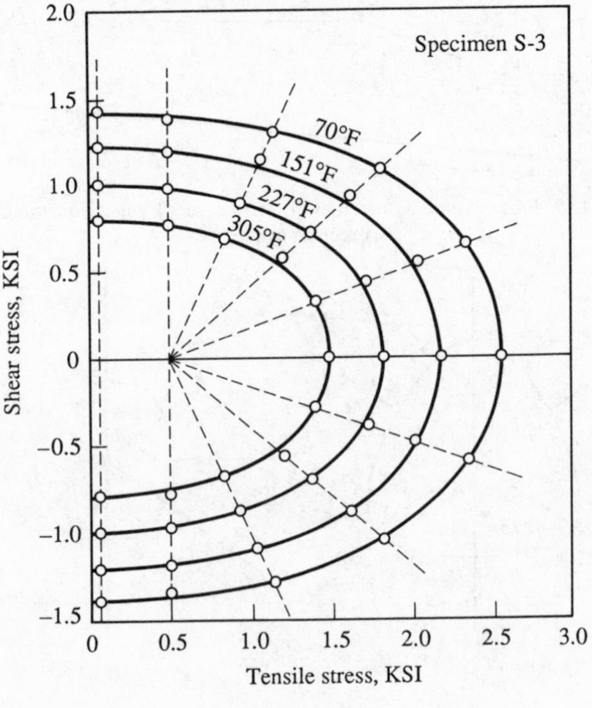

1.5 Examples of plastic-flow criteria

The Tresca criterion was experimentally obtained by H.E. Tresca (1872) and its mathematical formulation was given simultaneously by Barré de Saint-Venant in 1871. According to this criterion, plastic strain will appear at a certain point when the maximum tangential stress attains a certain value, $\tau_M = k$. It is also referred to as the *maximum-shear theory of yielding*. This is a criterion that yields good results in the case of pure metals. The obvious mathematical formulation is as follows (see also Problem 1.5):

Fig. 1.17. The effect of prestress, loading path and temperature on the yield surface of pure aluminium (after Phillips and Tang, 1972; reproduced in Bell, 1973, p. 684).

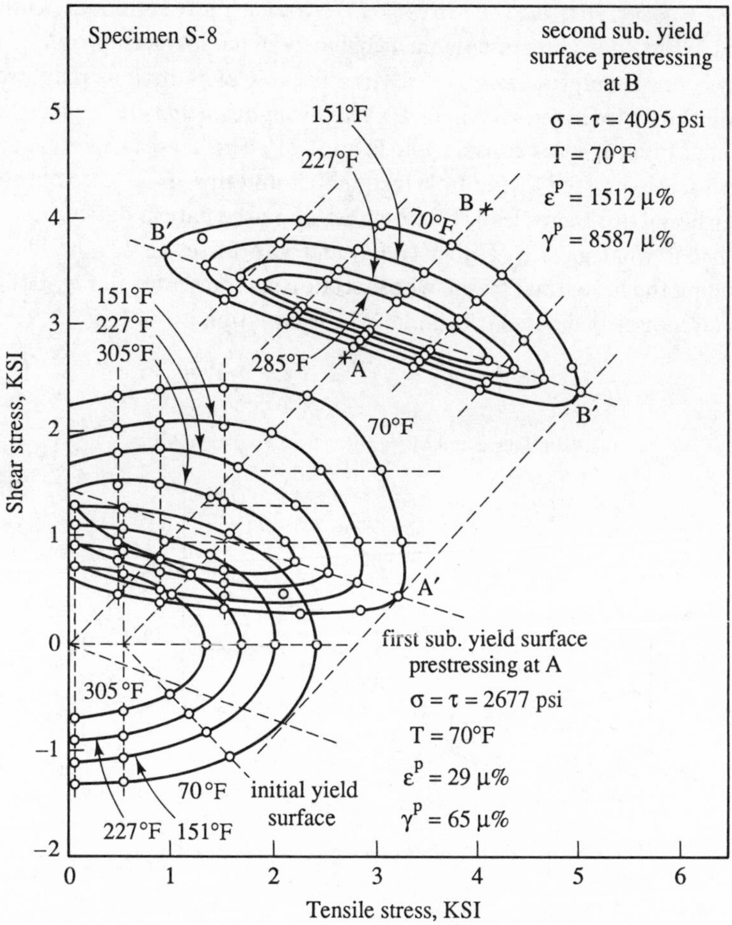

$$\operatorname*{Sup}_{\alpha,\beta} |\sigma_\alpha - \sigma_\beta| = 2k, \qquad \alpha, \beta = 1, 2, 3, \tag{1.20}$$

where the σ_α's are the main stresses, i.e.

$$\left.\begin{aligned} 2|\tau_1| &\equiv |\sigma_2 - \sigma_3| \leqslant 2k, \\ 2|\tau_2| &\equiv |\sigma_3 - \sigma_1| \leqslant 2k, \\ 2|\tau_3| &\equiv |\sigma_1 - \sigma_2| \leqslant 2k, \end{aligned}\right\} \tag{1.21}$$

where the conditions $\sigma_1 \geqslant \sigma_2 \geqslant \sigma_3$ may not be respected (otherwise we have always $2\tau_M = \sigma_1 - \sigma_3$). In the P_Δ plane representation with three axes at $\pi/3$, the elastic domain is limited by a hexagon whose sides are parallel to the axes. The corresponding flow curve is really convex but it presents six angular points (Fig. 1.18(a)).

The Huber (1904) and von Mises (1913) criterion, more commonly known as the *Mises criterion*, presents the peculiarity of having been proposed on a purely mathematical basis by von Mises because of its obvious simple representation. Moreover, it is, in fact, a *maximum-distortion-energy theory* of yielding. Indeed, let us consider the form (1.21). The Tresca–Saint-Venant criterion is expressed by multiple inequalities and this, in the case of three-dimensional problems, leads to a number of mathematical difficulties.

That is what gave to Huber (1904) and von Mises (1913) the idea of replacing the hexagonal prism, whose section is represented in Fig. 1.18(a), by a circumscribed circular cylinder whose equation is

$$(\sigma_1 - \sigma_2)^2 + (\sigma_2 - \sigma_3)^2 + (\sigma_3 - \sigma_1)^2 = 2k; \tag{1.22}$$

Fig. 1.18. Tresca and Mises criteria: (a) Tresca; (b) Mises.

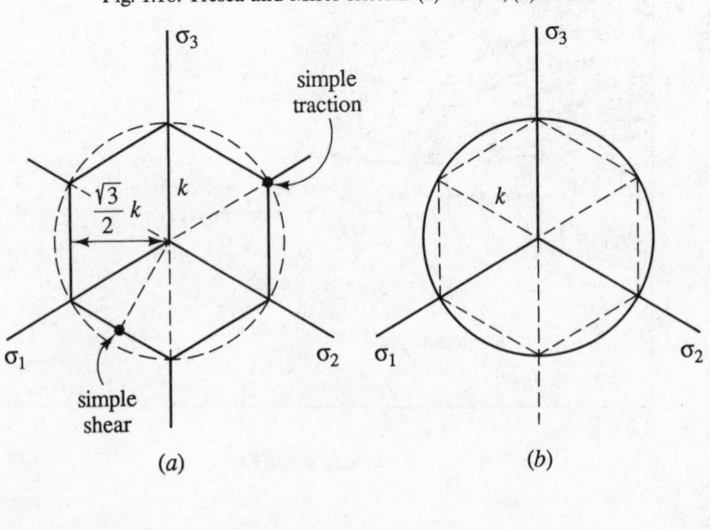

that is

$$I_2(\boldsymbol{\sigma}^d) = \mathrm{tr}(\boldsymbol{\sigma}^d)^2 = \tfrac{2}{3}k^2 \tag{1.23}$$

since

$$I_2(\boldsymbol{\sigma}^d) = \tfrac{1}{3}[(\sigma_1 - \sigma_2)^2 + (\sigma_2 - \sigma_3)^2 + (\sigma_3 - \sigma_1)^2]. \tag{1.24}$$

The intersection of the cylinder (1.22) with the plane $(\sigma_1, \sigma_2, \sigma_3)$ is a circle of radius k (Fig. 1.18(b)). The two Tresca and Mises criteria give equal resistances to simple traction and to simple compression (but different resistances in pure shear). In the case of metals, the experimental results are contained between these two criteria, although they are generally closer to the Mises criterion. This was shown in the well-known experiments by Taylor and Quinney (1931) (Fig. 1.19).

Other well-known criteria were proposed by Rankine (maximum-stress theory) and Beltrami (maximum-strain-energy theory) – see Timoshenko (1953). Of special notice also, although not pertaining to metals, are the Mohr criterion and the internal-friction theory of yielding (Coulomb). These look as follows.

The Mohr criterion (1900) or the intrinsic-curve criterion (Caquot), For bodies *other than metals*, Mohr (1900), working on the idea that plastic stress is generated by slip, proposed the generalization of the Tresca criterion in the form

$$|\tau| = K(\sigma_n), \tag{1.25}$$

Fig. 1.19. Experimental results of Taylor and Quinney (1931) for metals compared to Mises and Tresca criteria

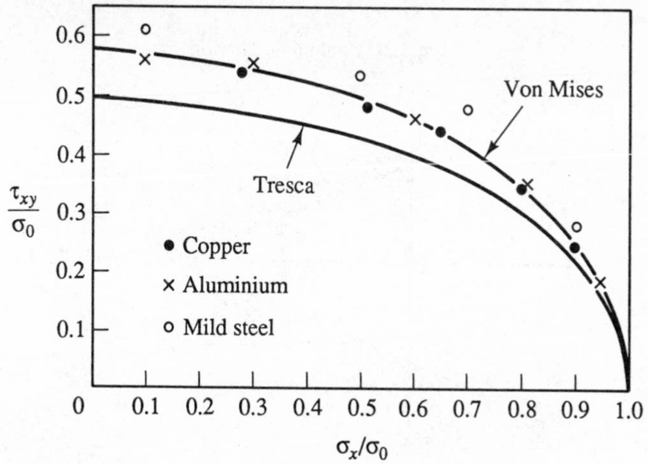

where τ is the shear on the slip facet and σ_n is the normal stress (see Fig. 1.20(a)). The curve (1.25) in the plane (σ_n, τ) is called the *intrinsic curve* (Caquot) and looks like the shape indicated in Fig. 1.20(b). This curve is said to be *intrinsic*, because, according to this curve, different materials differ only as to the position of the stress origin O with respect to the curve S, while the distance is great for very plastic bodies and small for brittle bodies (such as concrete, rocks, porcelain, cast iron). The criterion of the intrinsic curve is not very rigorous.

The Coulomb criterion (soil mechanics). This takes into account the internal friction and is written

$$f: |\tau| + \sigma_n \tan\phi \leqslant C, \tag{1.26}$$

Fig. 1.20. Mohr's criterion (intrinsic curve). $\sigma_n = n_i T_i = n_i \sigma_{ij} n_j$. $\tau_i = T_i - \sigma_n n_i = \sigma_{ij} n_j - \sigma_n n_i$.

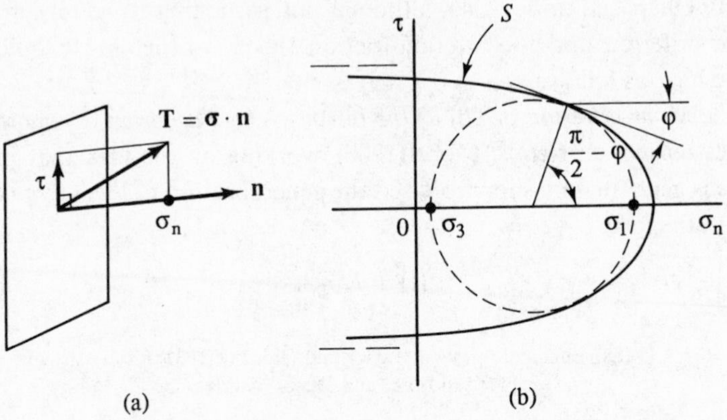

Fig. 1.21. Coulomb's criterion

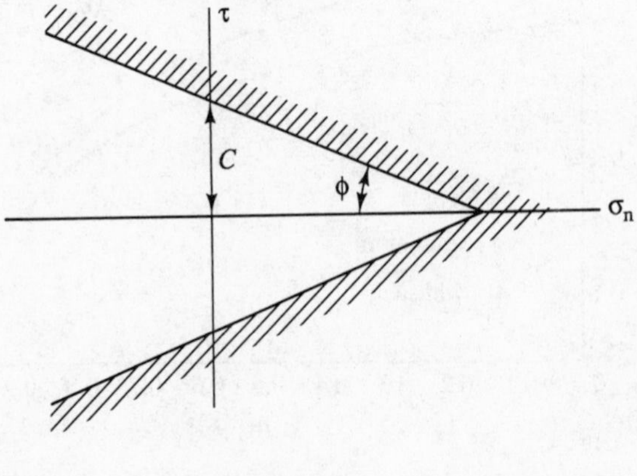

where ϕ is called the internal friction angle (Fig. 1.21). Some other criteria that generalize those of Mises and Tresca, are as follows:

(a) *the von Schleicher* (1926) *criterion*, such as

$$I_2(\boldsymbol{\sigma}^{\mathrm{d}}) = K(\sigma_m), \tag{1.27}$$

where

$$\sigma_m = \tfrac{1}{3}(\sigma_1 + \sigma_2 + \sigma_3) = \tfrac{1}{3}I_1(\boldsymbol{\sigma}) \tag{1.28}$$

is the so-called mean stress;

(b) *criteria* such as

$$\operatorname*{Sup}_{\alpha,\beta} |\sigma_\alpha - \sigma_\beta| = 2K(\sigma_m) \tag{1.29}$$

with, for rocks,

$$K(\sigma_m) = A|\sigma_m|^\alpha \tag{1.30}$$

where α is only slightly smaller than 1.

The criteria (1.27) and (1.29) emphasize the effect of mean stress on the yield surface. This mean stress acts practically like a *hardening* parameter (see below) for the Mises and Tresca criteria.

(c) *Gurson's* (1977) *criterion for isotropic porous media with cavities.* This reads (compare to eqn (1.27))

$$f(I_2, I_1, p) = \frac{(I_2(\sigma^{\mathrm{d}}))^{1/2}}{\sigma_0} - 1 - p\left[q\cosh\left(\frac{2}{3}\frac{\sigma_{\mathrm{m}}}{\sigma_*}\right) + p\right] = 0,$$

where σ_0 and σ_* are two material characteristic constants, p is the porosity (volume fraction of cavities), and q is a numerical factor (found by J.B. Leblond to be equal to 4/e, where e is the base of natural logarithms). For $p = 0$, we recover the Huber–Mises criterion with $\sigma_0^2 = 2k^2$. Gurson's criterion is well adapted to *ductile plasticity* accounting for the presence of evolving cavities.

To be complete, we also give an example of *anisotropic yield criterion* (due to R. Hill). It seems that recrystallized Zircalloy fits the proposed anisotropic criterion better than Mises' isotropic one (Fig. 1.22).

Hardening Equation (1.11) can also be written

$$f(\boldsymbol{\sigma}) = \text{const.} = K, \tag{1.31}$$

where K is a constant of the material, related to the yield limit. We may now pose the following question. Consider a body in the plastic state, characterized at a given moment by the stresses σ_{ij} to which we apply infinitely small increments $d\sigma_{ij}$ (complementary load). Can this complementary load lead to a complementary plastic strain? In other words, according to Fig. 1.23, the stress σ_{1M} would be somehow the *elastic limit* at a certain moment, depending upon the previous plastic strain and permitting the distinction between the beginning of the loading process (accompanied by plastic strain) and the unloading (which is purely elastic). This is the phenomenon of *hardening*. The flow surface is no longer fixed (as was the case in perfect plasticity (1.31). It dilates in any way whatever and is displaced while the hardening spreads. If the load (or yield) surface is uniformly dilating (in an *isotropic* way) then its equation, which replaces (1.31), takes the form

$$f(\boldsymbol{\sigma}) = \mathscr{F}(\beta) \tag{1.32}$$

where $\mathscr{F}$ is an *increasing* function of a certain positive parameter β characterizing the previous plastic strain.

Fig. 1.22. Anisotropic yield criterion (so-called Hill criterion) (after Lee and Zarvel, from Lemaître and Chaboche, 1985, p. 186)

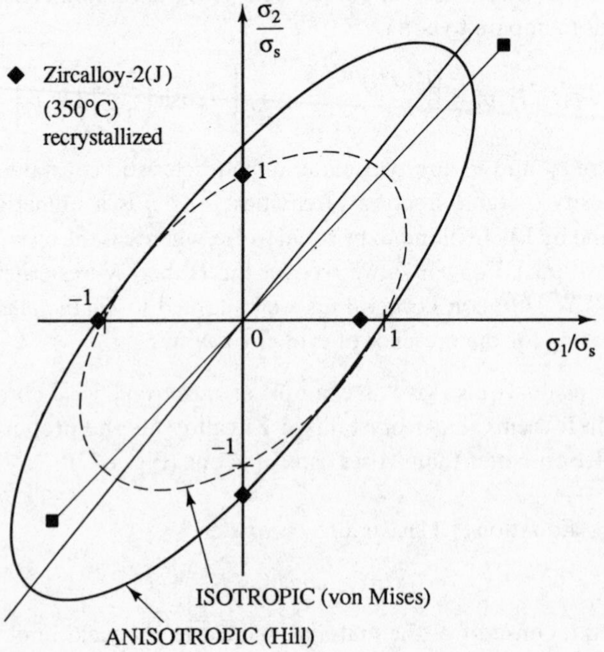

Examples of isotropic hardening measures (a) One possible hardening *scalar* measure accounting for past history is *the work of plastic strains* defined by

$$\beta \equiv W_{\mathrm{p}} = \int \sigma_{ij} \mathrm{d}\varepsilon_{ij}^{\mathrm{p}} = \int_0^t \sigma_{ij} \dot{\varepsilon}_{ij} \, \mathrm{d}t. \tag{1.33}$$

(b) Another possibility is provided by *the length of the path of the plastic strains* in the strain space $\boldsymbol{\varepsilon}^{\mathrm{p}}$,

Fig. 1.23. Hardening. ($\mathrm{d}\sigma_{ij}$ additional loading)

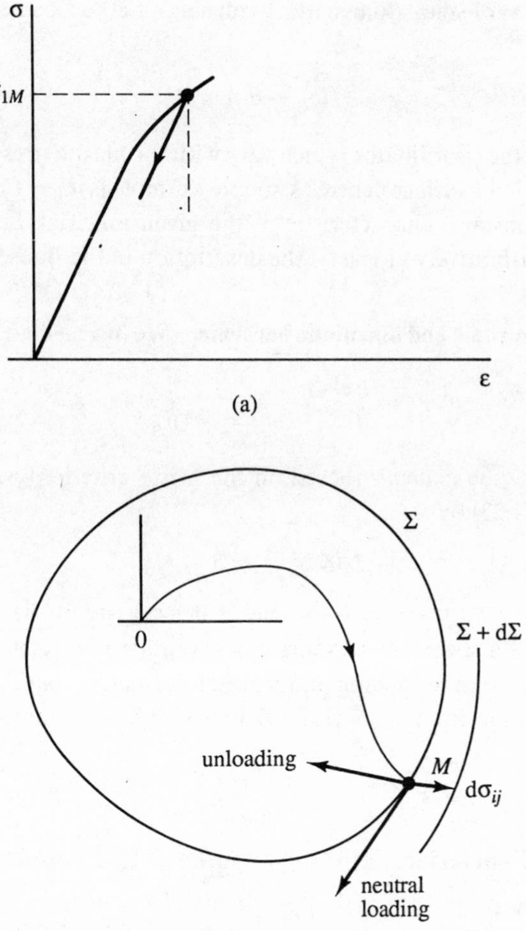

(a)

(b)

$$\beta = \oint \mathrm{d}\Gamma_{\mathrm{p}} = \int (\tfrac{2}{3}\mathrm{d}\varepsilon_{ij}^{\mathrm{p}}\,\mathrm{d}\varepsilon_{ij}^{\mathrm{p}})^{1/2} = \int_0^t (\tfrac{2}{3}\dot{\varepsilon}_{ij}^{\mathrm{p}}\dot{\varepsilon}_{ij}^{\mathrm{p}})^{1/2}\,\mathrm{d}t, \tag{1.34}$$

also called *cumulative plastic strain* or *Odqvist parameter*. Introduced by Odqvist in 1933, this parameter fits perfectly in a thermomechanical framework (see Chapter 5).

Actually, eqn(1.32) may include several hardening parameters and we can replace the scalar β by a vector $\boldsymbol{\beta}$ of n dimensions, in such a way that (1.32) is replaced by

$$f(\boldsymbol{\sigma}) = \mathscr{F}(\boldsymbol{\beta}). \tag{1.35}$$

Translation hardening (kinematic hardening) Let us consider the load surface

$$f(\sigma_{ij} - a_{ij}) = k, \tag{1.36}$$

where a_{ij} are the coordinates, which vary with the plastic stress, in the stress space, of the load surface centre. A simple example is $a_{ij} = C\varepsilon_{ij}^{\mathrm{p}}$, where C is a positive constant, characteristic of the given material. Such a scheme permits – qualitatively at least – the description of the Bauschinger effect.

Combined isotropic and kinematic hardening We have a load surface of the type

$$f(\sigma_{ij} - a_{ij}) = \mathscr{F}(\beta), \tag{1.37}$$

A simple criterion example (based on the Mises criterion) *with hardening*. We replace (1.23) by

$$[I_2(\boldsymbol{\sigma}^{\mathrm{d}})]^{1/2} + B \leqslant K_0, \tag{1.38}$$

where $B = -\partial W/\partial\beta$ and W is the energy density (see further on). Fig. 1.17 has presented a spectacular example of evolution of a yield surface under the action of strain hardening (influence of the loading path in the manner of both isotropic and kinematic hardening combined to anisotropy due to prestressing).

1.6 Conclusions: working hypotheses in elastoplasticity

From all the properties and experimental facts already given, we may conclude that the plasticity of *metals*, with or without hardening, at room or low temperature – with which we shall particularly deal in the following chapters – is

(H1) a thermodynamically *irreversible* phenomenon (existence of plastic strains),
(H2) which is practically *independent* of the strain rate,
(H3) and of volume changes,
(H4) implying the notion of a *threshold* or plastic-yield surface,
(H5) represented by a *convex* hypersurface in the stress space,
(H6) possibly presenting angular points (nonuniqueness of the exterior normal),
(H7) and possibly dependent upon the plastic-strain history (hardening).
(H8) Moreover, for an isotropic medium, with an identical response in traction or compression, the yield surface in the stress space admits a section with a symmetry of order 6 (hexagonal).

Every mathematical theory of metal plasticity must reflect or be based upon the working hypotheses (H1) to (H7). Let us observe that the mathematical and thermomechanical modelling problems that arise from them belong to a class that includes also those of the fracture and damage theories. This latter will not be entered into in great detail, except in what concerns its coupling with plasticity (Chapter 10). Fracture, however, constitutes an important part of this course.

Fracture and plasticity may be considered within the same context. The same is true of fracture and damage (see Bui (1982)).

For the experimental aspects of plasticity that we shall not discuss further here, the reader may consult Bell (1968, 1973) – the second of which will remain like a Bible to experimental solid mechanics – and more technologically oriented books than the present one, e.g., Bridgman (1952), Calladine (1985), Chen and Baladi (1985), Johnson and Mellor (1962, 1973), Mendelson (1968), Nadai (1950), and Vyalov (1986).

The main purpose of the problems accompanying this chapter is to familiarize the reader with the notion of yield surface and manipulating second-order tensors (stresses).

Problems for Chapter 1

1.1. Invariants of a second-order tensor

Consider a second-order symmetric tensor $\boldsymbol{\sigma}$ of Cartesian components σ_{ij}. We look for the eigenvalues σ_α of this tensor. They satisfy the characteristic equations

$$(\sigma_{ij} - \sigma_\alpha \delta_{ij}) d_j^\alpha = 0, \qquad (1)$$

where the $\mathbf{d}^\alpha$ are a set of eigenvectors.

(a) Show that the condition of solvability of (1) yields (for any $\sigma = \sigma_\alpha$, α fixed)

$$-\sigma^3 + \mathrm{I}_\sigma \sigma^2 - \mathrm{II}_\sigma \sigma + \mathrm{III}_\sigma = 0, \qquad (2)$$

where I_σ, II_σ and III_σ are the fundamental invariants of $\boldsymbol{\sigma}$. Give the expressions for these invariants.

(b) Show that the three invariants σ_I, σ_II and σ_III, defined as $\mathrm{tr}\,\boldsymbol{\sigma}^K$, $K = 1, 2, 3$, respectively, form an equivalent set of invariants for $\boldsymbol{\sigma}$. (tr = trace)

(c) Recall the expression of the Cayley – Hamilton Theorem for tensor $\boldsymbol{\sigma}$ (in terms of the first set).

[Answer: $\mathrm{I}_\sigma = \sigma^\mathrm{I}$, $\mathrm{II}_\sigma = \frac{1}{2}(\sigma_\mathrm{I}^2 - \sigma_\mathrm{II})$, $\mathrm{III}_\sigma = \det \boldsymbol{\sigma} = \frac{1}{3}(\sigma_\mathrm{III} - \frac{3}{2}\sigma_\mathrm{II}\sigma_\mathrm{I} + \frac{1}{2}\sigma_\mathrm{I}^3)$]

1.2. Isotropic scalar-valued function

(a) Let $f(\boldsymbol{\sigma})$ be a scalar-valued function of the second-order symmetric tensor $\boldsymbol{\sigma}$. The function f is said to be *isotropic* if and only if $f(\boldsymbol{\sigma}) = f(\mathbf{Q}\boldsymbol{\sigma}\mathbf{Q}^\mathrm{T})$, T = transpose, for any orthogonal transformation $\mathbf{Q}$ such that $\det \mathbf{Q} = +1$. Prove that this imposes the condition that $f = f(\sigma_\mathrm{I}, \sigma_\mathrm{II}, \sigma_\mathrm{III})$ only with notation as in 1.1(b).

(b) Let $f(\boldsymbol{\sigma}_1, \boldsymbol{\sigma}_2)$ be an *isotropic* scalar-valued function of two second-order symmetric tensors. Then prove that f depends on $\boldsymbol{\sigma}_1$ and $\boldsymbol{\sigma}_2$ only through ten scalar invariants: a possible set of such invariants is

$$\mathrm{tr}\,\boldsymbol{\sigma}_1,\ \mathrm{tr}\,\boldsymbol{\sigma}_1^2,\ \mathrm{tr}\,\boldsymbol{\sigma}_1^3,\ \mathrm{tr}\,\boldsymbol{\sigma}_2,\ \mathrm{tr}\,\boldsymbol{\sigma}_2^2,\ \mathrm{tr}\,\boldsymbol{\sigma}_2^3,\ \mathrm{tr}\,\boldsymbol{\sigma}_1\boldsymbol{\sigma}_2,\ \mathrm{tr}\,\boldsymbol{\sigma}_1^2\boldsymbol{\sigma}_2,\ \mathrm{tr}\,\boldsymbol{\sigma}_1\boldsymbol{\sigma}_2^2,\ \mathrm{tr}\,\boldsymbol{\sigma}_1^2\boldsymbol{\sigma}_2^2.$$

1.3. Isotropic plastic body

Prove that the yield condition for an *isotropic* plastic body is necessarily of the form ($\mathbf{s}$ = deviator of $\boldsymbol{\sigma}$)

$$f(\sigma_\mathrm{I}, (s_\mathrm{II})^{1/2}, (s_\mathrm{III})^{1/3}) = 0. \qquad (1)$$

For incompressibility show that this reduces to

$$f((s_\mathrm{II})^{1/2}, (s_\mathrm{III})^{1/3}) = 0. \qquad (2)$$

1.4. Yield function in terms of principal-stress differences

Show that the knowledge of $I_2(\boldsymbol{\sigma}^\mathrm{d})$ and $I_3(\boldsymbol{\sigma}^\mathrm{d})$ is equivalent to that of the differences between principal stresses. Are Tresca and Mises criteria admissible from that point of view?

1.5. Tresca criterion

Let $\boldsymbol{\sigma}$ be a deviatoric stress and σ_I, σ_II, σ_III be its invariants defined by $\sigma_\mathrm{I} = \mathrm{tr}\,\boldsymbol{\sigma} \equiv 0$, $\sigma_\mathrm{II} = \frac{1}{2}(\sigma_1^2 + \sigma_2^2 + \sigma_3^2)$, and $\sigma_\mathrm{III} = \frac{1}{3}(\sigma_1^3 + \sigma_2^3 + \sigma_3^3)$, where σ_1, σ_2 and σ_3 are principal stresses.

Show that Tresca's (maximum-shear) yield criterion can be written in the form

$$4\sigma_{\rm II}^3 - 27\sigma_{\rm III}^2 - 36k_{\rm II}^2\sigma_{\rm II}^2 + 96k^4\sigma_{\rm II} - 64k^2 = 0, \qquad (1)$$

while Mises' criterion simply reads $\sigma_{\rm II} = k^2$ (!). Obviously, this is an awkward formulation.

1.6. Anisotropic strain-hardening

If $\boldsymbol{\varepsilon}^{\rm p}$ is the decisive factor of hardening we may envisage a yield function for isotropic bodies as a scalar-valued function of the joint invariants of $\boldsymbol{\sigma}$ and $\boldsymbol{\varepsilon}^{\rm p}$: $f = f(I_\alpha)$, $\alpha = 1, 2, \ldots, 10$.

(a) Give the list of I_α's using the result of 1.2.

(b) For incompressible materials that are insensitive to isotropic pressure and neglecting higher-order invariants than those of order 2, show that we have the reduction

$$f(\operatorname{tr}(\mathbf{s}^2), \operatorname{tr}(\boldsymbol{\varepsilon}^{\rm p})^2, \operatorname{tr}(\boldsymbol{\sigma}\boldsymbol{\varepsilon}^{\rm p})) = 0,$$

Remark: This allows one to show that the principal axes of stresses and strain rate no longer coincide.

1.7. Anisotropic hardening (continued)

(a) Show that Prager's form

$$f = \operatorname{tr}(\mathbf{s} - C\boldsymbol{\varepsilon}^{\rm p})^2 - 2k^2 = 0 \qquad (1)$$

is a special polynomial form of the expression obtained in 1.6.

(b) Show that

$$2f = s_{ij}s_{ij} - 2s_{ij}\alpha_{ij} + A_0\alpha_{ij}\alpha_{rs}(s_{ij} - \alpha_{ij})(s_{rs} - \alpha_{rs}) - 2k^2 = 0, \qquad (2)$$

where $\alpha_{ij} = C\varepsilon_{ij}^{\rm p}$, A_0 and C constant, is the simplest form of the Huber–Mises condition generalized to the case of anisotropic hardening that takes into account translation, rotation and distortion of the yield surface due to plastic straining.

1.8. Kinematic strain-hardening

Show that a yield condition of the type

$$f(\boldsymbol{\sigma} - \boldsymbol{\alpha}) - k(\operatorname{tr}(\boldsymbol{\varepsilon}^{\rm p})^2, \operatorname{tr}(\boldsymbol{\varepsilon}^{\rm p})^3) = 0, \qquad \boldsymbol{\alpha} = \boldsymbol{\alpha}(\boldsymbol{\varepsilon}^{\rm p}), \qquad (1)$$

combines kinematic hardening and isotropic hardening rules, but it allows only for uniform growth and rigid-body *translation* of the yield surface in stress space and *not* for possible rotations and nonisotropic expansion of the surface.

1.9. Beam of rectangular cross-section

We call *interaction surface* (see Sawczuk (1989) the expression of the yield surface in terms of generalized stresses that is suitable for a particular type of structure. Here we consider beams of rectangular cross-section $2h \times b$. Let σ_0 be the yield point of the material in uniaxial tension or compression. According to the strength of materials the bending moment M and the axial force N are given by

$$M = \sigma_0 bh^2(1 - \zeta^2), \qquad N = 2\sigma_0 hb\zeta, \qquad (1)$$

where $\zeta = z/h$ is the reduced distance from the neutral axis. Then show that the yield condition $f = \sigma \mp \sigma_0 = 0$ translates to

$$\phi(M, N) = (M/M_0)^2 + (N/N_0)^2 - 1 = 0, \tag{2}$$

where M_0 and N_0 are the ultimate values of M and N (what are those values?). Represent (2) in $(M/M_0, N/N_0)$ space.

1.10. Huber–Mises plates

Let z be the thickness direction of the plate of thickness $2h$. Let the membrane force, moment and transverse shear force be defined by

$$N_{\alpha\beta} = \int_{-h}^{+h} \sigma_{\alpha\beta}\,\mathrm{d}z, \qquad M_{\alpha\beta} = \int_{-h}^{+h} \sigma_{\alpha\beta} z\mathrm{d}z, \qquad s_\alpha = \int_{-h}^{+h} \sigma_{z\alpha}\,\mathrm{d}z, \tag{1}$$

$\alpha = 1, 2$. M_0 and N_0 are ultimate values of M and N per unit width of the cross-section. We set $n_{\alpha\beta} = N_{\alpha\beta}/N_0$, $m_{\alpha\beta} = M_{\alpha\beta}/M_0$ and the rates $\lambda_{\alpha\beta} = L_{\alpha\beta}$, $\chi_{\alpha\beta} = (M_0/N_0)K_{\alpha\beta}$; what are these last two kinematic quantities? Discarding the effects of transverse shear, show that the energy dissipation per unit area of the middle surface of the plate reads

$$\phi = \int_{-h}^{+h} \sigma_{\alpha\beta}\dot{\varepsilon}_{\alpha\beta}\,\mathrm{d}z = N_{\alpha\beta}L_{\alpha\beta} + M_{\alpha\beta}K_{\alpha\beta} = N_0(n_{\alpha\beta}\lambda_{\alpha\beta} + \chi_{\alpha\beta}m_{\alpha\beta}) \tag{2}$$

The interaction surface will have the form (for further use)

$$\varphi(n_{\alpha\beta}, m_{\alpha\beta}) = 1 \tag{3}$$

1.11. Simplified interaction surface for shells

The yield condition is supposed to be satisfied in *integral form* over the entire cross-section of the wall of a shell. For instance, take an assumed stress distribution of the type ($\alpha = 1, 2$)

$$\sigma_{\alpha\beta} = \sigma_0(n_{\alpha\beta} + (\operatorname{sgn} z)m_{\alpha\beta}) \tag{1}$$

with $n_{\alpha\beta}$ and $m_{\alpha\beta}$ defined as in 1.10, and integrated Huber–Mises condition reading

$$\frac{1}{2h}\int_{-h}^{+h} (3\sigma_{\alpha\beta}\sigma_{\alpha\beta} - \sigma_{\alpha\alpha}\sigma_{\beta\beta})\,\mathrm{d}z = 2\sigma_0^2 \tag{2}$$

(a) Show that the interaction surface has for equation

$$3n_{\alpha\beta}n_{\alpha\beta} - n_{\alpha\alpha}n_{\beta\beta} + 3m_{\alpha\beta}m_{\alpha\beta} - m_{\alpha\alpha}m_{\beta\beta} = 2. \tag{3}$$

Such a surface is also obtained for sandwich plates.

(b) This result can be written as $N^2 + M^2 = 1$, a fact that emphasizes separately moment and membrane-force effects. What are N and M?

1.12. Cylindrical shells (see, e.g., Sawczuk (1989))

Make a literature search and describe a possible simplified interaction surface for a cylindrical shell subject to the maximum-normal-stress criterion.

1.13. Tresca shells (term paper)
Make a literature search and describe the bases (interaction surface, flow rule) of the theory of rotationally symmetric sandwich Tresca shells.

1.14. Fibre-reinforced structures (term paper)
Make a literature search and describe the interaction surface (yield criterion in generalized stress space) for shells with nonsymmetrically reinforced cross-section made of Huber–Mises material.

2

Thermomechanics of elastoviscoplastic continua

The purpose of this chapter Our purpose is to recall the essential elements of the thermomechanics of continuous media with applications in elastoplasticity, within the framework of the small-perturbation hypothesis (SPH). We are basing this on the general formulation given in an introductory course of *Thermomechanics.* (*See Appendix I*)

2.1 The small-perturbation hypothesis

The general motion and deformation of a continuum between a reference configuration $K_{\mathbf{R}}$, free of loads and strains (the so-called natural state), and the actual configuration K_t, at time t, after motion and deformation have taken place, are described by a space-time mapping (in fact a spatial diffeomorphism parametrized by the time) called the *motion* or transformation

$$\mathbf{x} = \chi(\mathbf{X}, t), \qquad \text{or} \qquad x_i = \chi_i(X_j, t), \tag{2.1}$$

when positions $\mathbf{X}$ and $\mathbf{x}$ at times t_0 and t here are referred to the same Cartesian frame. For the time being we are concerned only with 'small' transformations for which the displacement

$$u_i = x_i - X_i \tag{2.2}$$

remains small for all points of the body. In some sense $\mathbf{u}$ is only a small variation $\delta\mathbf{x}$ about the 'points' $\mathbf{X}$ that make up the body in its reference configuration (see Fig. 2.1, to be contrasted with Fig. 8.1 for 'finite' transformations). That is, we can also write

$$\mathbf{x} = \mathbf{X} + \delta\chi \tag*{(2.2)$_2$}$$

where $|\delta\chi|$ remains small as compared to a macroscopic length of reference for all $\mathbf{X}$'s: K_t *remains in a neighbourhood* of $K_{\mathbf{R}}$.

The material body is supposed to occupy a simply connected open set S_0 of Euclidean three-dimensional space in the reference configuration. Let ∂S_0 be the regular boundary of S_0. These become S and ∂S, respectively, after motion and deformation.

The *small-perturbation hypothesis* SPH consists in supposing that in ordinary conditions of use, most solids undergo only small transformations (2.2) and that the gradients of displacement,

$$\nabla \mathbf{u} : u_{i,j}(\mathbf{X}, t) = \frac{\partial u_i}{\partial X_j}, \tag{2.3}$$

are also small and, to fix ideas, do not exceed a few per cent. From standard courses in continuum mechanics and/or elasticity, the reader knows that SPH justifies the following approximations:

the domain S of the physical space occupied by the material body at a given moment t is assimilated to the domain S_0 occupied in the configuration of reference, and this holds for every t; as a consequence, boundary conditions are always applied at the unperturbed boundary ∂S_0 in SPH;

the relation between the strain tensor $\boldsymbol{\varepsilon} = \{\varepsilon_{ij}\}$ and the gradient of displacement is reduced to the *linear* relation

$$\boldsymbol{\varepsilon} = \boldsymbol{\varepsilon}(\mathbf{u}) = (\nabla \mathbf{u})_S, \qquad \text{i.e., } \varepsilon_{ij} = u_{(i,j)} = \tfrac{1}{2}(u_{i,j} + u_{j,i}), \tag{2.4}$$

where the subscript S indicates the symmetrization defined in the component equation;

according to the continuity equation, the mass per unit volume, $\rho(\mathbf{X}, t)$, is considered as a constant; this follows from the fact that $\rho(\mathbf{x}, t)$ in K_t and $\rho_0(\mathbf{X})$ in K_ρ are generally related by

$$\rho_0 = \rho \times [\text{Jacobian determinant of } \boldsymbol{\chi}(\mathbf{X}, t)];$$

Fig. 2.1. Small transformations. $\mathbf{X} = X_j \mathbf{e}_j$, $\mathbf{x} = x_i \mathbf{e}_i$, $\mathbf{e}_i \cdot \mathbf{e}_j = \delta_{ij}$.

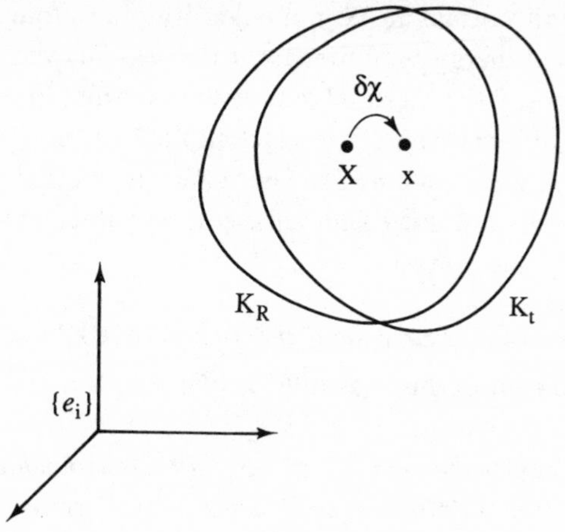

but the quantity within square brackets, denoted by J, is such that

$$J = 1 + O(|\nabla \mathbf{u}|);$$

as ρ will be in all instances already in order of magnitude at least $O(|\mathbf{u}|)$ or $O(|\nabla \mathbf{u}|)$, it can be replaced by ρ_0 in SPH;

the material derivative is replaced by the *partial derivative* $\partial/\partial t$ with respect to time;

the tensor of strain rates $\mathbf{D} = \{D_{ij} = D_{ji}\}$ is reduced to the partial time *derivative* of the deformation, $\dot{\boldsymbol{\varepsilon}}$, that is, in components,

$$D_{ij} = \frac{\partial}{\partial t}\varepsilon_{ij} = \dot{\varepsilon}_{ij} = \tfrac{1}{2}(\dot{u}_{i,j} + \dot{u}_{j,i}). \tag{2.5}$$

Let us recall that the relations (2.1) show that the *six* functions ε_{ij} cannot be independent, since, through integration, they must determine *three* scalar quantities, the displacement components. In order then that a strain be integrable the following conditions of *compatibility* must be satisfied:

$$\varepsilon_{ij,kl} + \varepsilon_{kl,ij} = \varepsilon_{ik,jl} + \varepsilon_{jl,ik}, \tag{2.6}$$

which represent *three* conditions, thus removing indeterminacy.

2.2 General principles of continuous-media thermomechanics

These are the principle of virtual power (PVP) and the first two principles (laws) of thermodynamics which are globally expressed for a material system S bounded by ∂S.

Let $\boldsymbol{\sigma} = \{\sigma_{ij} = \sigma_{ji}\}$ be the Cauchy stress tensor, $v_i = \partial\chi_i/\partial t = \dot{u}_i$ be the velocity field, ρ the mass density (considered here as constant), T_i the field of *tractions* (stress vectors) acting upon the surface ∂S of the material body, n_i the exterior unit normal at ∂S, f_i the density of external forces per unit mass, e the internal energy per unit mass, q_i the heat flux vector, h the source heat per unit mass, θ the thermodynamic temperature ($\theta > 0, \inf\theta = 0$), η the entropy per unit mass and Ψ the Helmholtz free energy per unit mass. We have the following statements. The reader is referred to Appendix 1 for a brief, but self-contained and sufficient, introduction to the thermodynamics of continuous media.

2.2.1 Principle of virtual power (PVP)

The verbal statement of this is as follows (see, e.g., Maugin, 1980).

PVP *In a Galilean frame and for an absolute Newtonian chronology, the virtual power of the inertial forces of a mechanical system S balances the* *virtual power of* all *other forces, internal or external, impressed on the system, for any virtual velocity fields.*

Mathematically this is translated into the formula

$$\mathscr{P}^*_{(a)} = \mathscr{P}^*_{(i)} + \mathscr{P}^*_{(v)} + \mathscr{P}^*_{(c)}, \tag{2.7}$$

where

$$\mathscr{P}^*_{(a)} = \int_S \rho \dot{v}_i v_i^* \, dv, \tag{2.8}$$

$$\mathscr{P}^*_{(i)} = -\int_S \sigma_{ij} v^*_{j,i} \, dv, \tag{2.9}$$

$$\mathscr{P}^*_{(v)} = \int_S \rho f_i v_i^* \, dv, \tag{2.10}$$

and

$$\mathscr{P}^*_{(c)} = \int_{\partial S} T_i v_i^* \, ds \tag{2.11}$$

are respectively, the virtual powers of inertial forces ($\rho\dot{\mathbf{v}}$), internal forces ($\boldsymbol{\sigma}$), volume (or at-a-distance, or 'body') forces ($\rho\mathbf{f}$), and contact forces ($\mathbf{T}$).

The expression (2.9) is essential in the statement (2.7) where the asterisk is used to denote a *virtual field* or the expression for a quantity in such a virtual field. Indeed, (2.9) can be written as

$$\mathscr{P}^*_{(i)} = -\int_S \sigma_{ij} D^*_{ij} \, dv = -\int_S \boldsymbol{\sigma} : \mathbf{D}^* \, dv. \tag{2.12}$$

But

$$\mathbf{D}(\mathbf{x}, t) = 0, \qquad \forall t, \tag{2.13}$$

defines, in differential (local) form, a so-called *rigid-body* motion (see, e.g., Germain, 1973, Maugin, 1980, 1988, pp. 78–9). By integration such a motion has as component equations for the velocity

$$v_i = \Omega_{ij} x_j + V_i, \tag{2.14}$$

where

$$\Omega_{ij} = -\Omega_{ji}, \qquad \Omega_{ij,k} = 0, \qquad V_{i,j} = 0. \tag{2.15}$$

There are two direct consequences of (2.12)–(2.13). First, in a virtual velocity field $\mathbf{v}^*$ which is that of a rigid-body motion, because of (2.13) – written with an asterisk –

$$\mathscr{P}^*_{(i)} \equiv 0. \tag{2.16}$$

This is sometimes referred to as the axiom of virtual power of internal forces:

Axiom of VPIF *The virtual power of forces internal to a system S vanishes for all rigid virtual motions of S considered at any time.*

The second consequence of this is that the *a priori* statement (2.7) (as a principle) in the manner of *d'Alembert* is entirely equivalent to the *Newtonian* mechanical-balance equations, the so-called conservation laws of linear and angular momenta. In effect, if we select for $\mathbf{v}^*$ in (2.7) a rigid-body velocity field such as (2.14) – with asterisks on $\mathbf{v}$, $\mathbf{\Omega}$ and $\mathbf{V}$ – $\mathbf{\Omega}^*$ and $\mathbf{V}^*$ can be taken out of the integrals because of eqns (2.15) and, for any such $\mathbf{\Omega}^*$ and $\mathbf{v}^*$, we shall obtain

the global balance law of linear momentum,

$$\frac{\mathrm{d}}{\mathrm{d}t}\int_S \rho\mathbf{v}\,\mathrm{d}v = \int_S \rho\mathbf{f}\,\mathrm{d}v + \int_{\partial S} \mathbf{T}\,\mathrm{d}s, \tag{2.17}$$

the global balance law of angular momentum,

$$\frac{\mathrm{d}}{\mathrm{d}t}\int_S \rho\mathbf{v}\times\mathbf{x}\,\mathrm{d}v = \int_S \rho\mathbf{f}\times\mathbf{x}\,\mathrm{d}v + \int_{\partial S} \mathbf{T}\times\mathbf{x}\,\mathrm{d}s, \tag{2.18}$$

where '$\times$' denotes the vector product, and we have taken account of the fact that $\mathbf{\Omega}^*$ is antisymmetric and, $\mathrm{d}/\mathrm{d}t$ denoting the material time derivative,

$$\int_S \rho\frac{\mathrm{d}A}{\mathrm{d}t}\mathrm{d}v \equiv \frac{\mathrm{d}}{\mathrm{d}t}\int_S \rho A\,\mathrm{d}v \tag{2.19}$$

holds true in continuum mechanics by virtue of the continuity equation.

The *local* mechanical-balance equations can be deduced either from (2.7) – d'Alembert's point of view – or from (2.17)–(2.18) – Newton's point of view. In the first case we simply apply the PVP statement by assuming that (2.7) holds true for any field $\mathbf{v}^*$, any volume element in S, and any surface element on ∂S, providing thus

the Euler–Cauchy equations of motion in a continuous medium,

$$\rho\dot{v}_i = \sigma_{ij,j} + \rho f_i \qquad \text{or} \qquad \rho\dot{\mathbf{v}} = \operatorname{div}\boldsymbol{\sigma} + \rho\mathbf{f} \quad \text{in } S, \tag{2.20}$$

with

$$\sigma_{ij} = \sigma_{ji} \qquad \text{or} \qquad \boldsymbol{\sigma} = \boldsymbol{\sigma}^{\mathrm{T}}, \tag{2.21}$$

and

the natural boundary conditions for stresses,

$$\sigma_{ij} n_j = T_i \qquad \text{or} \qquad \boldsymbol{\sigma}.\,\mathbf{n} = \mathbf{T} \quad \text{on } \partial S. \tag{2.22}$$

Had we started with eqns (2.17) and (2.18) as primary statements (where the notion of stress does not appear), we should first assume Cauchy's principle that **T** depends on the normal to ∂S and not on the geometry of ∂S at a higher order; the celebrated *tetrahedron* argument of Cauchy would yield the relation (2.22) *linear* in **n**. Finally, localization of eqns (2.17) and (2.18) on account of this would produce eqns (2.20) and (2.21), respectively – (2.20) being obtained on account of (2.21). This is not to prove the superiority of (2.7) over the 'Newtonian' point of view as they mathematically are strictly equivalent and it is a matter of preference. However, it appears that the formulation (2.7), global as it is, and clearly using the notion of test function such as $\mathbf{v}^*$, is very close to the modern spirit of analysis (formulation of *weak* solutions) and in fact is *directly* the formulation needed for *numerical computations* (e.g., in finite elements) and for any mathematical proofs relying on *energy* considerations. In some sense, while

> Nature, and Nature's Laws lay hid in Night
> God said – Let Newton be! and All was *Light*
> (Alexander Pope),

d'Alembert's point of view is closer to modern applied mathematics (more on this in Maugin, 1980).

For later use note the following particular cases of statement (2.7):

for a *real* velocity field **v** (no asterisk), the statement (2.7) produces the global equation

$$\dot{K} = \mathscr{P}_{(\mathrm{i})} + \mathscr{P}_{(\mathrm{ext})}, \tag{2.23}$$

where

$$K = \int_S \tfrac{1}{2}\rho \mathbf{v}^2 \,\mathrm{d}v \tag{2.24}$$

and

$$\mathscr{P}_{(\mathrm{ext})} = \int_S \rho \mathbf{f} \cdot \mathbf{v} \,\mathrm{d}v + \int_{\partial S} \mathbf{T} \cdot \mathbf{v} \,\mathrm{d}s \tag{2.25}$$

are respectively, the total kinetic energy and the power developed by external (body and surface) forces;

in *quasi-statics* where inertial forces are neglected, and SPH (2.7) takes on the simplified form

$$\int_S \boldsymbol{\sigma} : \dot{\boldsymbol{\varepsilon}}^* \, dv = \int_S \rho \mathbf{f} \cdot \dot{\mathbf{u}}^* \, dv + \int_{\partial S} \mathbf{T} . \dot{\mathbf{u}}^* \, ds. \tag{2.26}$$

This can also be written in terms of infinitesimal variations (*virtual work*) as

$$\int_{S_0} \boldsymbol{\sigma} : \delta \boldsymbol{\varepsilon} \, dv = \int_{S_0} \rho \mathbf{f} \cdot \delta \mathbf{u} \, dv + \int_{S_0} \mathbf{T} \cdot \delta \mathbf{u} \, ds. \tag{2.27}$$

2.2.2 Principles (laws) of thermodynamics for continuous media

First law

Verbal statement *The total time rate of change in energy (kinetic and internal) balances the total supply of energy through external forces and heat.*

That is, mathematically,

$$\dot{K} + \dot{E} = \mathscr{P}_{(\text{ext})} + \dot{Q}, \tag{2.28}$$

where

$$E = \int_S \rho e \, dv \tag{2.29}$$

and

$$\dot{Q} = \int_S \rho h \, dv - \int_{\partial S} q_i n_i \, ds \tag{2.30}$$

are the total internal energy and total energy rate of heat supply, respectively. Here e is the internal energy per unit mass, h is the source heat per unit mass and $\mathbf{q} = \{q_i\}$ is the *heat flux* vector. The power $\mathscr{P}_{(\text{ext})}$ has already been defined by (2.25). As a matter of fact, the combination of eqns (2.28) and (2.23) provides the *energy theorem*:

$$\dot{E} + \mathscr{P}_{(\text{i})} = \dot{Q}. \tag{2.31}$$

The local form of this at every regular point of S is

$$\rho \dot{e} = \sigma_{ij} D_{ij} - \nabla \cdot \mathbf{q} + \rho h, \tag{2.32}$$

where (2.19) has been used.

Second law

Verbal statement *The total time rate of change in entropy is never less than the supply of entropy through heat.*

That is, mathematically (see Appendix 1)

$$\dot{\mathcal{N}} \geqslant \int_S \frac{1}{\theta} \rho h \, \mathrm{d}v - \int_{\partial S} \frac{1}{\theta} q_i n_i \, \mathrm{d}s \tag{2.33}$$

where θ is the thermodynamic temperature and

$$\mathcal{N} = \int_S \rho \eta \, \mathrm{d}v \tag{2.34}$$

is the total entropy if η is the entropy per unit mass.

The local form of the *inequality* (2.33) at every regular point of S is

$$\rho \dot{\eta} - \frac{1}{\theta} \rho h + \nabla \cdot (\mathbf{q}/\theta) \geqslant 0 \tag{2.35}$$

or

$$\rho \theta \dot{\eta} - \rho h + \nabla \cdot \mathbf{q} - \frac{1}{\theta} (\mathbf{q} \cdot \nabla) \theta \geqslant 0. \tag{2.36}$$

Introducing the (Helmholtz) *free energy* per unit mass Ψ defined by

$$\Psi = e - \eta \theta, \tag{2.37}$$

we have

$$\rho \theta \dot{\eta} = \rho \dot{e} - \rho (\dot{\Psi} + \eta \dot{\theta}). \tag{2.38}$$

The elimination of $\dot{e}$ between (2.32) and (2.38) with the help of (2.35) leads to the *Clausius–Duhem* inequality

$$-\rho (\dot{\Psi} + \eta \dot{\theta}) + \sigma_{ij} D_{ij} - \frac{1}{\theta} \mathbf{q} \cdot \nabla(\theta) \geqslant 0, \tag{2.39}$$

or, since ρ is constant in *SPH*, with

$$W = \rho \Psi, \qquad S = \rho \eta \tag{2.40}$$

defining the free energy and entropy per unit volume,

$$-(\dot{W} + S\dot{\theta}) + \sigma_{ij} \dot{\varepsilon}_{ij} + \theta \mathbf{q} \cdot \nabla \left(\frac{1}{\theta} \right) \geqslant 0. \tag{2.41}$$

Thermodynamic admissibility is the requirement that any thermodynamic process accompanied by a strain and by a spatial change of temperature satisfy the thermodynamic *constraint* represented by the inequality (2.41).

2.3 Using the Clausius–Duhem inequality

Let us suppose that the infinitesimal strain is the sum of an elastic strain ε^e and a plastic strain ε^p (elastoplasticity; eventually certain viscosity phenomena may also be present):

$$\boldsymbol{\varepsilon} = \boldsymbol{\varepsilon}^e + \boldsymbol{\varepsilon}^p. \tag{2.42}$$

Let $\boldsymbol{\alpha}$ be an n-vector, of components α_k, of so-called internal variables that we introduce in order to account for our ignorance about certain complex phenomena that happen at a microscopic level and that are macroscopically evident in certain irreversibilities. Variables $\boldsymbol{\varepsilon}$ and θ are *observable* variables. Examples of internal variables will be given later on. Let us suppose then that

$$W = W(\boldsymbol{\varepsilon}^e, \boldsymbol{\alpha}, \theta), \qquad \mathscr{E}(\boldsymbol{\varepsilon}^e, \boldsymbol{\alpha}, S) = W + S\theta. \tag{2.43}$$

The entropy and the temperature are defined as in thermostatics (local state axiom; see Appendix 1):

$$S = -\frac{\partial W}{\partial \theta}, \qquad \theta = \frac{\partial \mathscr{E}(\boldsymbol{\varepsilon}^e, \boldsymbol{\alpha}, S)}{\partial S}, \tag{2.44}$$

where $\mathscr{E} = \rho e$ is *convex* in the variables $\boldsymbol{\varepsilon}^e$ and $\boldsymbol{\alpha}$ and S whereas W is *concave* in the variable θ and *convex* in the variables $\boldsymbol{\varepsilon}^e$ and $\boldsymbol{\alpha}$ (the properties of convexity of W are the result of the Legendre transformation (2.37); see Appendix 1 for this notion). We define

$$\sigma^e = \frac{\partial W}{\partial \boldsymbol{\varepsilon}^e}, \qquad \mathbf{A} = -\frac{\partial W}{\partial \boldsymbol{\alpha}} \tag{2.45}$$

and

$$\boldsymbol{\sigma}^v = \boldsymbol{\sigma} - \boldsymbol{\sigma}^e. \tag{2.46}$$

If we take into account (2.44), (2.45) and (2.46) the inequality (2.41) is written

$$\phi = \boldsymbol{\sigma}^v : \dot{\boldsymbol{\varepsilon}}^e + \boldsymbol{\sigma} : \dot{\boldsymbol{\varepsilon}}^p + \mathbf{A} \cdot \dot{\boldsymbol{\alpha}} + \theta \mathbf{q} \cdot \nabla(1/\theta) \geqslant 0, \tag{2.47}$$

which is nothing but the *dissipation inequality* where $\boldsymbol{\sigma}^v$, the thermodynamic dual of $\dot{\boldsymbol{\varepsilon}}^e$, is said to be the *viscous stress*. We will set

$$\left.\begin{aligned} \phi_v &= \boldsymbol{\sigma}^v : \dot{\boldsymbol{\varepsilon}}^e, \\ \phi_p &= \boldsymbol{\sigma} : \dot{\boldsymbol{\varepsilon}}^p + \mathbf{A} \cdot \dot{\boldsymbol{\alpha}}, \\ \phi_q &= \theta \mathbf{q} \cdot \nabla(1/\theta). \end{aligned}\right\} \tag{2.48}$$

The dissipation ϕ_q is said to be the *thermal dissipation.* The dissipation

$$\phi_{\text{intr}} := \phi_v + \phi_p \tag{2.49}$$

is said to be the *intrinsic dissipation.* We verify (when $h = 0$; exercise) that

$$\phi_{\text{intr}} = \theta \dot{S} + \nabla \cdot \mathbf{q}. \tag{2.50}$$

A more restrictive form than (2.47) of the second law is formulated by accepting the condition that ϕ_q and ϕ_{intr} be separately nonnegative:

$$\phi_{\text{intr}} \geqslant 0, \qquad \phi_q \geqslant 0, \tag{2.51}$$

where the second condition corresponds to the intuitive notion of heat propagation from a hot point to a cold point. If $\phi_{\text{intr}} = 0$, then the dissipation is purely thermal.

The total intrinsic power Φ_{intr} dissipated in any volume V is introduced by

$$\Phi_{\text{intr}} = \int_V \phi_{\text{intr}} \, \mathrm{d}v \tag{2.52}$$

We can then verify (exercise) the following results:

$$\forall V, \Phi_{\text{intr}} = \mathscr{P}_{(\text{ext})} - \frac{\mathrm{d}}{\mathrm{d}t} \int_V \left(\tfrac{1}{2}\rho \mathbf{v}^2 + W\right) \mathrm{d}v - \int_V S\dot{\theta} \, \mathrm{d}v \geqslant 0, \tag{2.53}$$

and, if the transformation is *isothermal* ($\dot{\theta} = 0$),

$$\forall V, \Phi_{\text{intr}} = \mathscr{P}_{(\text{ext})} - \frac{\mathrm{d}}{\mathrm{d}t} \int_V \left[\tfrac{1}{2}\rho \mathbf{v}^2 + W(\cdot, \theta_0)\right] \mathrm{d}v \geqslant 0, \tag{2.54}$$

whereas, if the transformation is *isentropic*, then

$$\forall V, \Phi_{\text{intr}} = \mathscr{P}_{(\text{ext})} - \frac{\mathrm{d}}{\mathrm{d}t} \int_V \left[\tfrac{1}{2}\rho \mathbf{v}^2 + \mathscr{E}(\cdot, S_0)\right] \mathrm{d}v \geqslant 0. \tag{2.55}$$

Let it be noted that the splitting of dissipation has no meaning unless the functions $\theta \dot{S}$ and $\mathbf{q} \cdot \nabla(1/\theta)$ are well defined. In the case of propagation of shock waves where θ and S are discontinuous across the wave, we must come back to the complete expression of the dissipation.

Thermal equation (or equation of heat 'propagation') This equation arises from the definition (2.50), so that

$$\nabla \cdot \mathbf{q} + \theta \dot{S} - \phi_{\text{intr}} = 0, \tag{2.56}$$

where S is given by (2.44). If we accept the Fourier conduction law that satisfies $\phi_q \geqslant 0$,

$$\mathbf{q} = -k\,\nabla\theta, \qquad k \geqslant 0, \tag{2.57}$$

then (2.56) gives

$$\theta \dot{S} - \phi_{\text{intr}} = k\,\Delta\theta. \tag{2.58}$$

We have

$$\begin{aligned} \dot{S} = \frac{\partial}{\partial t}\left(-\frac{\partial W}{\partial \theta}\right) &= -\frac{\partial^2 W}{\partial \theta^2}\dot{\theta} - \frac{\partial^2 W}{\partial \theta \partial \boldsymbol{\varepsilon}^{\mathrm{e}}} : \dot{\boldsymbol{\varepsilon}}^{\mathrm{e}} - \frac{\partial^2 W}{\partial \theta \partial \boldsymbol{\alpha}} \cdot \dot{\boldsymbol{\alpha}} \\ &= \frac{\mathscr{C}}{\theta_0}\dot{\theta} - \frac{\partial \boldsymbol{\sigma}^{\mathrm{e}}}{\partial \theta} : \dot{\boldsymbol{\varepsilon}}^{\mathrm{e}} + \frac{\partial \mathbf{A}}{\partial \theta} \cdot \dot{\boldsymbol{\alpha}}, \end{aligned} \tag{2.59}$$

with (because of the concavity of W with respect to θ)

$$\mathscr{C} = -\theta_0 \frac{\partial^2 W}{\partial \theta^2} \geqslant 0. \tag{2.60}$$

Eqn (2.58) then takes the form

$$\mathscr{C}\dot{\theta} - \left(\phi_{\text{intr}} + \theta_0 \frac{\partial \boldsymbol{\sigma}^{\mathrm{e}}}{\partial \theta} : \dot{\boldsymbol{\varepsilon}}^{\mathrm{e}} - \theta_0 \frac{\partial \mathbf{A}}{\partial \theta} \cdot \dot{\boldsymbol{\alpha}}\right) = k\,\Delta\theta, \tag{2.61}$$

where the contribution between parentheses represents the thermomechanical coupling at the level of the thermal equation (this will be developed in Chapter 12). Without this coupling, eqn (2.61) is reduced to the classical *parabolic* heat equation.

2.4 Particular cases of solid media

By assuming the presence or absence of $\boldsymbol{\varepsilon}^{\mathrm{p}}$ and by selecting the n-vector $\boldsymbol{\alpha}$, we may reproduce large classes of solid media. Here are some examples.

2.4.1 Neither plastic strain nor associated phenomena

We set

$$\boldsymbol{\varepsilon}^{\mathrm{p}} = \mathbf{0}, \qquad \boldsymbol{\alpha} = \mathbf{0}. \tag{2.62}$$

There remain

$$\boldsymbol{\varepsilon}^{\mathrm{e}} = \boldsymbol{\varepsilon}, \qquad \boldsymbol{\sigma}^{\mathrm{e}} = \frac{\partial W}{\partial \boldsymbol{\varepsilon}}, \qquad \boldsymbol{\sigma}^{\mathrm{v}} = \boldsymbol{\sigma} - \boldsymbol{\sigma}^{\mathrm{e}}, \tag{2.63}$$

and the inequality of dissipation is reduced to

$$\phi = \boldsymbol{\sigma}^{\mathrm{v}} : \dot{\boldsymbol{\varepsilon}} + \theta \mathbf{q} \cdot \boldsymbol{\nabla}\left(\frac{1}{\theta}\right) \geqslant 0. \tag{2.64}$$

This is *thermoviscoelasticity*. In the particular case where $\boldsymbol{\sigma}^{\mathrm{v}}$ is linear in $\dot{\boldsymbol{\varepsilon}}$ and the thermal conduction law is of the Fourier type, we have the Kelvin–Voigt thermoviscoelasticity. In one dimension, discarding temperature effects, we thus have $\sigma^{\mathrm{v}} = \eta\dot{\varepsilon}$ while $\sigma^{\mathrm{e}} = E\varepsilon$.

From the last of eqns (2.63) there follows the model

$$\sigma = \sigma^{\mathrm{e}} + \sigma^{\mathrm{v}} = E\varepsilon + \eta\dot{\varepsilon} = E\left(\varepsilon + \frac{1}{\tau_{\varepsilon}}\dot{\varepsilon}\right). \tag{2.65}$$

This can be represented by the rheological model in Fig. 2.2, where a spring (Hookean) element H and a dashpot (Newtonian viscous body) element N are set in *parallel*.

2.4.2 Maxwell's viscoelasticity

We set

$$\boldsymbol{\varepsilon}^{\mathrm{p}} = \mathbf{0}, \qquad \alpha_{ij}\ (\text{tensor}) \neq 0. \tag{2.66}$$

If we ignore thermal conduction, the dissipation inequality is written as

$$\phi = \boldsymbol{\sigma}^{\mathrm{v}} : \dot{\boldsymbol{\varepsilon}} + \mathbf{A} : \dot{\boldsymbol{\alpha}} \geqslant 0. \tag{2.67}$$

Let us suppose that $\boldsymbol{\sigma}^{\mathrm{v}} = \mathbf{0}$, and that α_{ij} is $\varepsilon^{\mathrm{v}}_{ij}$, a deformation, said to be viscoelastic, and that W depends upon $\boldsymbol{\varepsilon}$ only through $\boldsymbol{\varepsilon} - \boldsymbol{\varepsilon}^{\mathrm{v}}$ as intermediate. We get then

Fig. 2.2. Kelvin–Voigt (KV) rheological model (KV = H ∥ N)

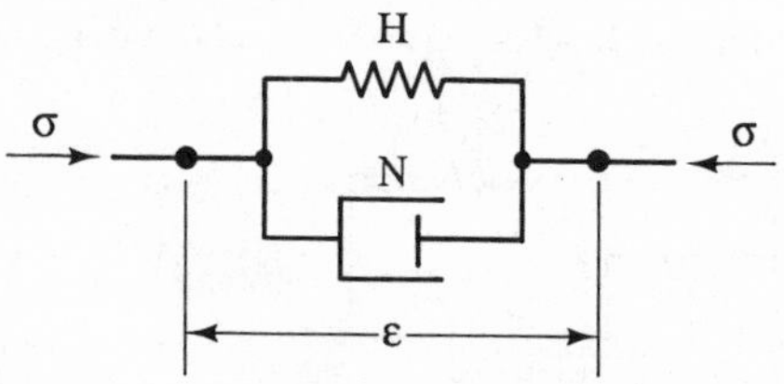

For example,

$$W = \tfrac{1}{2}E_{ijkl}(\varepsilon_{ij} - \varepsilon_{ij}^{\mathrm{v}})(\varepsilon_{kl} - \varepsilon_{kl}^{\mathrm{v}}). \tag{2.69}$$

This is certainly *Maxwell's viscoelasticity.* In fact, for a one-dimensional model we have $\sigma = E(\varepsilon - \varepsilon^{\mathrm{v}})$, which may also be written as $\dot{\sigma} = E(\dot{\varepsilon} - \dot{\varepsilon}^{\mathrm{v}})$. According to (2.68), we may take $\dot{\boldsymbol{\varepsilon}}^{\mathrm{v}} = (E\tau)^{-1}\boldsymbol{\sigma}$ where $\tau > 0$ is a relaxation time, in such a way as to obtain Maxwell's rheological model,

$$\dot{\sigma} + \frac{1}{\tau}\sigma = E\dot{\varepsilon}. \tag{2.70}$$

This is pictured in Fig. 2.3 by means of a spring (Hookean) element H and a Newtonian viscous-body element N set *in series*, so that stretches ε add up while the *same* σ is transmitted to the two elements. While the model (2.65) exhibits so-called *after-effects*, the Maxwell (M) model M = H—N exhibits *relaxation*.

It is not difficult to build generalizations of the simple models (2.65) and (2.70) that account for *several* characteristic times. For instance, the so-called *standard* (S) model very common in practical rheology is obtained by S = KV—H_0 = (H_1 ∥ N)—H_0 as in Fig. 2.4. Its governing equation is of the type

$$\sigma + \frac{1}{\tau_\sigma}\dot{\sigma} = E\left(\varepsilon + \frac{1}{\tau_\varepsilon}\dot{\varepsilon}\right). \tag{2.71}$$

Other models can be constructed by using the following 'recipes'.

the Poynting – Thomson (PT) model PT = H ∥ M;

the Lethersich (L) model L = N—KV;

Fig. 2.3. Maxwell (M) model (M = H—N)

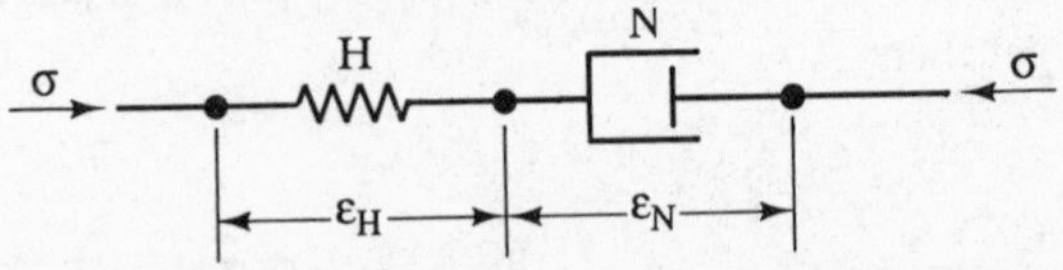

Fig. 2.4. Standard (S) model (S = (KV)—H_0)

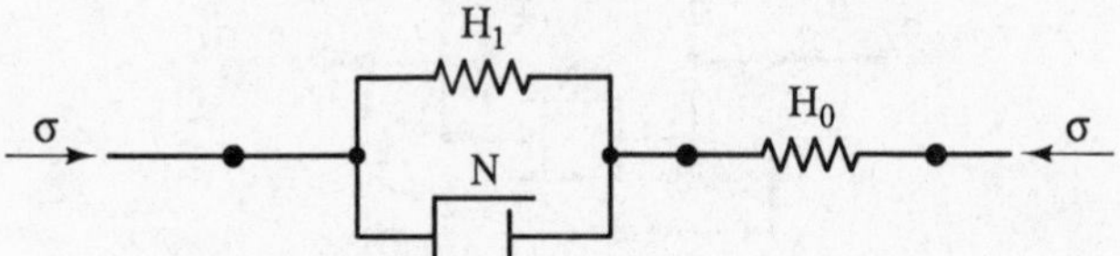

the Jeffreys (J) model J = N ∥ M;

the Burgers (Bu) model Bu = M—KV.

For example, the J-model used to model the Earth's crust has formula

$$\sigma + \frac{1}{\tau_\sigma}\dot{\sigma} = \eta\left(\dot{\varepsilon} + \frac{1}{\tau_J}\ddot{\varepsilon}\right). \tag{2.72}$$

In a general manner multi-element models built from H and N elements yield stress – strain relations of the general form

$$\alpha_0\sigma + \alpha_1\dot{\sigma} + \alpha_2\ddot{\sigma} + \cdots + \alpha_n\sigma^{(n)} = \beta_0\varepsilon + \beta_1\dot{\varepsilon} + \beta_2\ddot{\varepsilon} + \cdots + \beta_m\varepsilon^{(m)}, \tag{2.73}$$

where (n) denotes the nth-order time derivative, and it is to be noted that (m) is *not* necessarily equal to (n) – this is illustrated by the J-model. All these models are *viscoelastic* ones as they present characteristic times but no stress threshold.

2.4.3 Thermoelasticity

We set

$$\boldsymbol{\varepsilon}^{\mathrm{p}} = \mathbf{0}, \qquad \boldsymbol{\alpha} = \mathbf{0}. \tag{2.74}$$

Then

$$\boldsymbol{\sigma} = \boldsymbol{\sigma}^{\mathrm{e}} = \frac{\partial W}{\partial \boldsymbol{\varepsilon}}(\boldsymbol{\varepsilon}, \theta), \qquad \phi = \phi_q \geqslant 0, \tag{2.75}$$

with $\tilde{\theta} = \theta - \theta_0$, $|\tilde{\theta}| \ll \theta_0$, with

$$\left.\begin{aligned} W &= \frac{1}{2}E_{ijkl}\varepsilon_{ij}\varepsilon_{kl} - \tilde{\theta}v_{ij}\varepsilon_{ij} - \frac{1}{2\theta_0}\mathscr{C}\tilde{\theta}^2, \\ q_i &= -k_{ij}\theta_{,j}, \end{aligned}\right\} \tag{2.76}$$

This is *Duhamel's thermoelasticity* of anisotropic linear media where the v_{ij}'s are the thermoelastic coupling coefficients. The reader is referred to the many monographs on thermoelasticity for a deeper insight into this theory and its applications (e.g., Nowacki, 1986).

2.4.4 The difference between viscous and plastic phenomena

We consider the split hypothesis (2.51) and in particular the inequality

$$\phi_{\mathrm{intr}} = \phi_{\mathrm{v}} + \phi_{\mathrm{p}} \geqslant 0, \tag{2.77}$$

where

$$\phi_{\mathrm{v}} = \boldsymbol{\sigma}^{\mathrm{v}} : \dot{\boldsymbol{\varepsilon}}^{\mathrm{e}}, \qquad \phi_{\mathrm{p}} = \boldsymbol{\sigma} : \dot{\boldsymbol{\varepsilon}}^{\mathrm{p}} + \mathbf{A}\cdot\dot{\boldsymbol{\alpha}}, \tag{2.78}$$

where the variables $\boldsymbol{\alpha}$, contrary to the cases (2.66)–(2.67), also represent plastic-type phenomena.

In the *viscoelastic* cases, the viscosity phenomena imply the influence of a *characteristic time* through a dependence on the strain rate. In the elastoplastic case for metals at room temperature, there is no such influence. We admit then that $\boldsymbol{\sigma}^{v} = \mathbf{0}$ and $\boldsymbol{\alpha}$'s are of the plastic type. Consequently ϕ_{intr} is reduced to

$$\phi_{\text{intr}} = \boldsymbol{\sigma} : \dot{\boldsymbol{\varepsilon}}^{p} + \mathbf{A} \cdot \dot{\boldsymbol{\alpha}} \geqslant 0. \tag{2.79}$$

We can rewrite this in the form of a scalar product in a space of appropriate dimension (e.g., $6 + n$ dimensions) as

$$\phi(\dot{\mathbf{X}}) := \mathbf{Y} \cdot \dot{\mathbf{X}} \geqslant 0, \tag{2.80}$$

where

$$\mathbf{Y} = (\boldsymbol{\sigma}, \mathbf{A}), \qquad \mathbf{X} = (\boldsymbol{\varepsilon}^{p}, \boldsymbol{\alpha}). \tag{2.81}$$

Owing to the existence of a threshold, the generalized forces $\mathbf{Y}$ are not unbounded; they are restrained inside a *convex* domain C, which is called the elasticity domain and contains the origin O. Mathematically speaking, dissipation is expressed then in terms of $\dot{\mathbf{X}}$ in the form (see Appendix 2)

$$\phi(\dot{\mathbf{X}}) = \operatorname*{Sup}_{\mathbf{Y}^* \in C} \mathbf{Y}^* \cdot \dot{\mathbf{X}} \geqslant 0. \tag{2.82}$$

ϕ then is a *positively homogeneous* function of degree 1 of the velocity $\dot{\mathbf{X}}$, that is, $\phi(k\dot{\mathbf{X}}) = k\phi(\dot{\mathbf{X}})$ if $k \geqslant 0$.

We notice that this last behaviour is completely different from the viscoelastic case where ϕ, although convex in $\dot{\mathbf{X}}$, is typically *homogeneous* of degree 2 in $\dot{\mathbf{X}}$ (i.e. quadratic).

The dissipation definition (2.82) is equivalent to the inequality

$$(\mathbf{Y} - \mathbf{Y}^*) \cdot \dot{\mathbf{X}} \geqslant 0, \qquad \forall \mathbf{Y}^* \in C, \tag{2.83}$$

called the Hill – Mandel *maximal-dissipation* principle (Hill, 1948), expressing the following *normality* property: velocity $\dot{\mathbf{X}}$ is a vector belonging to the cone of the outward normals in Y, at the elasticity domain C, or

$$\dot{\mathbf{X}} \in N_C(\mathbf{Y}), \tag{2.84}$$

which may be considered as an evolution law for the variables $\mathbf{X}$ (illustration in Fig. 2.5)

If $\mathbf{Y}$ is interior to C, $\dot{\mathbf{X}} = \mathbf{0}$ and $\phi_{\text{intr}} = 0$, then the response is purely reversible, that is, purely elastic. The parameters $\mathbf{X}$ cannot evolve unless the state of generalized forces $\mathbf{Y}$ attains a threshold corresponding to the boundary of the convex domain C. Condition (2.83) corresponds to the fact that the 'angle' between $\mathbf{YY}^*$ and $\dot{\mathbf{X}}$ is obtuse ($>\pi/2$) We may notice that the function of $\dot{\mathbf{X}}$, $\mathscr{D}(\dot{\mathbf{X}})$ such that $\mathscr{D} = \phi$, is differentiable for $\dot{\mathbf{X}} \neq 0$ and that we simply obtain (Euler's identity)

$$\mathbf{Y} = \frac{\partial \mathscr{D}}{\partial \dot{\mathbf{X}}}. \tag{2.85}$$

The *dissipation potential* $\mathscr{D}$, if associated to the elastoplastic model, is a convex function, positively homogeneous of degree 1. The notion of a convex-function *subgradient* may be introduced by

$$\partial\mathscr{D}(\dot{\mathbf{X}}) = \{\mathbf{Y} | \mathscr{D}(\dot{\mathbf{X}}^*) \geqslant \mathscr{D}(\dot{\mathbf{X}}) + \mathbf{Y}\cdot(\dot{\mathbf{X}}^* - \dot{\mathbf{X}}), \qquad \forall \dot{\mathbf{X}}^*\} \tag{2.85}$$

in such a manner that, *even for* $\dot{\mathbf{X}} = \mathbf{0}$ we might write

$$\mathbf{Y} \in \partial\mathscr{D}(\dot{\mathbf{X}}), \tag{2.87}$$

which extends the notion of gradient to nondifferentiable convex functions.

We may observe that the mathematical elastoplastic model that we very formally and summarily developed above – we shall re-examine this in Chapter 3 – satisfies the working hypotheses (H1)–(H7) in Chapter 1, provided that we take into consideration only the deviator $\mathbf{e}$ of $\boldsymbol{\varepsilon}$. We may also notice how important the notions related to *convexity* are. Before going in greater detail into the elastoplasticity theory, with or without hardening, any reader somewhat uncertain about his knowledge of convexity should look up Appendix 2.

Fig. 2.5. Convex of plasticity. (a) General case (smooth yield surface). (b) Plasticity without hardening (yield surface with noncontinuous tangent plane).

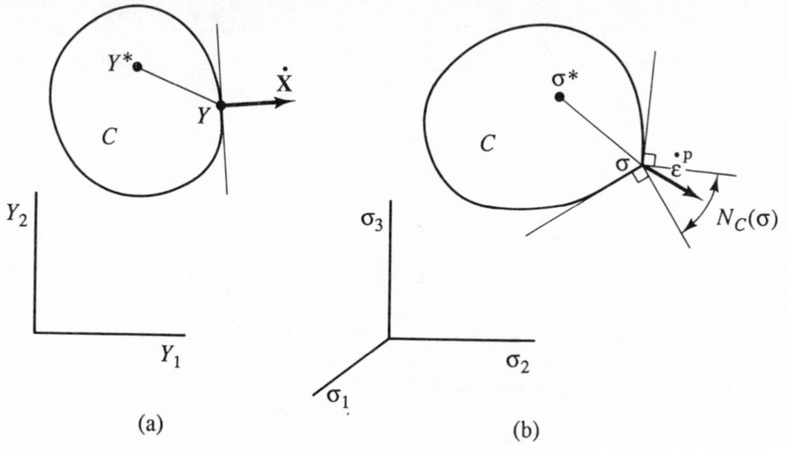

To conclude this chapter we mention the existence of *viscoplastic* rheological models. We already know (eqn (1.8), Fig. 1.8) that a Saint-Venant body (SV) or dry-friction element adequately represents the notion of elastic limit or plasticity threshold. The model of one-dimensional perfect elastoplasticity illustrated in Fig. 1.8 is the so-called *Prandtl* (Pr) *model.* As no N element is present, the Pr-model does not have any characteristic time. The *Bingham* (Bi) and *Schwedoff* (SW) *models* with recipes

$$\text{Bi} = (\text{N} \parallel \text{SV})\text{—H}, \qquad \text{SW} = (\text{M} \parallel \text{SV})\text{—H}$$

and illustrated in Fig. 2.6 *do* include N elements and SV elements simultaneously and, therefore, are *viscoplastic* models. To be complete we should note that rheological models used in the description of the behaviour of *soils* also admit dashpots with viscous sinking through perforated pistons (e.g., Terzaghi, Taylor, Tan, Gibson–Lo models; see Vyalov, 1986, p. 229), and there even exist models with *structural-change* elements (e.g., the so-called Budin model.)

The accompanying problems have for object to introduce the reader to examples of plasticity dissipation and flow rules (more on this in Chapter 3 and the accompanying problems).

Problems for Chapter 2

2.1. Local intrinsic dissipation

Check that the intrinsic dissipation per unit volume in a general continuum is given by eqn. (2.50).

Fig. 2.6. Bingham (a) and Schwedoff (b) viscoplastic rheological models

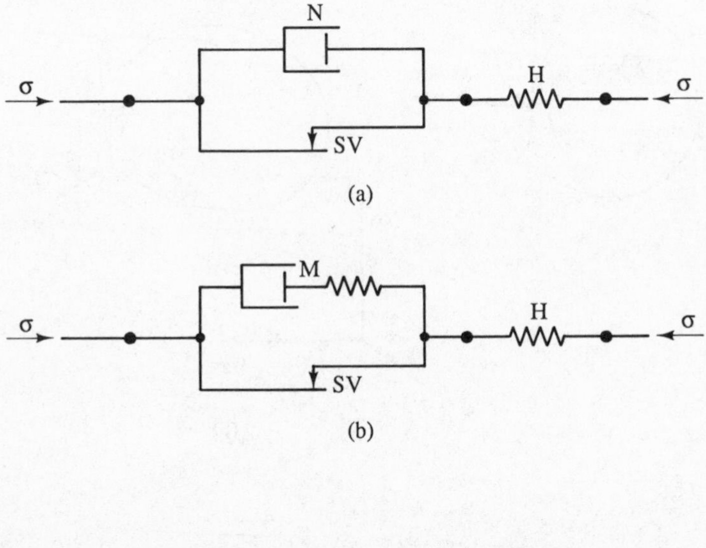

2.2. Global intrinsic dissipation

Prove that the total intrinsic dissipation of a material body occupying the volume V of $\mathbb{E}^3$ at time t is given by eqn (2.54) when the thermodynamic evolution is *isothermal*, and by eqn (2.55) when the transformation is *isentropic* (find first the general expression in the absence of these simplifying hypotheses).

2.3. Surface of constant dissipation

The plastic dissipation is given by $\phi = \boldsymbol{\sigma} : \dot{\boldsymbol{\varepsilon}}^{\mathrm{p}}$ with $\boldsymbol{\sigma} = \partial\phi/\partial\dot{\boldsymbol{\varepsilon}}^{\mathrm{p}}$. The surfaces $\phi = \text{const.}$ are called surfaces of constant dissipation. If $f(\boldsymbol{\sigma}) = 0$ is convex in stress space, show that $\phi = \text{const.}$ is convex in strain-rate space. Determine the polygon of constant dissipation in the plane-stress case of a material obeying Tresca's criterion.

2.4. Flow rule for a rigid–plastic material

(a) Prove that for isotropic bodies the most general form of an isotropic tensor-valued stress tensor $\boldsymbol{\sigma}(\mathbf{D})$, where $\mathbf{D}$ is the strain-rate tensor, reads

$$\boldsymbol{\sigma} = \alpha_0 \mathbf{I} + \alpha_1 \mathbf{D} + \alpha_2 \mathbf{D}^2, \tag{1}$$

where α_0, α_1 and α_2 are scalar functions of the fundamental invariants of $\mathbf{D}$

(b) If (1) is homogeneous of degree zero (strain-rate-independent theory), then show that

$$\frac{\partial \boldsymbol{\sigma}}{\partial \mathbf{D}} : \mathbf{D} = \mathbf{0} \qquad \text{for } \partial\boldsymbol{\sigma}/\partial\mathbf{D} \neq \mathbf{0}. \tag{2}$$

Eq (1) and (2) define a plastic body. *Why* does there exist then a yield condition $f(\boldsymbol{\sigma}) = 0$?

(c) Show that $\mathbf{D} = \dot{\lambda}\mathbf{G}(\boldsymbol{\sigma})$, where $\dot{\lambda}$ is a scalar multiplier. This is the plasticity flow rule.

2.5. Rigid–plastic materials (continued)

Let $\mathbf{s}$ and $\hat{\mathbf{D}}$ be the deviatoric parts of $\boldsymbol{\sigma}$ and $\mathbf{D}$, respectively. Let s_α and $\hat{D}_\alpha$, $\alpha = \mathrm{I, II, III}$, be the fundamental invariants of $\mathbf{s}$ and $\mathbf{D}$ defined by $s_\alpha = \operatorname{tr} \mathbf{s}^\alpha$, etc. We set

$$s \equiv (\hat{D}_{\mathrm{III}})^{1/3}/(\hat{D}_{\mathrm{II}})^{1/2}, \qquad t \equiv D_{\mathrm{I}}/\hat{D}_{\mathrm{II}}. \tag{1}$$

Then prove that the plasticity flow rule of 2.4 can be written as

$$\hat{\mathbf{D}} = \dot{\lambda}\left\{\mathbf{s} - \frac{6C}{6B^2 - C^2}\left[\mathbf{s}^2 - \frac{1}{3}s_{\mathrm{III}}\mathbf{I}\right]\right\} \tag{2}$$

with

$$\dot{\lambda} = \sqrt{\hat{D}_{\mathrm{II}}}\Big/\left[B - \frac{2C^2}{6B^2 - C^2}(B - Cs^3)\right],$$

where C and B are functions of s and t only.

2.6. Lévy–Mises laws

If $C = 0$, show that (2) of 2.5 reduces to the Lévy–Mises law

$$\hat{\mathbf{D}} = \dot{\lambda}\mathbf{s} = \frac{\sqrt{\hat{D}_{\mathrm{II}}}}{B}\mathbf{s} \tag{1}$$

with, also,

$$\dot{\lambda} = \frac{\mathrm{tr}(\mathbf{s}\hat{\mathbf{D}})}{s_{\mathrm{II}}} > 0. \tag{2}$$

2.7. Homogeneous functions

Show that the functions

$$f(x, y) = \frac{\alpha x + \beta y}{(Ax^2 + Bxy)^{1/2}}, \; g(x, y) = Ax^2 + Bxy + Cy^2$$

are homogeneous of degree zero and 2, respectively.

2.8. Simplified flow law

Redo 2.4 but starting from

$$\boldsymbol{\sigma} = \alpha_0 \mathbf{I} + \boldsymbol{\alpha}_1 \mathbf{D} \tag{1}$$

by neglecting second-order tensor terms in $\mathbf{D}$. Then show that

$$\sqrt{s_{\mathrm{II}}} \equiv s = \phi(\hat{D}_{\mathrm{II}}, \hat{D}_{\mathrm{III}})\hat{D}_{\mathrm{II}} > 0 \tag{2}$$

satisfies the differential equation

$$\frac{\mathrm{d}\phi}{\mathrm{d}x} + \phi = 0, \qquad x = \log \hat{D}_{\mathrm{II}}, \qquad y = \log \hat{D}_{\mathrm{III}}/\hat{D}_{\mathrm{II}}, \tag{3}$$

of which the solution is $\phi = g(y)\exp(-x)$. Deduce from this that

$$\hat{\mathbf{D}} = \dot{\lambda}\mathbf{s}, \qquad \dot{\lambda} = \frac{\sqrt{\hat{D}_{\mathrm{II}}}}{F(s)}. \tag{4}$$

2.9. Flow rule for a beam of rectangular cross-section

Let L be the *rate* elongation and K the rate of curvature in Problem **1.9.** Show that $\dot{\varepsilon} = L + zK$, and that the flow rule reads

$$(L, K) = \nu\left(\frac{\partial \phi}{\partial N}, \frac{\partial \phi}{\partial M}\right), \qquad \nu \geqslant 0$$

where ϕ is given by eqn (2) in Problem **1.9**.

2.10. Flow rule for Huber–Mises plates

Complete the results of Problem **1.10** by establishing the flow rule

$$(\lambda_{\alpha\beta}, \chi_{\alpha\beta}) = \nu\left(\frac{\partial \varphi}{\partial n_{\alpha\beta}}, \frac{\partial \varphi}{\partial m_{\alpha\beta}}\right), \nu \geqslant 0,$$

as, if normals to the middle surface are assumed to remain straight and normal to this surface,

$$\dot{\varepsilon}_{\alpha\beta} = \lambda_{\alpha\beta} + 2\zeta\chi_{\alpha_\beta},$$

where $\zeta = z/h$.

2.11. Tresca material

Show that, if in a Tresca material the principal stress directions are parallel to the spatial axes, then we can consider the dissipation potential

$$\mathscr{D} = \frac{k}{2}\left(\sum_{\alpha} |D_{\alpha}^{\mathrm{p}}|\right),$$

where the D_{α}^{p} satisfy jointly the incompressibility condition $\sum_{\alpha} D_{\alpha}^{\mathrm{p}} = 0$. [*Note*: The Tresca criterion is to be used only if one knows the principal directions.]

2.12. Rheological models

Write the differential (constitutive) equations of the Bingham and Schwedoff rheological models illustrated in Fig. 2.6.

2.13. Bingham viscoplasticity

Make a literature search and write a short report on Bingham's model of viscoplastic 'fluids'. By way of illustration, treat Poiseuille's flow problem in a cylindrical pipe for such a fluid in terms of the pressure gradient.

2.14. Complex rheological models

Make a literature search and write a ten-page report on complex rheological models (used to model soils – see e.g. Vyalov, 1986) and their applications in soil mechanics including porosity and structural changes.

3

Small-strain elastoplasticity

The object of the chapter In this chapter we examine more explicitly the thermomechanical bases of the model presented in outline at the end of Chapter 2 and we specify the incremental nature of the laws of plasticity as well as the most usual forms that these laws take, emphasizing in particular the Prandtl–Reuss relations as they appear in practical problems, whether in variational form or not.

3.1 Reminder of the thermomechanical formulation

Let us recall the dissipation inequality (2.80) or

$$\phi(\dot{\mathbf{X}}) = \mathbf{Y} \cdot \dot{\mathbf{X}} \geqslant 0, \tag{3.1}$$

where

$$\mathbf{Y} = (\boldsymbol{\sigma}, \mathbf{A}), \qquad \mathbf{X} = (\boldsymbol{\varepsilon}^{\mathrm{p}}, \boldsymbol{\alpha}) \tag{3.2}$$

and

$$\boldsymbol{\sigma} = \partial W / \partial \boldsymbol{\varepsilon}^{\mathrm{e}}, \qquad \mathbf{A} = -\partial W / \partial \boldsymbol{\alpha}, \qquad S = -\partial W / \partial \theta. \tag{3.3}$$

Eqs (3.3) are *state equations* or *state laws*, whereas (3.1) is the dissipation inequality which will permit us to formulate the *complementary laws* that govern dissipative processes (plasticity, hardening). The latter belong to a class of mechanisms that we are going to specify. The relative generality of these laws allows us to present in a unified way a number of dissipative phenomena (other than those in plasticity, e.g., damage, fracture).

3.1.1 The notion of normal dissipative mechanism

We accept that the scalar product $\mathbf{Y} \cdot \dot{\mathbf{X}}$ which is present in (3.1) can be expressed in the form of a function $\mathscr{D}$, which depends only upon the values of the variables $\dot{\mathbf{X}}$ and upon the thermodynamic state variables at the time considered. We accept then that, for every $\dot{\mathbf{X}}$, this function $\mathscr{D}$ is defined, and that it is continuous, nonnegative, convex, quasi-homogeneous and zero for $\dot{\mathbf{X}} = \mathbf{0}$. We say then that such a function $\mathscr{D}(\dot{\mathbf{X}})$ determines a *normal dissipative mechanism.*

Actually, this quasi-homogeneity corresponds to the fact that $\mathscr{D}(\dot{\mathbf{X}})$ may be defined by (see Germain, 1973, p. 150)

$$\mathscr{D}(\dot{\mathbf{X}}) = b(\lambda), \qquad \lambda = h(\dot{\mathbf{X}}), \qquad b(0) = 0, \qquad b(1) = 1. \tag{3.4}$$

Since $\mathscr{D}(\dot{\mathbf{X}})$ is convex, $b(\lambda)$ is itself convex and $\lambda^{-1}\, b(\lambda)$ is a nonnegative and nondecreasing function of λ. We suppose that it is increasing, zero for $\lambda = 0$, and that it takes arbitarily large values when λ increases indefinitely.

As a matter of fact, a *normal dissipative mechanism* is defined by a convex and quasi-homogeneous dissipation function $\mathscr{D}(\dot{\mathbf{X}})$ and by the *orthogonality axiom* which specifies the property which must be satisfied by any vector $\mathbf{Y}$ associated to a given vector $\dot{\mathbf{X}}$, through the application of the complementary laws. We may then state the following.

Orthogonality axiom *Any vector* $\mathbf{Y}$ *associated to a given vector* $\dot{\mathbf{X}}$ *is a normal vector, exterior to the domain* Δ *whose boundary* $\partial\Delta$ *passes through* $\dot{\mathbf{X}}$

The vector $\mathbf{Y}$ is necessarily collinear to **grad** $\mathscr{D}$ or **grad** h. We may then write

$$\mathbf{Y} = c\,\mathbf{grad}\, h, \qquad Y_\alpha = c\frac{\partial h}{\partial \dot{X}_\alpha}, \tag{3.5}$$

where c is a scalar. If we calculate the product $\mathbf{Y}\cdot\dot{\mathbf{X}}$ we have then

$$\mathbf{Y}\cdot\dot{\mathbf{X}} = ch(\dot{\mathbf{X}}) = c\lambda = \mathscr{D}(\dot{\mathbf{X}}) = b(\lambda) \tag{3.6}$$

because $\mathbf{Y}\cdot\dot{\mathbf{X}} = \mathscr{D}(\dot{\mathbf{X}})$, and h is homogeneous of order 1. We may then write

$$\mathbf{Y} = \frac{b(\lambda)}{\lambda}\,\mathbf{grad}\, h. \tag{3.7}$$

But we can also write

$$\mathbf{grad}\,\mathscr{D} = (\mathbf{grad}\, h)\frac{\mathrm{d}b}{\mathrm{d}\lambda}$$

and

$$\dot{\mathbf{X}}\cdot\mathbf{grad}\,\mathscr{D} = \lambda\frac{\mathrm{d}b}{\mathrm{d}\lambda} = f(\mathscr{D}), \tag{3.8}$$

where $f(\mathscr{D})$ is a function of $\mathscr{D}$ at least equal to $\mathscr{D}$, since $\lambda^{-1}b(\lambda) \leqslant \mathrm{d}b/\mathrm{d}\lambda$, owing to the convexity properties of the function $b(\lambda)$. Eqn (3.7) then may also be written as

$$\mathbf{Y} = \frac{\mathscr{D}}{f(\mathscr{D})}\,\mathbf{grad}\,\mathscr{D} = \frac{\mathscr{D}}{(\dot{\mathbf{X}}\cdot\mathbf{grad}\,\mathscr{D})}\,\mathbf{grad}\,\mathscr{D}. \tag{3.9}$$

3.1.2 Dissipation pseudo-potential

Let $\varphi(\dot{\mathbf{X}})$ be the quasi-homogeneous function, called the dissipation pseudo-potential, defined by

$$\lambda = h(\dot{\mathbf{X}}), \qquad \varphi(\dot{\mathbf{X}}) = a(\lambda) = \int_0^\lambda b(s)\frac{\mathrm{d}s}{s} = \int_0^1 b(\lambda t)\frac{\mathrm{d}t}{t}, \tag{3.10}$$

or, alternatively,

$$\varphi(\dot{\mathbf{X}}) = \int_0^1 \mathscr{D}(t,\dot{\mathbf{X}})\frac{dt}{t}. \tag{3.11}$$

We verify that $\varphi(\dot{\mathbf{X}})$ is convex and we note that

$$\mathbf{grad}\,\varphi = \frac{\mathrm{d}a}{\mathrm{d}\lambda}\,\mathbf{grad}\,h = \frac{b(\lambda)}{\lambda}\,\mathbf{grad}\,h. \tag{3.12}$$

Consequently, (3.7) may be written as

$$\mathbf{Y} = \mathbf{grad}\,\varphi, \tag{3.13}$$

which justifies the name of φ. We observe then the following.

(a) If $\mathscr{D}(\dot{\mathbf{X}})$ does not admit any gradient at $\dot{\mathbf{X}}$, then we introduce the notion of subgradient and (3.13) is replaced by (see Appendix 2)

$$\mathbf{Y} \in \partial\varphi, \tag{3.14}$$

(b) If $\mathscr{D}(\dot{\mathbf{X}})$ is *positively homogeneous* of order p (particular case of quasi-homogeneous function) we have

$$b(\lambda) = \lambda^p, \qquad a(\lambda) = \frac{\lambda^p}{p}, \qquad f(\mathscr{D}) = p\mathscr{D}. \tag{3.15}$$

In this case, if $\mathscr{D}(\dot{\mathbf{X}})$ is continuously differentiable, we have then

$$\mathbf{Y} = \frac{1}{p}\,\mathbf{grad}\,\mathscr{D}. \tag{3.16}$$

(c) If the pseudo-potential φ is introduced, then the famous Onsager symmetry relations of the theory of irreversible processes may be generalized to the nonlinear cases too. In fact, from (3.13) we may deduce that

$$Y_\alpha = \frac{\partial\varphi}{\partial\dot{\mathbf{X}}_\alpha}, \qquad Y_\beta = \frac{\partial\varphi}{\partial\dot{\mathbf{X}}_\beta}$$

in such a way that (Ziegler, 1968, Edelen, 1973)

$$\frac{\partial Y_\alpha}{\partial \dot{\mathbf{X}}_\beta} = \frac{\partial Y_\beta}{\partial \dot{\mathbf{X}}_\alpha}. \tag{3.17}$$

Dual properties The question we may ask now is the following: for a given $\mathbf{Y}$, try to find which are the $\dot{\mathbf{X}}$ such that that $\mathbf{Y}$ might be associated to one among them. Given the properties of convexity it is sufficient to introduce the notion of the *Legendre–Fenchel* transform of φ by (see Appendix 2)

$$\varphi^*(\mathbf{Y}) = \operatorname*{Sup}_{\dot{\mathbf{X}}} [\mathbf{Y} \cdot \dot{\mathbf{X}} - \varphi(\dot{\mathbf{X}})]. \tag{3.18}$$

We get immediately

$$\varphi^*(\mathbf{Z}) - \varphi(\mathbf{Y}) - \dot{\mathbf{X}} \cdot (\mathbf{Z} - \mathbf{Y}) \geqslant 0. \tag{3.19}$$

Consequently (see the definition (2.86)), $\dot{\mathbf{X}}$ is a subgradient of φ^* at the point $\mathbf{Y}$:

$$\dot{\mathbf{X}} \in \partial\varphi^*, \tag{3.20}$$

or therefore, if φ^* is continuously differentiable,

$$\dot{\mathbf{X}} = \frac{\partial \varphi^*}{\partial \mathbf{Y}}. \tag{3.21}$$

The properties of convexity and of *homogeneity* are preserved in the transformation (3.18). Eqn (3.20) is the dual expression to (3.14) and φ^* may be said to be the *dual dissipation pseudopotential.* Eqn (3.20) or (3.21) is an *evolution equation*, generally nonlinear, for $\dot{\mathbf{X}}$.

3.1.3 Positively homogeneous dissipation functions of degree 1

Let us suppose that $b(\lambda)/\lambda$ is constant and equal to 1. $\varphi(\dot{\mathbf{X}}) = \mathscr{D}(\dot{\mathbf{X}})$, then, is a positively homogeneous function of degree 1. If $\dot{\mathbf{X}}_\alpha$ are the orthonormal coordinates in a Euclidean space $\mathbb{E}_m$ and the Y_α are the corresponding coordinates in the dual space $\mathbb{E}_m^*$, then we have (Germain, 1973)

$$\left.\begin{aligned} &\dot{\mathbf{X}} = \mathbf{O} \Rightarrow \mathbf{Y} \in \mathscr{S} + \partial\mathscr{S}, \\ &\dot{\mathbf{X}} \neq \mathbf{O} \Rightarrow \mathbf{Y} = \mathbf{grad}\,\mathscr{D}, \quad \mathbf{Y} \in \partial\varphi, \end{aligned}\right\} \tag{3.22}$$

where $\partial\mathscr{S}$ is a *convex closed surface* surrounding the origin in the space $\mathbb{E}_m^*$. If we examine now the dual function $\varphi^*(\mathbf{Y})$ defined by (3.18), we notice that it is zero if $\mathbf{Y} \in \mathscr{S} + \partial\mathscr{S}$, and infinite if $\mathbf{Y}$ is in the interior of $\mathscr{S}$. This function φ^* is only lower semi-continuous (l.s.c., see Appendix 2). According to Appendix 2, φ^* is nothing but the indicator function of the convex

$\mathscr{S}$, or

$$\varphi^* = I_{\mathscr{S}}(\mathbf{Y}). \tag{3.23}$$

If $\partial\mathscr{S}$ is a surface with a continuous tangent plane and if $\varphi(\mathbf{Y})$ is a function continuously differentiable with respect to $\mathbf{Y}$, zero on $\partial\mathscr{S}$ and negative in $\mathscr{S}$, then we can write

$$\left.\begin{array}{l} \mathbf{Y} \in \mathscr{S}, \quad \dot{\mathbf{X}} = 0, \\ \mathbf{Y} \in \partial\mathscr{S}, \quad \dot{\mathbf{X}} = \dot{\lambda}\,\mathbf{grad}\, f, \quad \dot{\mathbf{X}}_\alpha = \dot{\lambda}\dfrac{\partial f}{\partial Y_\alpha}, \quad \dot{\lambda} \geqslant 0. \end{array}\right\} \tag{3.24}$$

If φ^* and f are irregular, we introduce the notions of *cone of outward normals* and of *subgradient*. The scalar coefficient $\dot{\lambda}$ introduced has nothing to do with the λ used, e.g., in eqns (3.10)–(3.15).

We see that the correspondence established between $\mathbf{Y}$ and $\dot{\mathbf{X}}$ is clearly *singular*. But it is directly applicable in the case of *plasticity* where the flow surface $f = 0$ may be assimilated to the plastic potential (so-called 'associated' plasticity) and $\dot{\lambda}$ to the plastic multiplier. We see that in plasticity the singularity of the representation (3.24) of the complementary laws is the consequence of the fact that the dissipation is homogeneous of degree 1, as a result of the experimental fact that plasticity depends hardly at all upon strain rate. If there is *viscosity*, the complementary law is much more regular. We may also observe that in plasticity the *dissipation function* and the *plastic potential* are proportional. Finally, the plastic multiplier $\dot{\lambda}$ should also be determined.

The plastic multiplier is denoted here by $\dot{\lambda}$ (and not λ as is the case in other works; this is to show more vividly that, contrary to what happens in viscoelasticity and viscoplasticity, here the evolution law (3.24) *does not* imply the intervention of a characteristic time.

3.2. Perfect plasticity equations in SPH

Perfect plasticity in SPH is characterized by three statements of existence, as follows.

(i) the existence of a strain energy $W(\boldsymbol{\varepsilon}^e)$ such that

$$\boldsymbol{\sigma} = \frac{\partial W}{\partial \boldsymbol{\varepsilon}^e}. \tag{3.25}$$

This corresponds to the elastic part of the mechanical behaviour.

(ii) the existence of a convex elasticity domain C, in the space of (deviatoric) stresses $\boldsymbol{\sigma}$ whence the criterion of plasticity

$$f(\boldsymbol{\sigma}) = 0 \Rightarrow C = \{\boldsymbol{\sigma} | f(\boldsymbol{\sigma}) \leqslant 0\}. \tag{3.26}$$

(iii) The existence of a *normality law*

$$\dot{\boldsymbol{\varepsilon}}^{\mathrm{p}} \in N_C(\boldsymbol{\sigma}) \qquad \text{or} \qquad \dot{\boldsymbol{\varepsilon}}^{\mathrm{p}} = \dot{\lambda}\frac{\partial f}{\partial \boldsymbol{\sigma}}. \tag{3.27}$$

The latter can also be written in an equivalent variational form, the so-called *Hill–Mandel maximal-dissipation principle*:

$$(\boldsymbol{\sigma} - \boldsymbol{\sigma}^*) : \dot{\boldsymbol{\varepsilon}}^{\mathrm{p}} \geqslant 0, \qquad \forall \boldsymbol{\sigma}^* \in C \tag{3.28}$$

(see eqns (2.82) to (2.84) and Fig. 2.5).

Example Let us suppose the *following linear anisotropic* elastic behaviour:

$$W = \tfrac{1}{2}\boldsymbol{\varepsilon}^{\mathrm{e}} : \mathbf{E} : \boldsymbol{\varepsilon}^{\mathrm{e}} = \tfrac{1}{2}\varepsilon^{\mathrm{e}}_{ij} E_{ijkl} \varepsilon^{\mathrm{e}}_{kl}, \tag{3.29}$$

where E_{ijkl} is an elasticity (or rigidity) modulus tensor, satisfying the symmetries

$$E_{ijkl} = E_{jikl} = E_{ijlk} = E_{klij} \tag{3.30}$$

in such a way the E_{ijkl} might be represented by a symmetric table in $\mathbb{R}^6$. (Consequently, prove as an exercise that E_{ijkl} has, at most, *twenty-one* independent elements; this reduces to *two*, the celebrated Lamé coefficients, for an isotropic body).

We also admit the *positivity* condition

$$E_{ijkl} A_{ij} A_{kl} \geqslant m A_{ij} A_{ij}, \quad m > 0, \quad \forall A_{ij} = A_{ji} \text{ real}. \tag{3.31}$$

Moreover, we have the decomposition

$$\varepsilon_{ij} = \tfrac{1}{2}(u_{i,j} + u_{j,i}) = \varepsilon^{\mathrm{e}}_{ij} + \varepsilon^{\mathrm{p}}_{ij}. \tag{3.32}$$

Let S_{ijkl} be the tensor of *compliances* such that $\mathbf{S} : \mathbf{E} = \mathbf{I}$ in $\mathbb{R}^6$, that is, in ordinary tensorial language

$$S_{ijkl} E_{klpq} = \tfrac{1}{2}(\delta_{ip}\delta_{jq} + \delta_{jp}\delta_{iq}). \tag{3.33}$$

Then

$$\sigma_{ij} = E_{ijkl}\varepsilon^{\mathrm{e}}_{kl} \qquad \text{and} \qquad \varepsilon^{\mathrm{e}}_{kl} = S_{klpq}\sigma_{pq}. \tag{3.34}$$

In Voigt's notation in $\mathbb{R}^6$, eqns (3.33) and (3.34) read ($\alpha, \beta = 1, \ldots, 6$)

$$\sigma_\alpha = E_{\alpha\beta}\varepsilon^{\mathrm{e}}_\beta, \qquad \varepsilon^{\mathrm{e}}_\beta = S_{\beta\gamma}\sigma_\gamma, \qquad S_{\alpha\beta}E_{\beta\gamma} = \delta_{\alpha\gamma},$$

where $\delta_{\alpha\gamma}$ is the Kronecker delta in $\mathbb{R}^6$.

By using (3.32) and taking the derivative with respect to time, we obtain

$$\dot{\varepsilon}_{ij} = \dot{\varepsilon}^{\mathrm{p}}_{ij} + S_{ijpq}\dot{\sigma}_{pq}. \tag{3.35}$$

Given (3.27) we have then

$$\dot{\varepsilon}_{ij} = S_{ijpq}\dot{\sigma}_{pq} + \dot{\lambda}\frac{\partial f}{\partial \sigma_{ij}}. \tag{3.36}$$

Eqns (3.34), (3.27) and (3.26) define adequately a constitutive equation, but the correspondence

(strain history) → (stress history)

is much more complex than that of elasticity, because

$$\left.\begin{aligned} \dot{\lambda} &\geqslant 0 \quad \text{if } f(\boldsymbol{\sigma}) = 0, \\ \dot{\lambda} &= 0 \quad \text{if } f(\boldsymbol{\sigma}) < 0. \end{aligned}\right\} \tag{3.37}$$

We have then the following.

Proposition *The equations*

$$\left.\begin{aligned} &\boldsymbol{\varepsilon} = \boldsymbol{\varepsilon}^{\mathrm{e}} + \boldsymbol{\varepsilon}^{\mathrm{p}}, \\ &\boldsymbol{\sigma} = \mathbf{E} : \boldsymbol{\varepsilon}^{\mathrm{e}}, \\ &\dot{\boldsymbol{\varepsilon}}^{\mathrm{p}} = \dot{\lambda}\partial f/\partial\boldsymbol{\sigma}, \quad \dot{\lambda} \geqslant 0 \text{ if } f = 0, \quad \dot{\lambda} = 0 \text{ if } f < 0, \end{aligned}\right\} \tag{3.38}$$

are characteristic of the material's response.

By this proposition we mean the following. If $\boldsymbol{\varepsilon}(t)$ is known for the time interval $[0, T]$, we may then, starting from an initial state $\boldsymbol{\sigma}(0)$, $\boldsymbol{\varepsilon}^{\mathrm{p}}(0)$, define the stress path $\boldsymbol{\sigma}(t)$ and the associated plastic strains, $\dot{\boldsymbol{\varepsilon}}^{\mathrm{p}}(t)$.

Proof We have

$$\left.\begin{aligned} \dot{\boldsymbol{\varepsilon}}^{\mathrm{p}} &\in N_C(\boldsymbol{\sigma}) = N_C\left(\frac{\partial W}{\partial \boldsymbol{\varepsilon}^{\mathrm{e}}}(\boldsymbol{\varepsilon}, \boldsymbol{\varepsilon}^{\mathrm{p}})\right), \\ \boldsymbol{\varepsilon}^{\mathrm{p}}(0) &= \boldsymbol{\varepsilon}^{\mathrm{p}}_0. \end{aligned}\right\} \tag{3.39}$$

This constitutes an *evolution problem* since $\boldsymbol{\varepsilon}(t)$ is known, but N_C is a non-one-to-one operator. The problem (3.39) has an equivalent variational formulation in the form of a variational *inequation*.

Indeed, according to (3.28)

$$(\boldsymbol{\sigma} - \boldsymbol{\sigma}^*) : \dot{\boldsymbol{\varepsilon}}^{\mathrm{p}} \geqslant 0,$$

so that

$$\left.\begin{aligned} \dot{\boldsymbol{\varepsilon}}^{\mathrm{p}} : \left(\frac{\partial W}{\partial \boldsymbol{\varepsilon}^{\mathrm{e}}}(\boldsymbol{\varepsilon}, \boldsymbol{\varepsilon}^{\mathrm{p}}) - \boldsymbol{\sigma}^* \right) \geqslant 0, \quad \forall \boldsymbol{\sigma}^* \in C, \\ \boldsymbol{\varepsilon}^{\mathrm{p}}(0) = \boldsymbol{\varepsilon}^{\mathrm{p}}_0 \end{aligned}\right\} \qquad (3.40)_{(\varepsilon)}$$

If W is quadratic (the elastic part being linear) we have the problem

$$\left.\begin{aligned} \dot{\boldsymbol{\varepsilon}}^{\mathrm{p}} : [\mathbf{E} : (\boldsymbol{\varepsilon} - \boldsymbol{\varepsilon}^{\mathrm{p}}) - \boldsymbol{\sigma}^*] \geqslant 0, \quad \forall \boldsymbol{\sigma}^* \in C, \\ \boldsymbol{\varepsilon}^{\mathrm{p}}(0) = \boldsymbol{\varepsilon}^{\mathrm{p}}_0, \end{aligned}\right\} \qquad (3.40)_{(\varepsilon')}$$

i.e.,

$$\dot{\boldsymbol{\varepsilon}}^{\mathrm{p}} \in N_C(\mathbf{E} : (\boldsymbol{\varepsilon} - \boldsymbol{\varepsilon}^{\mathrm{p}})).$$

The mathematical study of problems of the type of (3.40) has been carried on by H. Brézis and J.J. Moreau, to whom the reader is referred. As a result, we get the *existence* and the *uniqueness* of the path $\boldsymbol{\varepsilon}^{\mathrm{p}}(t)$ and $\boldsymbol{\sigma}(t)$.

Let us notice that we also have an evolution problem for $\boldsymbol{\sigma}$; actually, according to (3.28) and in the case of (3.34) we have

$$(\dot{\boldsymbol{\varepsilon}} - \mathbf{S} : \dot{\boldsymbol{\sigma}}) : (\boldsymbol{\sigma} - \boldsymbol{\sigma}^*) \geqslant 0,$$

hence the problem (where $\boldsymbol{\varepsilon}(t)$ is known)

$$\left.\begin{aligned} \dot{\boldsymbol{\sigma}} : \mathbf{S} : (\boldsymbol{\sigma} - \boldsymbol{\sigma}^*) - \dot{\boldsymbol{\varepsilon}} : (\boldsymbol{\sigma} - \boldsymbol{\sigma}^*) \geqslant 0, \quad \forall \boldsymbol{\sigma}^* \in C, \\ \boldsymbol{\sigma}(0) = \boldsymbol{\sigma}_0. \end{aligned}\right\} \qquad (3.40)_{(\sigma)}$$

3.3 Incremental nature of the elastoplasticity laws

Following the previous proposition, the knowledge of $\boldsymbol{\varepsilon}(t)$ in the interval $[0, T]$ and of $\boldsymbol{\sigma}(0)$ and $\boldsymbol{\varepsilon}^{\mathrm{p}}(0)$ implies also that of $\boldsymbol{\sigma}(t)$ and $\boldsymbol{\varepsilon}^{\mathrm{p}}(t)$. We may now pose the following problem. If, at a given moment t, we know $\dot{\boldsymbol{\varepsilon}}(t)$, $\boldsymbol{\varepsilon}^{\mathrm{p}}(t)$ and $\boldsymbol{\sigma}(t)$, and if we know $\dot{\boldsymbol{\varepsilon}}$, is it possible to determine the associated expressions $\dot{\boldsymbol{\varepsilon}}^{\mathrm{p}}$ and $\dot{\boldsymbol{\sigma}}$, that is, *increment* the solution? As a matter of fact, if we can determine $\dot{\boldsymbol{\varepsilon}}^{\mathrm{p}}$ and $\dot{\boldsymbol{\sigma}}$, then, by integrating in time, we may construct by successive steps the stress path associated to a given strain path $\boldsymbol{\varepsilon}(t)$. It so happens that we may obtain an explicit relation

$$\dot{\boldsymbol{\sigma}} = \dot{\hat{\boldsymbol{\sigma}}}(\boldsymbol{\sigma}, \dot{\boldsymbol{\varepsilon}}) \tag{3.41}$$

in the following manner and in four steps.

(a) *Orthogonality property.* Let us consider (Fig. 3.1) an evolution $\boldsymbol{\sigma}(t)$ only differentiable to *the right* with respect to time (this is what we need in order to evolve with increasing time). Let us apply (3.28) with a derivative to *the right* at a regular moment t.

We have

$$\dot{\boldsymbol{\varepsilon}}^{\mathrm{p}}_{(t')} : (\boldsymbol{\sigma}_{(t')} - \boldsymbol{\sigma}^*) \geqslant 0. \tag{3.42}$$

We choose

$$\boldsymbol{\sigma}^* = \boldsymbol{\sigma}_{(t'+h)}$$

and we divide by h, that is

$$\dot{\boldsymbol{\varepsilon}}^{\mathrm{p}}_{(t')} : \left(\frac{\boldsymbol{\sigma}_{(t')} - \boldsymbol{\sigma}_{(t'+h)}}{h} \right) \geqslant 0. \tag{3.43}$$

If we take the limit $h \to 0$ we obtain

$$\left. \begin{aligned} \dot{\boldsymbol{\varepsilon}}^{\mathrm{p}}_{(t')} : \dot{\boldsymbol{\sigma}}_{(t')} &\geqslant 0 \quad \text{for } h > 0, \\ \dot{\boldsymbol{\varepsilon}}^{\mathrm{p}}_{(t')} : \dot{\boldsymbol{\sigma}}_{(t')} &\leqslant 0 \quad \text{for } h < 0, \end{aligned} \right\} \tag{3.44}$$

in such a manner that we can only conclude that

$$\dot{\boldsymbol{\sigma}}_{(t')} : \dot{\boldsymbol{\varepsilon}}^{\mathrm{p}}_{(t')} = 0.$$

If then we let t' tend towards t, we obtain for *all* ts

$$\dot{\boldsymbol{\sigma}} : \dot{\boldsymbol{\varepsilon}}^{\mathrm{p}} = 0, \tag{3.45}$$

where all derivatives are understood as to the right.

Fig. 3.1. Evolution $\sigma(t)$

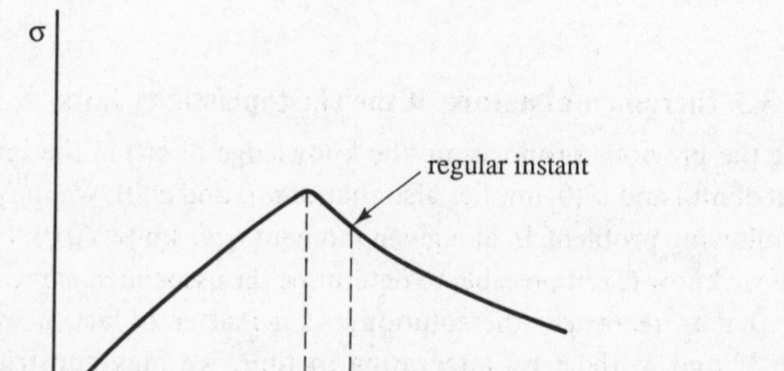

Eqn (3.45) is an *orthogonality* relation between the stress rate (to the right) and the plastic strain rate.

(b) Let us show that we can now re-write (3.38) in the form

$$\left.\begin{aligned}&\boldsymbol{\varepsilon} = \boldsymbol{\varepsilon}^{e} + \boldsymbol{\varepsilon}^{p},\\ &\boldsymbol{\sigma} = \mathbf{E} : \boldsymbol{\varepsilon}^{e},\\ &\dot{\boldsymbol{\varepsilon}}^{p} = \dot{\lambda}\frac{\partial f}{\partial \boldsymbol{\sigma}} \quad \text{with} \begin{cases}\dot{\lambda} \geqslant 0 & \text{if } f = 0 \text{ } \textit{and} \text{ } \dot{f} = 0,\\ \dot{\lambda} = 0 & \text{if } f < 0, \textit{or } f = 0 \text{ and } \dot{f} < 0.\end{cases}\end{aligned}\right\} \tag{3.46}$$

In fact, from (3.45) it follows that

$$\dot{\lambda}\frac{\partial f}{\partial \boldsymbol{\sigma}} : \dot{\boldsymbol{\sigma}} = 0$$

or

$$\dot{\lambda}\dot{f} = 0, \tag{3.47}$$

whence the precisions indicated in the third line of (3.46).

(c) *Determination of the plastic multiplier* $\dot{\lambda}$. Eqn (3.45) may be re-written in the following successive forms:

$$\left.\begin{aligned}\dot{\boldsymbol{\sigma}} : \dot{\boldsymbol{\varepsilon}}^{p} &= 0,\\ \dot{\boldsymbol{\sigma}} : \dot{\lambda}\frac{\partial f}{\partial \boldsymbol{\sigma}} &= 0,\\ (\dot{\boldsymbol{\varepsilon}} - \dot{\boldsymbol{\varepsilon}}^{p}) : \mathbf{E} : \dot{\lambda}\frac{\partial f}{\partial \boldsymbol{\sigma}} &= 0,\\ \left(\dot{\boldsymbol{\varepsilon}} - \dot{\lambda}\frac{\partial f}{\partial \boldsymbol{\sigma}}\right) : \mathbf{E} : \dot{\lambda}\frac{\partial f}{\partial \boldsymbol{\sigma}} &= 0.\end{aligned}\right\} \tag{3.48}$$

We have then the following alternatives. If $f < 0$, then $\dot{\lambda} = 0$. If $f = 0$ and $\dot{\lambda} \geqslant 0$ the last of eqns (3.48) gives

$$\dot{\lambda} = \frac{\left\langle \dfrac{\partial f}{\partial \boldsymbol{\sigma}} : \mathbf{E} : \dot{\boldsymbol{\varepsilon}} \right\rangle}{\left(\dfrac{\partial f}{\partial \boldsymbol{\sigma}} : \mathbf{E} : \dfrac{\partial f}{\partial \boldsymbol{\sigma}}\right)}, \tag{3.49}$$

where the symbol $\langle \ldots \rangle$ indicates the 'positive part', that is

$$\langle x \rangle = \begin{cases} x & \text{if } x \geqslant 0,\\ 0 & \text{if } x < 0.\end{cases} \tag{3.50}$$

(d) *Expression of* $\dot{\boldsymbol{\sigma}}(\dot{\boldsymbol{\varepsilon}})$. If we take into account (3.49) we may write

$$\dot{\boldsymbol{\sigma}} = E : \dot{\boldsymbol{\varepsilon}}^{e} = \mathbf{E} : (\dot{\boldsymbol{\varepsilon}} - \dot{\boldsymbol{\varepsilon}}^{p})$$

or

$$\dot{\boldsymbol{\sigma}} = \mathbf{E} : \dot{\boldsymbol{\varepsilon}} - \frac{\left\langle \dfrac{\partial f}{\partial \boldsymbol{\sigma}} : \mathbf{E} : \dot{\boldsymbol{\varepsilon}} \right\rangle}{\left(\dfrac{\partial f}{\partial \boldsymbol{\sigma}} : \mathbf{E} : \dfrac{\partial f}{\partial \boldsymbol{\sigma}} \right)} \mathbf{E} : \frac{\partial f}{\partial \boldsymbol{\sigma}}. \tag{3.51}$$

We immediately verify that this is the equivalent of the formula

$$\dot{\boldsymbol{\sigma}} = \frac{\partial \Psi^{\sigma}(\dot{\boldsymbol{\varepsilon}})}{\partial \dot{\boldsymbol{\varepsilon}}}, \tag{3.52}$$

where Ψ^{σ} is the *strain-rate potential* defined by

$$\Psi^{\sigma}(\dot{\boldsymbol{\varepsilon}}) = \frac{1}{2} \dot{\boldsymbol{\varepsilon}} : \mathbf{E} : \dot{\boldsymbol{\varepsilon}} - \frac{1}{2} \frac{\left\langle \dfrac{\partial f}{\partial \boldsymbol{\sigma}} : \mathbf{E} : \dot{\boldsymbol{\varepsilon}} \right\rangle^{2}}{\left(\dfrac{\partial f}{\partial \boldsymbol{\sigma}} : \mathbf{E} : \dfrac{\partial f}{\partial \boldsymbol{\sigma}} \right)} \tag{3.53}$$

or

$$\Psi^{\sigma} = \Psi_1^{\sigma} - \Psi_2^{\sigma}, \tag{3.54}$$

where Ψ_1^{σ} is the 'elastic' part. Ψ_2^{σ} is quadratic in $\dot{\boldsymbol{\varepsilon}}$ for $\dot{\boldsymbol{\varepsilon}} > 0$ and zero for $\dot{\boldsymbol{\varepsilon}} < 0$. Since we must subtract Ψ_2^{σ} from Ψ_1^{σ} (which is quadratic), the result is *not strictly* quadratic and the potential Ψ^{σ} is *not* strictly convex (see Fig. 3.2) Fig. 3.3 illustrates in a more systematic way the law obtained in rates

Fig. 3.2. Strain-rate potential

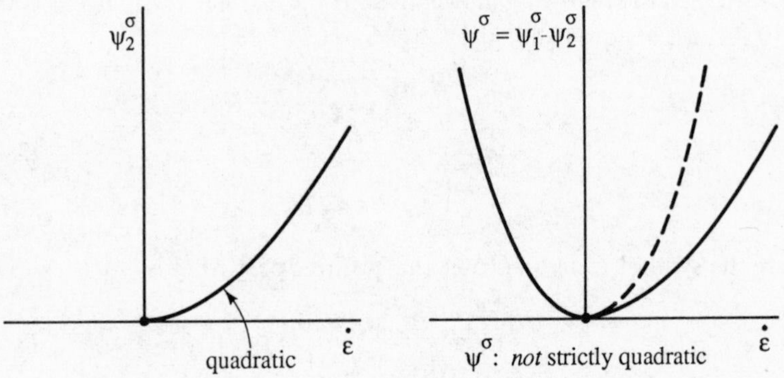

(for illustration purposes we take $E = 1$; $\dot{\boldsymbol{\sigma}}$ is then the projection of $\dot{\boldsymbol{\sigma}}$ upon the tangent plane at $\boldsymbol{\sigma}$ to the elasticity domain, since $\dot{\boldsymbol{\sigma}} : \dot{\boldsymbol{\varepsilon}}^{\mathrm{p}} = 0$).

Since Ψ^{σ} is convex we may consider its Legendre–Fenchel transform $\Psi^*(\dot{\boldsymbol{\sigma}})$ such that (see Appendix 2)

$$\Psi^*(\dot{\boldsymbol{\sigma}}) = \operatorname*{Sup}_{\dot{\boldsymbol{\varepsilon}}} [\dot{\boldsymbol{\sigma}} : \dot{\boldsymbol{\varepsilon}} - \Psi^{\sigma}(\dot{\boldsymbol{\varepsilon}})], \qquad \dot{\boldsymbol{\varepsilon}} = \frac{\partial \Psi^*}{\partial \boldsymbol{\sigma}}. \tag{3.55}$$

Given (3.53) we prove that

$$\Psi^*(\dot{\boldsymbol{\sigma}}) = \begin{Bmatrix} \frac{1}{2}\dot{\boldsymbol{\sigma}} : \mathbf{S} : \dot{\boldsymbol{\sigma}} & \text{if} \frac{\partial f}{\partial \boldsymbol{\sigma}} : \dot{\boldsymbol{\sigma}} \leqslant 0, \\ +\infty & \text{otherwise.} \end{Bmatrix} \tag{3.56}$$

It is remarkable that eqn (3.52), if we ignore the superimposed dots, is similar to the basic constitutive equation of elasticity (1.1) except that Ψ^{σ} is *not* strictly quadratic. If unloading does not occur, then eqn (3.52) is a *nonlinear elastic* constitutive equation, but in terms of *velocities* of strains and stresses. The expressions (3.53) and (3.56), however, are fully equipped to treat both loading *and* unloading cases.

3.4 Remarks

3.4.1 Energy aspect

Let Ω be the volume occupied by the material; following (2.53)–(2.54) in a quasi-static state, in the absence of volume forces and in isothermal evolution, we have

Fig. 3.3. Law in velocities

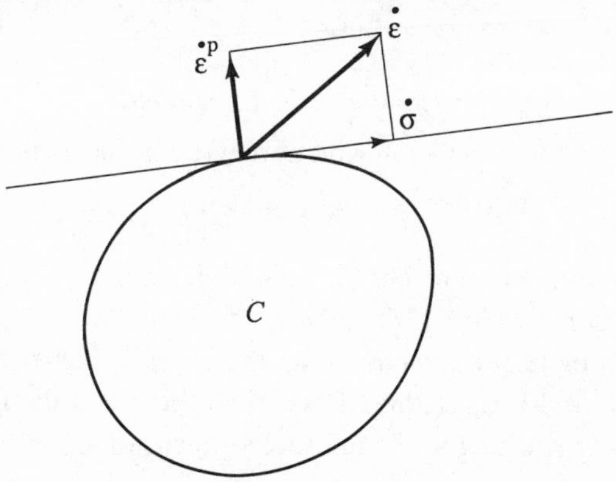

$$\Phi_{\text{intr}} = \int_{\partial\Omega} \mathbf{T}\cdot\mathbf{v}\,\mathrm{d}s - \frac{\mathrm{d}}{\mathrm{d}t}\int_{\Omega} W(\boldsymbol{\varepsilon}^{\mathrm{e}})\,\mathrm{d}v$$

$$= \int_{\Omega} (\boldsymbol{\sigma} : \dot{\boldsymbol{\varepsilon}} - \dot{W})\,\mathrm{d}v,$$

and, with

$$\operatorname{div}\boldsymbol{\sigma} = \mathbf{O} \quad \text{in } \Omega, \qquad \boldsymbol{\sigma} = \frac{\partial W}{\partial \boldsymbol{\varepsilon}^{\mathrm{e}}}, \qquad \boldsymbol{\varepsilon} = \boldsymbol{\varepsilon}^{\mathrm{e}} + \boldsymbol{\varepsilon}^{\mathrm{p}},$$

it follows that

$$\Phi_{\text{intr}} = \int_{\Omega} \boldsymbol{\sigma} : \dot{\boldsymbol{\varepsilon}}^{\mathrm{p}}\,\mathrm{d}v \geqslant 0. \tag{3.57}$$

3.4.2 Thermodynamic restriction on the convex C

The convex C must contain the origin ($0 \in C$) in stress space. Its boundary $f = 0$ is none other than the plastic potential. This fulfils the wish expressed by G.I. Taylor (1947) for the description of the plastic behaviour of metals where identification of the yield locus and plastic potential allows one to derive the flow rule from an experimental determination of the yield locus. When this identification holds good, we say that we deal with '*plasticity with associated flow rule*'. If this appears to be very much valid for metals, it is not true in general, and there exist plasticity formulations with *non associated* flow rules (see Drucker, 1988). In this book we consider only 'associated' plasticity.

3.4.3 Regularity

Our previous premise was that everything was regular (that is, that it varies sufficiently slowly in space). But in the presence of discontinuity *surfaces* we have to take certain precautions.

3.4.4 The Prandtl–Reuss relations

We consider the relation (3.36), which we write in the form

$$\dot{\boldsymbol{\varepsilon}} = (\mathbf{S} + \mathscr{H}) : \dot{\boldsymbol{\sigma}}, \tag{3.58}$$

where $\mathscr{H}$ is a function of $\boldsymbol{\sigma}$. The variational formulation of this is nothing but the evolution problem $(3.40)_\sigma$ for $\boldsymbol{\sigma}$.

Eqns (3.36) or (3.58) are known as the *Prandtl–Reuss equations* after Prandtl (1924) and Reuss (1939). Prandtl considered only the plane problem while Reuss considered the full three-dimensional case.

3.4.5 The Lévy–Mises relations

This is the form taken by the Prandtl–Reuss equations in the absence of an elastic contribution. The body is then said to be *rigid–plastic*. Historically, the Lévy (1871) – Mises (1913) relations came before the Prandtl–Reuss relations. As a matter of fact, as both Lévy and Mises used the *total* strain increment (and not only the *plastic* strain increment) in their relationship between ratios of the components of strain increments and stress ratios, their theory is strictly valid only when elastic strains are zero, hence the *rigid*–plastic framework.

3.4.6 The Hencky–Nadai relations

If *all* strains remain very small, one is tempted to use directly the *strain*, instead of the strain increment, and *stress*, instead of stress increment, in the elastoplastic incremental constitutive equation. Thus Hencky's relations mimic (3.58) simply by

$$\boldsymbol{\varepsilon} = (\mathbf{S} + \boldsymbol{\Psi}) : \boldsymbol{\sigma},$$

where $\boldsymbol{\Psi}$ is essentially 'positive' during continued loading and zero during unloading. This simplified relationship was used by Nadai, Ilyushin, and many Russian engineers in the 1940s and 1950s. No need to say that replacing the derivative of a function, dy/dx, by the ratio of the function itself to the argument, y/x, is strictly valid only when y/x is a constant! The Hencky–Nadai relations belong to the small-strain 'deformation' theory of elastoplasticity, which we shall not consider further, except in problems (see Problems 5.1 and 5.2 – also 5.11).

3.5 Viscoplasticity

Generally speaking, plastic strains are not instantaneous; they are slowed down by viscosity. In this case, the stress tensor $\boldsymbol{\sigma}$ is no longer restricted to the interior of the convex C. When $\boldsymbol{\sigma}$ is inside C, the behaviour is elastic and, when $\boldsymbol{\sigma}$ is outside C, the rate of the plastic strain is proportional to the distance between $\boldsymbol{\sigma}$ and the convex C. For example, let $\eta > 0$ be a viscosity coefficient. We have

$$\boldsymbol{\sigma} = \mathbf{E} : \boldsymbol{\varepsilon}^e, \qquad \boldsymbol{\varepsilon} = \boldsymbol{\varepsilon}^e + \boldsymbol{\varepsilon}^p, \qquad \dot{\boldsymbol{\varepsilon}} = \mathbf{S} : \dot{\boldsymbol{\sigma}} + \dot{\boldsymbol{\varepsilon}}^p$$

and

$$\dot{\boldsymbol{\varepsilon}}^p = \frac{1}{2\eta}(\boldsymbol{\sigma} - \Pi_C \boldsymbol{\sigma}), \tag{3.59}$$

where Π_C is the projection operator on the convex C. Since $\boldsymbol{\sigma} = \Pi_C \boldsymbol{\sigma}$ for $\boldsymbol{\sigma} \in C$ we have $\dot{\boldsymbol{\varepsilon}}^p = \mathbf{0}$ for $\boldsymbol{\sigma} \in C$ (see Fig. 3.4) This is Perzyna's (1966) model in which $\dot{\boldsymbol{\varepsilon}}^p$ is derived from the potential of dissipation

$$\Omega = \frac{1}{4\eta} \|\boldsymbol{\sigma} - \Pi_C \boldsymbol{\sigma}\|^2 \tag{3.60}$$

by

$$\dot{\boldsymbol{\varepsilon}}^p = \partial \Omega / \partial \boldsymbol{\sigma}. \tag{3.61}$$

No plastic multiplier appears in eqn (3.59) which becomes much more regular than a plasticity evolution equation.

Viscosity plays a *regularizing* role in diminishing the singular character of the relations (3.24). We may then consider the question of the *transition*, within certain limits, from viscoplastic behaviour to plastic behaviour. When we consider the case (3.59) with the corresponding viscoplastic potential Ω, the elastoplastic case is obtained once more at the limit $\eta \to 0$, where Ω becomes the *indicator* of the convex C according to the definition (see Appendix 2)

$$I_C(\boldsymbol{\sigma}) = \begin{cases} 0 & \text{if } \boldsymbol{\sigma} \in C, \\ +\infty & \text{if } \boldsymbol{\sigma} \notin C. \end{cases} \tag{3.62}$$

Metal viscoplasticity presents its own specific interest. But the formulation of a viscoplastic type is also interesting, since a great number of mathematical proofs can be more easily carried out within this framework; the results of the plastic case are then obtained by the limit process of an asymptotically zero viscosity.

We may also visualize more vividly the question in the following manner (Fig. 3.4(b) and (c)). Let us consider a mechanism of plastic flow and let us

Fig. 3.4. Viscoplasticity

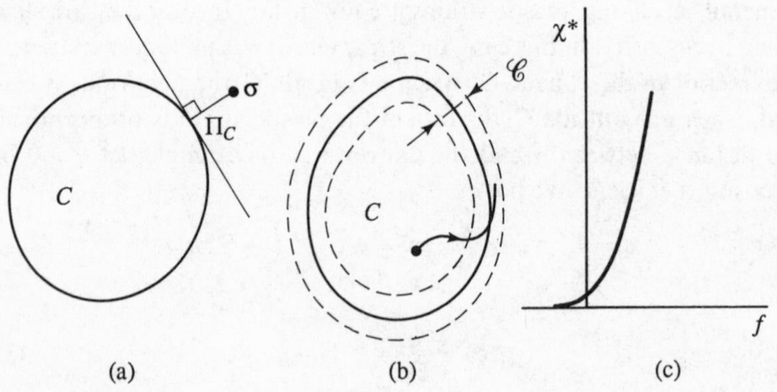

introduce viscosity into it, through the following process. We suppose that the material (e.g., a metal) presents a viscous internal friction which is particularly activated when the representative state point in the stress space tends to cross the edge of the convex C. Let us introduce a rather thin layer, in the neighbourhood of $S = \partial C$, such that, when we enter this layer, some kind of internal friction mechanism, strongly opposed to the crossing of the said layer, comes into play (we have then a stress path similar to that represented in Fig. 3.4(b)). This can be translated into mathematical terms by introducing a function $\chi(\boldsymbol{\sigma})$ endowed with the property of varying very fast in the layer $\mathscr{C}$ so that, outside $\mathscr{C}$, the value of $\boldsymbol{\sigma} \in C$ is almost zero, becoming very great when $\boldsymbol{\sigma} \in \mathscr{C}$ (see Fig. 3.4(c). Let $f(\boldsymbol{\sigma})$ be the function such that S is $f = 0$.

We define χ by

$$\chi(\boldsymbol{\sigma}) = \chi^*(f(\boldsymbol{\sigma})) \tag{3.63}$$

and we formulate the law of plastic flow by

$$\dot{\boldsymbol{\varepsilon}}^{\mathrm{p}} = \frac{\partial \chi}{\partial \boldsymbol{\sigma}} = \dot{\lambda}\frac{\partial f}{\partial \boldsymbol{\sigma}}, \tag{3.64}$$

where we have set

$$\dot{\lambda} = \mathrm{d}\chi^*/\mathrm{d}f. \tag{3.65}$$

The *purely plastic* kind of flow law is the law (3.64) in which we forget the definition (3.65) while $\dot{\lambda}$ becomes one of the unknowns of the problem. If we let the thickness of the layer $\mathscr{C}$ go towards zero, the function χ^*, in accordance with (3.62), becomes the indicator of the convex C. We may also, following Kestin and Rice (1970), say that plasticity may be considered as a limiting case where the flow potential (Ω for (3.59)) is reduced to a *singular* surface or, even, that the plasticity's flow surface (independent of the strain rate) may be considered as a *singular sequential limit* of surfaces of constant flow potential. This is the last time we discuss the elastoviscoplasticity case. Perzyna (1966) is still a useful reference, and Critescu and Suliciu (1976) is an excellent introduction to mathematical problems in elastoviscoplasticity.

Problems for Chapter 3

3.1. Huber–Mises criterion for plane stresses

Write first the Huber–Mises criterion in terms of principal stresses, select $\sigma_3 = 0$ and show that

$$f(\boldsymbol{\sigma}) = \sigma_1^2 - \sigma_1\sigma_2 + \sigma_2^2 - \sigma_0^2 = 0. \tag{1}$$

Show that $\dot{\varepsilon}_3^p$ is not necessarily zero. What is its value? What is the shape of the criterion curve?

3.2. Plane flow of a Huber–Mises material

Use principal strains and stresses. Select $\dot{\varepsilon}_3 = 0$ and show that the criterion reduces to

$$f(\boldsymbol{\sigma}) = (\sigma_1 - \sigma_2)^2 - \tfrac{4}{3}\sigma_0^2 = 0.$$

What is the shape of the criterion curve?

3.3. Locking materials

Locking is defined by the condition

$$\hat{\mathbf{D}} = \hat{\mathbf{D}}(\boldsymbol{\sigma}) \quad \text{if} \frac{\partial \hat{\mathbf{D}}}{\partial \boldsymbol{\sigma}} : \boldsymbol{\sigma} = 0.$$

What is the precise locking condition assuming that $\hat{\mathbf{D}}$ corresponds to small strains?

3.4. Lévy–Mises model

Show that

$$\dot{\lambda} = \left(\frac{\operatorname{tr}\mathbf{D}^2}{\operatorname{tr}\boldsymbol{\sigma}^2}\right)^{1/2}.$$

With the Huber–Mises criterion show that

$$\dot{\lambda}^2 = \frac{\operatorname{tr}\mathbf{D}^2}{2k^2} = g(\mathbf{D}) \quad \text{only.}$$

3.5. Prandtl–Reuss theory for small strains

Assume that the plastic part of the shear strain obeys the Lévy–Mises relation. Then prove that

$$\dot{e}_{ij} = \frac{1}{2\mu}\dot{s}_{ij} + \dot{\lambda}s_{ij}, \qquad \dot{\lambda} > 0, \text{ for } f(\sigma_{ij}) = 0,$$

and

$$\dot{e}_{ij} = \frac{1}{2\mu}\dot{s}_{ij} \qquad \text{when } \dot{f}(\sigma_{ij}) < 0.$$

Show that

$$\dot{\lambda} = \left[\frac{\operatorname{tr}(\dot{\mathbf{e}}^{\mathrm{p}})^2}{\operatorname{tr}\mathbf{s}^2}\right]^{1/2} = \frac{\dot{W}_{\mathrm{p}}}{\operatorname{tr}\mathbf{s}^2},$$

where $\dot{W}_{\mathrm{p}}$ is the rate of dissipation of energy. If (Huber–Mises) $f(\boldsymbol{\sigma}) = \operatorname{tr}\mathbf{s}^2 = 2k^2 = \text{const.}$, then show that $\dot{\lambda} = \dot{W}_{\mathrm{p}}/2k^2$.

3.6. Prandtl–Reuss theory for plane stresses

Derive the equations of the Prandtl–Reuss theory for *plane* stresses.

3.7. Hypoelasticity and plasticity

Make a literature search and write a report on the relationship between *hypoelasticity* (to be defined in the process) and the incremental formulation of elastoplasticity. [Hint: study works by C.A. Truesdell and T. Tokuoka in the years 1955–1975.]

3.8. Complementary stress-rate potential

Prove that $\Psi^*(\dot{\boldsymbol{\sigma}})$ has the expression (3.56) if the strain-rate potential $\Psi^\alpha(\dot{\boldsymbol{\varepsilon}})$ is expressed by means of (3.53).

3.9. Multiple plastic potential

Show that if the plasticity threshold is defined by several inequalities

$$f_\alpha(\boldsymbol{\sigma}) \leqslant 0, \qquad \alpha = 1, 2, \ldots, N, \tag{1}$$

then the Prandtl – Reuss relation takes on the form

$$\dot{\boldsymbol{\varepsilon}} = \mathbf{S} : \dot{\boldsymbol{\sigma}} + \sum_{\alpha=1}^{N} \dot{\lambda}_\alpha \frac{\partial f_\alpha}{\partial \boldsymbol{\sigma}}, \tag{2}$$

where the numbers $\dot{\lambda}_\alpha$ are necessarily nonnegative. The representation (1) is used when the plasticity threshold exhibits edges or apices (e.g., Tresca criterion).

3.10. Flexural analysis of plates

Let m_1 and m_2 be the principal bending moments in a plate. We define ω and Ψ by

$$2\omega = m_1 + m_2, \qquad 2\Psi = m_2 - m_1. \tag{1}$$

(a) Show that the interaction surface $F(m_x, m_y, m_{xy}) - 1 = 0$ can be rewritten as

$$\Psi - f(\omega) = 0. \tag{2}$$

This is a closed curve in the plane of principal moments. Note that

$$m_x = \omega + \Psi\cos 2\theta, \qquad m_y = \omega - \Psi\cos 2\theta, \qquad m_{xy} = \Psi\sin 2\theta,$$

(b) Then, while (1) is identically satisfied by the new variables, find the expression of the equilibrium equations in terms of f, ω, θ, p and a potential Φ of transverse shear introduced by

$$\Phi_{,y} = s_x - -p_x, \qquad \Phi_{,x} = -s_y + p_y.$$

(c) Show that the kinematics of the plate during plastic yielding are given by

$$w_{,xx} = \frac{v}{2}(f' - \cos 2\theta), \qquad w_{,xy} = -\frac{v}{2}\sin 2\theta, \qquad w_{,yy} = \frac{v}{2}(f' + \cos 2\theta), \tag{4}$$

where w is the deflection rate and $f' = \partial f/\partial\omega$. Then show that f and w

satisfy the equation

$$(1 + f')w_{,xx} + (1 - f')w_{,yy} = 0. \tag{5}$$

This shows that the deflection rate w depends on the yield criterion. In particular, this equation may be of elliptic, hyperbolic or parabolic type, depending on the value of f'.

3.11. Flexural analysis of plates (continued from P3.10)

Show that w satisfies the PDE

$$\frac{\partial^2 w}{\partial x^2}\frac{\partial^2 w}{\partial y^2} - \frac{\partial^2 w}{\partial x \partial y} = \frac{v^2}{4}(f'^2 - 1). \tag{1}$$

If $f' = \pm 1$ the right-hand side vanishes. Conclude that the state of stress is parabolic and that the surface of deflection is *developable.*

3.12. Yield-line theory (Hencky, 1923)

Starting from the premises of **3.11** make a literature search and present the essentials of the *yield-line theory for rigid–plastic materials* in a ten-page-long report. This theory was mostly developed by Hencky in the 1920s. [See Hill (1950) for an excellent introduction; in a simpler form Mendelson (1968).]

3.13 Yield-line method

Make a literature search and present the essentials of the 'yield-line method' (yield lines joining plastic hinges) together with a worked-out example (simply supported – on the four edges – and uniformly loaded rectangular plate) in a ten-page-long report.

3.14. Cylindrical shells

Make a literature search and describe the theory (interaction surface, flow rule) of plastic cylindrical shells made of a Huber–Mises material.

3.15. Sandwich shells

Make a literature search and describe the theory (interaction surface, flow rule) of plastic sandwich shells made of a Huber–Mises material.

4

Problems in perfect elastoplasticity

The object of the chapter In this chapter we examine the essential mathematical problems (evolution in time, asymptotic behaviour) of the solutions of *perfect* elastoplasticity problems, that is, in the absence of strain hardening. The analytical solution of specific elastoplasticity problems is given in Appendix 3. Let us notice that what leads to the study of problems of evolution in time, in the form of problems *in terms* of velocities – even if these problems are considered as if in a quasi-static state (inertia forces being left out) – is the expression in this form of the plasticity laws.

4.1 Reminder of the perfect elastoplasticity equations

We shall limit ourselves to the case in which the *elastic* response is *linear*, while it may also be anisotropic. We obtain the following constitutive equations (Chapter 2 or Chapter 3)

$$\left.\begin{aligned} &\boldsymbol{\varepsilon} = \boldsymbol{\varepsilon}^{e} + \boldsymbol{\varepsilon}^{p}, \\ &\boldsymbol{\sigma} = \mathbf{E} : \boldsymbol{\varepsilon}^{e}, \\ &\dot{\boldsymbol{\varepsilon}}^{p} = \dot{\lambda}\frac{\partial f}{\partial \boldsymbol{\sigma}}, \quad \begin{cases} \dot{\lambda} \geqslant 0 & \text{if } f = 0 \text{ and } \dot{f} = 0, \\ \dot{\lambda} = 0 & \text{if } f < 0, \text{ or } f = 0 \text{ and } \dot{f} < 0. \end{cases} \end{aligned}\right\} \tag{4.1}$$

The last relation (normality) could also be written more succinctly as

$$\dot{\boldsymbol{\varepsilon}}^{p} \in N_C(\boldsymbol{\sigma}).$$

Evidently, eqns (4.1) are much more adequate for formulating the mathematical problems of elastoplasticity *in terms of velocities*. Actually, in order to establish the quasi-static response of an elastoplastic structure subject to any load path, it is necessary to follow the evolution of the material step by step. The *incremental* formulation of the plastic constitutive equations is very well suited to this approach. Here is the usual formulation of the problem

Problem Starting from a given initial point ($\mathbf{u}(0), \boldsymbol{\sigma}(0), \boldsymbol{\varepsilon}^p(0)$), determine the response of the system ($\mathbf{u}(t), \boldsymbol{\sigma}(t)$), according to the data, in terms of forces and displacement ($\mathbf{T}^d(t), \mathbf{g}(t), \mathbf{u}^d(t)$).

We have (Fig. 4.1)

$$\mathbf{u}^d(t) \text{ prescribed on } \partial\Omega_u,$$

$$\mathbf{T}^d(t) \text{ prescribed on } \partial\Omega_T,$$

$$\mathbf{g}(t) \text{ prescribed in } \Omega,$$

with the initial data

$$\mathbf{u}(\mathbf{x},0) = \mathbf{u}_0(\mathbf{x}), \quad \forall \mathbf{x} \in \Omega,$$

$$\boldsymbol{\sigma}(\mathbf{x},0) = \boldsymbol{\sigma}_0(\mathbf{x}), \quad \forall \mathbf{x} \in \Omega,$$

$$\boldsymbol{\varepsilon}^p(\mathbf{x},0) = \boldsymbol{\varepsilon}_0^p(\mathbf{x}), \quad \forall \mathbf{x} \in \Omega.$$

The equations of the problem (in velocities) are as follows:

the constitutive equation $\dot{\boldsymbol{\sigma}}(\dot{\boldsymbol{\varepsilon}}), \dot{\boldsymbol{\varepsilon}}^p(\dot{\boldsymbol{\varepsilon}}), \boldsymbol{\sigma}(\mathbf{x}, t) \in C$;

the conditions of kinematical type

$$\left.\begin{aligned} \dot{\mathbf{u}} &= \dot{\mathbf{u}}^d(t) \qquad \text{on } \partial\Omega_u, \\ \dot{\boldsymbol{\varepsilon}}_{ij} &= \tfrac{1}{2}(\dot{\mathbf{u}}_{i,j} + \dot{\mathbf{u}}_{j,i}), \end{aligned}\right\} \tag{4.2}$$

the equations of statics (inertia terms are left out)

$$\begin{aligned} \operatorname{div} \boldsymbol{\sigma} + \mathbf{g} &= 0 \qquad \text{in } \Omega, \\ \boldsymbol{\sigma} \cdot \mathbf{n} &= \mathbf{T}^d \quad \text{on } \partial\Omega_T. \end{aligned} \tag{4.3}$$

The latter may be written in variational form (PVP, see eqn (2.27))

Fig. 4.1. Boundary-value initial-value problem

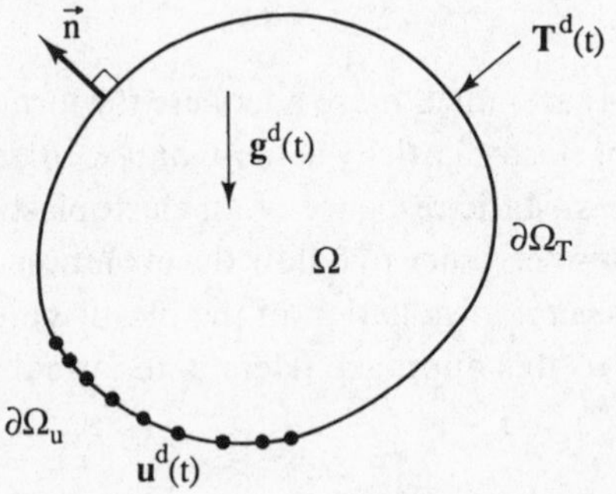

$$\int_{\Omega} \sigma_{ij}\delta\varepsilon_{ij}\,\mathrm{d}\Omega = \int_{\Omega} \mathbf{g}\cdot\delta\mathbf{u}\,\mathrm{d}\Omega + \int_{\partial\Omega} \mathbf{T}^{\mathrm{d}}\cdot\delta\mathbf{u}\,\mathrm{d}s. \tag{4.4}$$

Actually, if we take the derivative of (4.3) with respect to time, we can also write the problem in the following form:

$$\left.\begin{aligned} \operatorname{div}\dot{\boldsymbol{\sigma}} + \dot{\mathbf{g}} = \mathbf{0} \quad & \text{in } \Omega, \\ \dot{\boldsymbol{\sigma}}\cdot\mathbf{n} = \dot{\mathbf{T}}^{\mathrm{d}} \quad & \text{on } \partial\Omega_T, \\ \dot{\mathbf{u}} = \dot{\mathbf{u}}^{\mathrm{d}}(t) \quad & \text{on } \partial\Omega_u, \end{aligned}\right\} \tag{4.5}$$

with the constitutive equations

$$\left.\begin{aligned} \dot{\boldsymbol{\sigma}} = \mathbf{E}:\dot{\boldsymbol{\varepsilon}} \quad & \text{if } \mathrm{f}(\boldsymbol{\sigma}) < 0, \\ \dot{\boldsymbol{\sigma}} = \frac{\partial\Psi^{\sigma}(\dot{\boldsymbol{\varepsilon}})}{\partial\dot{\boldsymbol{\varepsilon}}} \quad & \text{if } f(\boldsymbol{\sigma}) = 0, \end{aligned}\right\} \tag{4.6}$$

with

$$\dot{\boldsymbol{\varepsilon}} = (\nabla\dot{\mathbf{u}})_{\mathrm{S}} \quad (\mathrm{S} = \text{symmetric part}), \tag{4.7}$$

Eqns (4.5)–(4.7) constitute a set of equations of the *nonlinear elastic* type! So we have reduced the elastoplastic problem *in velocities* to a *nonlinear* problem with partial derivatives, similar to a nonlinear elasticity problem! In order to study this problem we need to specify the few definitions that follow immediately.

Definitions (a) *Statically admissible (SA) stress field.* These are the *stress fields* compatible with the data in *forces* i.e., $(\mathbf{g}, \mathbf{T}^{\mathrm{d}})$. By definition such fields satisfy eqn (4.3) and the set of these fields, SA fields, is denoted by

$$S(t) = \{\mathbf{s}(t) | \operatorname{div}\mathbf{s} + \mathbf{g} = \mathbf{0} \text{ in } \Omega, \mathbf{s}\cdot\mathbf{n} = \mathbf{T}^{\mathrm{d}} \text{ on } \partial\Omega_T\}, \tag{4.8}$$

where $\mathbf{s}$ is a stress $\boldsymbol{\sigma}$ statically admissible at time t.

(b) *Kinematically admissible (KA) displacement fields.* These are the displacement fields compatible with the data in *displacement*, i.e., they satisfy the condition $\mathbf{u} = \mathbf{u}^{\mathrm{d}}$ on $\partial\Omega_u$, so that the set of KA fields, $\mathscr{U}$, is introduced by

$$\mathscr{U}(t) = \{\mathbf{u}^* | \mathbf{u}^* \text{ is KA}\}. \tag{4.9}$$

(c) *Plastically admissible (PA) stress fields.* These stress fields must belong to the convex C, hence the set is

$$P = \{\mathbf{s} | \mathbf{s}(\mathbf{x}) \in C, \forall\mathbf{x}\}, \tag{4.10}$$

where $\mathbf{s}$ is a stress tensor.

4.2 Problem in terms of velocities

4.2.1 The intuitive view point

Since we know $\dot{\boldsymbol{\varepsilon}}$ we can determine $\dot{\boldsymbol{\sigma}}(\dot{\boldsymbol{\varepsilon}})$ and $\dot{\boldsymbol{\varepsilon}}^p(\dot{\boldsymbol{\varepsilon}})$. If at $t = 0$ we know $\dot{\mathbf{g}}(0)$, $\dot{\mathbf{T}}^d(0)$ and $\dot{\mathbf{u}}^d(0)$, then we know the 'tendency' of the problem's evolution with $\dot{\mathbf{u}}(0)$, $\dot{\boldsymbol{\sigma}}(0)$ and $\dot{\boldsymbol{\varepsilon}}^p(0)$. After a lapse of time Δt we have the following *incremental* problem:

$$\left.\begin{aligned} \boldsymbol{\sigma}(0 + \Delta t) &= \boldsymbol{\sigma}(0) + \dot{\boldsymbol{\sigma}}(0)\Delta t, \\ \mathbf{u}(0 + \Delta t) &= \mathbf{u}(0) + \dot{\mathbf{u}}(0)\Delta t. \end{aligned}\right\} \tag{4.11}$$

This means that we can progress in time and that we know how to describe the evolution, step by step. Let us then assume that we know the *actual* state ($\mathbf{u}(t)$, $\boldsymbol{\sigma}(t)$, $\boldsymbol{\varepsilon}^p(t)$) at time t. From this moment, t, on, the data vary by $\dot{\mathbf{g}}(t)$, $\dot{\mathbf{T}}^d(t)$ and $\dot{\mathbf{u}}^d(t)$. The problem is to calculate the rates of the unknown fields of the problem ($\dot{\mathbf{u}}(t), \dot{\boldsymbol{\sigma}}(t), \dot{\boldsymbol{\varepsilon}}^p(t)$), which are associated to these variations. It is a *problem in velocities* which satisfies eqns (4.5)–(4.7) and these equations present the structure of a *nonlinear* elasticity problem (because the potential Ψ^σ is *not strictly* quadratic). The problem of the corresponding variational formulation is posed then once more (adapted to the numerical computations). There are *three* possibilities, depending upon whether this variational formulation is made:

in terms of velocities of displacement alone ($\dot{\mathbf{u}}$);

in terms of velocities of stress alone ($\dot{\boldsymbol{\sigma}}$);

in a combined formulation ($\dot{\mathbf{u}}, \dot{\boldsymbol{\sigma}}$), said to be a *two*-field one.

Let us recall (see appendix to this chapter) that in nonlinear elasticity, the existence of a convex elastic potential $W(\boldsymbol{\varepsilon})$, $\boldsymbol{\sigma} = \partial W/\partial \boldsymbol{\varepsilon}$, leads to the *principle of the minimum total potential energy*, and of the *complementary* energy. Since Ψ^σ *is convex* in $\dot{\boldsymbol{\varepsilon}}$, we can use the similarity of the equations of the problem in velocities with those of nonlinear elasticity to prove in the same way that the *elastoplastic problem in velocity* is associated to two minimum principles of the same type.

4.2.2 The Greenberg minimum principle

In *elasticity* the total potential energy is defined in terms of the elastic energy (expressed by means of strains) and the data in *forces*, i.e., ($\mathbf{g}, \mathbf{T}^d$), so that this potential energy $I(\boldsymbol{\xi})$ reads (see appendix for the case of *linear* elasticity)

$$I(\boldsymbol{\xi}) = \int_\Omega W(\boldsymbol{\varepsilon}(\boldsymbol{\xi}))\,\mathrm{d}\Omega - \int_\Omega \mathbf{g}\cdot\boldsymbol{\xi}\,\mathrm{d}\Omega - \int_{\partial\Omega_T} \mathbf{T}^{\mathrm{d}}\cdot\boldsymbol{\xi}\,\mathrm{d}s, \tag{4.12}$$

where $\boldsymbol{\xi}$ is a displacement.

Then we have the following statement.

Minimum total potential energy theorem in elasticity *The actual solution* $\mathbf{u}$ *in displacement is the one that minimizes the functional*

$$I(\boldsymbol{\xi})$$

among kinematically admissible fields, i.e.,

$$I(\mathbf{u}) = \min_{\boldsymbol{\xi}\in\mathscr{U}} I(\boldsymbol{\xi}). \tag{4.13}$$

The proof is given in the appendix to this chapter for *linear* elasticity. This result still holds for elastic energies that are *convex* but not necessarily quadratic. This is sufficient, by using the above-noted analogy, to state directly the Greenberg minimum principle (Greenberg, 1949) in elasto-plasticity in SPH.

Greenberg minimum principle *The velocity* $\dot{\mathbf{u}}$ *minimizes the functional*

$$A(\dot{\mathbf{u}}) = \int_\Omega \Psi^\sigma(\dot{\boldsymbol{\varepsilon}})\,\mathrm{d}\Omega - \int_\Omega \dot{\mathbf{g}}\cdot\dot{\mathbf{u}}\,\mathrm{d}\Omega - \int_{\partial\Omega_T} \dot{\mathbf{T}}^{\mathrm{d}}\cdot\dot{\mathbf{u}}\,\mathrm{d}s, \tag{4.14}$$

among the velocities $\dot{\mathbf{u}}^*$ *compatible with the kinematical conditions, i.e.,* $\dot{\mathbf{u}}^*|_{\partial\Omega_u} = \dot{\mathbf{u}}^{\mathrm{d}}$.

4.2.3 The Hodge–Prager minimum principle

In classical elasticity, the pointwise definition of the *complementary* elastic energy $W^*(\boldsymbol{\sigma})$, a function of stresses, is given by the *Legendre* transformation

$$W^*(\boldsymbol{\sigma}) = \boldsymbol{\sigma} : \boldsymbol{\varepsilon} - W(\boldsymbol{\varepsilon})$$

in such a way that

$$\boldsymbol{\varepsilon} = \partial W^*/\partial\boldsymbol{\sigma}, \qquad \boldsymbol{\sigma} = \partial W/\partial\boldsymbol{\varepsilon}.$$

This is illustrated in (linear *or* nonlinear) one-dimensional elasticity in Fig. 4.2, where W corresponds to the area *below* the curve and W^* to the area to the *left* of the curve (hatched area). In both linear and nonlinear elasticity the total *complementary* energy is defined in terms of $W^*(\boldsymbol{\sigma})$ and the data in *displacements* (i.e., $\mathbf{u}^d$ on $\partial\Omega$), i.e., this complementary energy $J(\mathbf{s})$ is the functional

$$J(\mathbf{s}) = \int_{\Omega} W^*(\mathbf{s})\,d\Omega - \int_{\partial\Omega_u} \mathbf{u}^d \cdot (\mathbf{s}\cdot\mathbf{n})\,ds, \tag{4.15}$$

where $\mathbf{s}$ is a stress tensor

The statement of the minimum total complementary energy theorem is the following.

Minimum total complementary energy in elasticity *The solution in stress $\boldsymbol{\sigma}$ minimizes the total complementary energy among the statically admissible fields of stresses, i.e.*

$$J(\boldsymbol{\sigma}) = \min_{\mathbf{s}\in S} J(\mathbf{s}). \tag{4.16}$$

The proof of this in linear elasticity is given in the appendix to this chapter. It holds also in nonlinear elasticity by using the *convexity* property of $W(\boldsymbol{\varepsilon})$. (Note, however, that some nonlinear elasticity theories admit *nonconvex* strain energies (e.g., Maugin and Cadet, 1991) in order to explain the twinning of certain elastic crystals, but the strain energy remains convex for a large range of temperatures.)

The Hodge–Prager minimum principle in elastoplasticity is the equivalent of the *minimum total complementary energy, for stress rates.* It is the dual of the Greenberg principle.

Fig. 4.2. Notion of complementary energy ($W + W^* = \boldsymbol{\sigma} : \boldsymbol{\varepsilon}$)

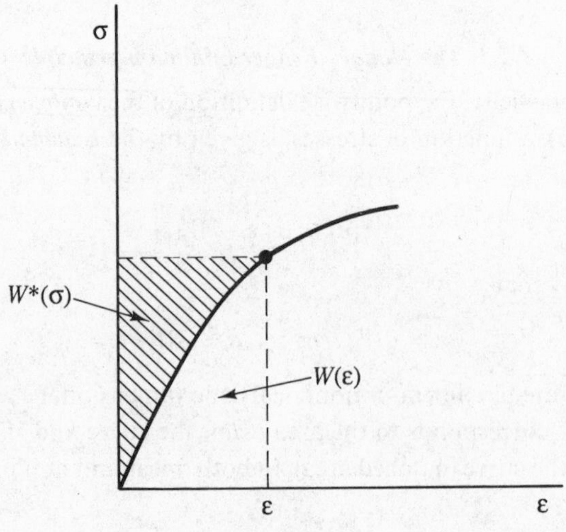

We recall that (see eqns (3.55)–(3.56))

$$\Psi^*(\dot{\boldsymbol{\sigma}}) = \left\{ \begin{array}{ll} \frac{1}{2}\dot{\boldsymbol{\sigma}} : \mathbf{S} : \dot{\boldsymbol{\sigma}} & \text{if} \frac{\partial f}{\partial \boldsymbol{\sigma}} : \dot{\boldsymbol{\sigma}} \leqslant 0, \\ +\infty & \text{if} \frac{\partial f}{\partial \boldsymbol{\sigma}} : \dot{\boldsymbol{\sigma}} > 0. \end{array} \right\} \tag{4.17}$$

If we only take into account the stress rates $\dot{\boldsymbol{\sigma}}^*$ compatible with the plasticity condition

$$\frac{\partial f}{\partial \boldsymbol{\sigma}} : \dot{\sigma} \leqslant 0 \qquad \text{if } f(\boldsymbol{\sigma}) = 0,$$

we have then the following statement of the Hodge–Prager minimum principle:

The rate of stress $\dot{\boldsymbol{\sigma}}$ minimizes the functional

$$B(\dot{\boldsymbol{\sigma}}^*) = \int_\Omega \frac{1}{2} \dot{\boldsymbol{\sigma}}^* : \mathbf{S} : \dot{\boldsymbol{\sigma}}^* \, \mathrm{d}\Omega - \int_{\partial\Omega} \dot{\mathbf{u}}^{\mathrm{d}} \cdot (\dot{\boldsymbol{\sigma}}^* \cdot \mathbf{n}) \, \mathrm{d}s, \tag{4.18}$$

among the rates $\dot{\boldsymbol{\sigma}}^*$ compatible with the plasticity conditions and the static conditions.

Remarks (i) The Greenberg principle allows us to introduce naturally the *weak*, that is, discontinuous, solutions.

(ii) Since $\Psi^\sigma(\dot{\boldsymbol{\varepsilon}})$ is convex but *not strictly convex*, usually there is *non-uniqueness* of the velocity $\dot{\mathbf{u}}$.

(iii) The mixed, two-field (($\mathbf{u}, \boldsymbol{\sigma}$) in elasticity, ($\dot{\mathbf{u}}, \dot{\boldsymbol{\sigma}}$) in velocities in elasto-plasticity) formulations call for variational principles of the *Hellinger–Reissner* type (after Hellinger, 1914, and Reissner, 1953). For this material, Washizu's book (1982) is of great interest. Such variational principles are dealt with by way of problems in Chapter 11, as they are most relevant to numerical implementations.

(iv) We notice that $B(\dot{\boldsymbol{\sigma}})$ defined in (4.18) is *quadratic*, not *strictly* convex on a set K such as

$$K = \left\{ \mathbf{s} \mid \frac{\partial f}{\partial \boldsymbol{\sigma}} : \dot{\mathbf{s}} \leqslant 0 \text{ if } f(\boldsymbol{\sigma}) = 0 \right\}. \tag{4.19}$$

Consequently K is convex and the Hodge–Prager principle is written as

$$\min_{\mathbf{s} \in K \text{ convex}} B(\mathbf{s}), \tag{4.20}$$

for which there is a rather regular *unique* solution (note: this a very difficult mathematical problem which is *not* examined here).

Uniqueness of response in stress (elementary demonstration) The uniqueness of response *in stress* of an elastoplastic material is obvious on a schematic (σ, ε) plot as only one value of stress corresponds to any prescribed value of ε either during the elastic regime or during plastic flow (on the plastic plateau in Fig. 1.8). Obviously the converse is *not* true in perfect elastoplasticity as the whole elastic limit or plastic threshold corresponds to a set of values of ε. An elementary 'elastic-energy evolution' argument allows one to prove the first statement in the full three-dimensional case. This goes as follows.

Let $\boldsymbol{\sigma}_1$ and $\boldsymbol{\sigma}_2$ be the two eventual solutions in stress. The solution $\Delta\boldsymbol{\sigma} = \boldsymbol{\sigma}_1 - \boldsymbol{\sigma}_2$ satisfies

$$\left.\begin{aligned} \operatorname{div}(\Delta\boldsymbol{\sigma}) &= \mathbf{0} \qquad \text{in } \Omega, \\ (\Delta\boldsymbol{\sigma})\cdot\mathbf{n} &= \mathbf{0} \qquad \text{on } \partial\Omega_T \end{aligned}\right\} \tag{4.21}$$

whereas $\Delta\dot{\boldsymbol{\varepsilon}} = \dot{\boldsymbol{\varepsilon}}_1 - \dot{\boldsymbol{\varepsilon}}_2$ satisfies

$$\left.\begin{aligned} \Delta\dot{\boldsymbol{\varepsilon}} &= [\nabla(\Delta\dot{\mathbf{u}})]_S, \\ \Delta\dot{\mathbf{u}} &= \mathbf{0} \qquad \text{on } \partial\Omega_u. \end{aligned}\right\} \tag{4.22}$$

According to the virtual-power principle (2.26) it follows that

$$\int_\Omega \Delta\boldsymbol{\sigma} : \Delta\dot{\boldsymbol{\varepsilon}}\, d\Omega = 0 = \int_\Omega (\Delta\boldsymbol{\sigma}) : \mathbf{S} : (\Delta\dot{\boldsymbol{\sigma}})\, d\Omega + \int_\Omega \Delta\boldsymbol{\sigma} : \Delta\dot{\boldsymbol{\varepsilon}}^{\mathrm{p}}\, d\Omega. \tag{4.23}$$

The last contribution in (4.23) is greater than or equal to zero, so that

$$\int_\Omega (\Delta\boldsymbol{\sigma}) : \mathbf{S} : (\Delta\dot{\boldsymbol{\sigma}})\, d\Omega \leqslant 0$$

and thus

$$\frac{d}{dt} \int_\Omega \frac{1}{2} (\Delta\boldsymbol{\sigma}) : \mathbf{S} : (\Delta\boldsymbol{\sigma})\, d\Omega \leqslant 0. \tag{4.24}$$

If follows that

$$I(t) = \frac{1}{2} \int_\Omega \Delta\boldsymbol{\sigma}(t) : \mathbf{S} : \Delta\boldsymbol{\sigma}(t)\, d\Omega \geqslant 0$$

decreases in time, with $I(0) = 0$. Necessarily, then, we have $\Delta\boldsymbol{\sigma} = 0$. QED.

4.2.4 Mathematical analysis of quasi-static evolution

The *perfectly*-plastic–elastic scheme is the most difficult to study when we wish to establish its general evolution results. The viscoplastic or visco-elastic model is more simple thanks to the regularizing effect of viscosity. The introduction of a strictly positive strain-hardening also regularizes the plastic problem.

The uniqueness of the response *in stress* is guaranteed from the moment that the force data are compatible with the plasticity threshold. On the contrary, the determination of the *displacements* can pose a problem of regularity since the solutions often exhibit some surfaces of discontinuity. The most complete results concerning the *existence* and the regularity in displacement are due to the work of several authors, among them more particularly C. Johnson (1976), Temam and Strang (1980), Temam (1988) and P. Suquet (1979, 1981a). The clue lies in the use of the functional space of functions of *bounded variation* (BD functional spaces). This allows the representation of the observed discontinuities (slip surfaces, plastic hinges). We refer the mathematically inclined reader to the authors cited. Practically a century has elapsed between embryonic formulations of plasticity and these mathematical proofs. This is a rather clear consequence of the singular nature of *perfect* elastoplasticity. Another manner to proceed is to apply the method of *viscoplastic regularization*, i.e., account for the constitutive equation (3.59) and, at the end of the procedure, pass to the limit of vanishing viscosity.

4.2.5 Evolution in stresses

Let $\boldsymbol{\sigma}(t) \in S(t)$ and $\boldsymbol{\sigma}(t) \in P$. We assume that $P \cap S(t) \neq \{\varnothing\}$. $\mathbf{g}(t)$ and $\mathbf{T}^d(t)$ are not arbitrary and their characterization is the subject of a course about fracture (see Chapter 7) and collapse design (see Chapter 6).

Purely elastic response Let us consider the same data $\{\mathbf{g}(t), \mathbf{T}^d(t), \mathbf{u}^d(t)\}$ and the same structure Ω, but this time with a *purely elastic* behaviour. In this case we must look for $\mathbf{u}^E \in \mathscr{U}(t)$ and $\boldsymbol{\sigma}^E \in S(t)$ such that

$$\boldsymbol{\sigma}^E = \mathbf{E} : \boldsymbol{\varepsilon}^E, \qquad \boldsymbol{\varepsilon}^E = (\nabla \mathbf{u}^E)_S. \tag{4.25}$$

This means that we must study a quasi-static evolution problem for $(\boldsymbol{\sigma}^E, \mathbf{u}^E)$ for a linear solid. Variational functional techniques for this case are on solid ground (see, e.g., Duvaut and Lions, 1972, Chap. 3) so that this problem is theoretically solved.

Elastoplastic problem We set

$$\left.\begin{aligned}\boldsymbol{\sigma}(t) &= \boldsymbol{\sigma}^{\mathrm{E}}(t) + \boldsymbol{\sigma}^{\mathrm{r}}(t) = \mathbf{E} : \boldsymbol{\varepsilon}^{\mathrm{e}}(t),\\ \mathbf{u}(t) &= \mathbf{u}^{\mathrm{E}}(t) + \mathbf{u}^{\mathrm{r}}(t),\end{aligned}\right\} \tag{4.26}$$

where $\boldsymbol{\sigma}^{\mathrm{r}}$ is a field of *residual* stresses, which satisfies

$$\left.\begin{aligned}\mathbf{u}^{\mathrm{r}} &\in \mathscr{U}_0 = \{\mathbf{u} | \mathbf{u} = \mathbf{0} \text{ on } \partial\Omega_u\},\\ \boldsymbol{\sigma}^{\mathrm{r}} &\in S_0 = \{\mathbf{s} | \operatorname{div} \mathbf{s} = \mathbf{0} \text{ in } \Omega, \mathbf{s}\cdot\mathbf{n} = \mathbf{0} \text{ on } \partial\Omega_T\},\end{aligned}\right\} \tag{4.27}$$

$$\boldsymbol{\sigma}^{\mathrm{r}} = \mathbf{E} : (\boldsymbol{\varepsilon}^{\mathrm{r}} - \boldsymbol{\varepsilon}^{\mathrm{p}}), \qquad \boldsymbol{\varepsilon}^{\mathrm{r}} = (\nabla \mathbf{u}^{\mathrm{r}})_{\mathrm{S}}. \tag{4.28}$$

The set S_0 is called the set of *self-equilibrated stress* fields (there are no prescribed forces in volume and at surfaces).

The representations (4.28) follow from consideration of the following argument. Inverting (4.26), we have $\boldsymbol{\varepsilon}^{\mathrm{e}} = \mathbf{S} : (\boldsymbol{\sigma}^{\mathrm{E}} + \boldsymbol{\sigma}^{\mathrm{r}})$, so that $\boldsymbol{\varepsilon} = \boldsymbol{\varepsilon}^{\mathrm{e}} + \boldsymbol{\varepsilon}^{\mathrm{p}} = \mathbf{S} : \boldsymbol{\sigma}^{\mathrm{E}} + (\mathbf{S} : \boldsymbol{\sigma}^{\mathrm{r}} + \boldsymbol{\varepsilon}^{\mathrm{p}}) = (\nabla \mathbf{u})_{\mathrm{S}}$. But $\mathbf{S} : \boldsymbol{\sigma}^{\mathrm{E}} \equiv \varepsilon^{\mathrm{E}} = (\nabla \mathbf{u}^{\mathrm{E}})_{\mathrm{S}}$ as $\boldsymbol{\sigma}^{\mathrm{E}}$ corresponds to a purely elastic response. Hence $\mathbf{S} : \boldsymbol{\sigma}^{\mathrm{r}} + \boldsymbol{\varepsilon}^{\mathrm{p}} \equiv \varepsilon^{\mathrm{r}} = (\nabla \mathbf{u}^{\mathrm{r}})_{\mathrm{S}}$ as both $\boldsymbol{\varepsilon}$ and $\boldsymbol{\varepsilon}^{\mathrm{E}}$ are derivable from a displacement. There follows (4.28) immediately. Considering two eventual fields of residual stresses, $\boldsymbol{\sigma}_1^{\mathrm{r}}$ and $\boldsymbol{\sigma}_2^{\mathrm{r}}$, for the same elastoplastic problem – both $\boldsymbol{\sigma}_1^{\mathrm{r}}$ and $\boldsymbol{\sigma}_2^{\mathrm{r}}$ are self-equilibrated – and applying the principle of virtual work allow one to show that

$$\int_\Omega \Delta\boldsymbol{\sigma}^{\mathrm{r}} : \mathrm{S} : \Delta\boldsymbol{\sigma}^{\mathrm{r}} \,\mathrm{d}\Omega = 0, \qquad \Delta\boldsymbol{\sigma}^{\mathrm{r}} \equiv \boldsymbol{\sigma}_1^{\mathrm{r}} - \boldsymbol{\sigma}_2^{\mathrm{r}},$$

so that $\Delta\boldsymbol{\sigma}^{\mathrm{r}} = 0$ by virtue of the nonnegativess of the linear operator $\mathbf{S}$. Whence the *unique* determination of $\boldsymbol{\sigma}^{\mathrm{r}}$ from $\boldsymbol{\varepsilon}^{\mathrm{p}}$ – see eqn (4.40) below.

Let us consider now the evolution problem for $\boldsymbol{\sigma}(t)$. We shall show that $\boldsymbol{\sigma}$ is the solution of a simple variational problem. So we note the following.

(a) The normality law (3.28) integrated over Ω, on the assumption that all fields are sufficiently regular, is written

$$\int_\Omega (\boldsymbol{\sigma} - \boldsymbol{\sigma}^*) : \dot{\boldsymbol{\varepsilon}}^{\mathrm{p}} \,\mathrm{d}\Omega \geqslant 0, \tag{4.29}$$

(b) The integrated equation of equilibrium (PVP; see eqn (2.27)) yields

$$\int_\Omega \boldsymbol{\sigma} : \delta\boldsymbol{\varepsilon} \,\mathrm{d}\Omega = \int_\Omega \mathbf{g} \cdot \delta\mathbf{u} \,\mathrm{d}\Omega + \int_{\partial\Omega} \mathbf{T}^{\mathrm{d}} \cdot \delta\mathbf{u} \,\mathrm{d}s, \qquad \forall \delta\mathbf{u} \text{ with } \delta\mathbf{u}|_{\partial\Omega_u} = 0, \tag{4.30}$$

(c) As a particular case in the application of (4.30) let us take

$$\delta\mathbf{u} = \dot{\mathbf{u}} - \dot{\mathbf{u}}^{\mathrm{E}}. \tag{4.31}$$

Consequently, (4.30) is written as

$$\int_\Omega \boldsymbol{\sigma} : (\dot{\boldsymbol{\varepsilon}} - \dot{\boldsymbol{\varepsilon}}^{\mathrm{E}})\,\mathrm{d}\Omega = \int_\Omega \mathbf{g}\cdot(\dot{\mathbf{u}} - \dot{\mathbf{u}}^{\mathrm{E}})\,\mathrm{d}\Omega + \int_{\partial\Omega} \mathbf{T}^{\mathrm{d}}\cdot(\dot{\mathbf{u}} - \dot{\mathbf{u}}^{\mathrm{E}})\,\mathrm{d}s. \quad (4.32)$$

Let us take now $\boldsymbol{\sigma}$ such that $\boldsymbol{\sigma}^* \in S(t)$. It gives

$$\int_\Omega \boldsymbol{\sigma}^* : (\dot{\boldsymbol{\varepsilon}} - \dot{\boldsymbol{\varepsilon}}^{\mathrm{E}})\,\mathrm{d}\Omega = \int_\Omega \mathbf{g}\cdot(\dot{\mathbf{u}} - \dot{\mathbf{u}}^{\mathrm{E}})\,\mathrm{d}\Omega + \int_{\partial\Omega} \mathbf{T}^{\mathrm{d}}\cdot(\dot{\mathbf{u}} - \dot{\mathbf{u}}^{\mathrm{E}})\,\mathrm{d}s. \quad (4.33)$$

Substracting (4.33) from (4.32), we obtain

$$\int_\Omega (\boldsymbol{\sigma} - \boldsymbol{\sigma}^*) : (\dot{\boldsymbol{\varepsilon}} - \dot{\boldsymbol{\varepsilon}}^{\mathrm{E}})\,\mathrm{d}\Omega = 0 \quad (4.34)$$

or

$$\int_\Omega (\boldsymbol{\sigma} - \boldsymbol{\sigma}^*) : (\dot{\boldsymbol{\varepsilon}}^{\mathrm{e}} + \dot{\boldsymbol{\varepsilon}}^{\mathrm{p}} - \dot{\boldsymbol{\varepsilon}}^{\mathrm{E}})\,\mathrm{d}\Omega = 0. \quad (4.35)$$

But $\dot{\boldsymbol{\varepsilon}}^{\mathrm{e}} = \mathbf{S} : \dot{\boldsymbol{\sigma}}$, and if we also remember the inequality (4.29), the result arising from (4.35) is that

$$\int_\Omega \dot{\boldsymbol{\sigma}} : \mathbf{S} : (\boldsymbol{\sigma}^* - \boldsymbol{\sigma})\,\mathrm{d}\Omega - \int_\Omega (\boldsymbol{\sigma}^* - \boldsymbol{\sigma}) : \dot{\boldsymbol{\varepsilon}}^{\mathrm{E}}\,\mathrm{d}\Omega \geqslant 0 \qquad \text{with } \boldsymbol{\sigma}(t) \in P \cap S(t),$$

$$\forall t, \forall \boldsymbol{\sigma}^* \in P \cap S(t), \quad (4.36)$$

to which we should join the initial condition $\boldsymbol{\sigma}(0) = \boldsymbol{\sigma}_0$.

Eqn (4.36) is a *variational equation of evolution* of a type often studied in mathematics. Let us define

$$\boldsymbol{\sigma} \in X = \{\boldsymbol{\sigma} \text{ defined in } \Omega | \langle \boldsymbol{\sigma}, \boldsymbol{\sigma}^* \rangle = \int_\Omega \boldsymbol{\sigma} : \mathbf{S} : \boldsymbol{\sigma}^*\,\mathrm{d}\Omega\}. \quad (4.37)$$

where the scalar product is that of the *energy*. We can rewrite (4.36) simply as

$$\langle \dot{\boldsymbol{\sigma}} - \dot{\boldsymbol{\sigma}}^{\mathrm{E}}, \boldsymbol{\sigma}^* - \boldsymbol{\sigma} \rangle \geqslant 0, \qquad \forall \boldsymbol{\sigma}^* \in P \cap S(t), \quad (4.38)$$

because $\dot{\boldsymbol{\varepsilon}}^{\mathrm{E}} = \mathbf{S} : \dot{\boldsymbol{\sigma}}$. We notice the similarity between (4.38) and the normality law $\dot{\boldsymbol{\varepsilon}}^{\mathrm{p}} : (\boldsymbol{\sigma} - \boldsymbol{\sigma}^*) \geqslant 0$, which is also written $\dot{\boldsymbol{\varepsilon}}^{\mathrm{p}} \in N_C(\boldsymbol{\sigma})$. Through this analogy, eqn (4.38) may also be written in the form

$$(\dot{\boldsymbol{\sigma}}^{\mathrm{E}} - \dot{\boldsymbol{\sigma}}) \in N_{P \cap S(t)}(\boldsymbol{\sigma}), \; \boldsymbol{\sigma}(0) = \boldsymbol{\sigma}_0, \quad (4.39)$$

where $P \cap S(t)$ is a ***time-dependent*** *convex*. We can say that (4.39) is a *Hill law* or *normality law*. This general equation has been studied by J.J. Moreau

(1977), for any *moving* convex (such as we meet in econometrics) and we get the following particular result.

Uniqueness theorem *The response in stress is unique.*

Indeed, let $\boldsymbol{\sigma}_1$ and $\boldsymbol{\sigma}_2$ be two of the possible responses. We have

$$\langle \dot{\boldsymbol{\sigma}}_1 - \dot{\boldsymbol{\sigma}}^{\mathrm{E}}, \boldsymbol{\sigma}_2 - \boldsymbol{\sigma}_1 \rangle \geqslant 0 \qquad \text{with } \boldsymbol{\sigma} = \boldsymbol{\sigma}_1, \boldsymbol{\sigma}^* = \boldsymbol{\sigma}_2,$$

$$\langle \dot{\boldsymbol{\sigma}}_2 - \dot{\boldsymbol{\sigma}}^{\mathrm{E}}, \boldsymbol{\sigma}_1 - \boldsymbol{\sigma}_2 \rangle \geqslant 0 \qquad \text{with } \boldsymbol{\sigma} = \boldsymbol{\sigma}_2, \boldsymbol{\sigma}^* = \boldsymbol{\sigma}_1,$$

whence

$$\langle \dot{\boldsymbol{\sigma}}_1 - \dot{\boldsymbol{\sigma}}_2, \boldsymbol{\sigma}_2 - \boldsymbol{\sigma}_1 \rangle \geqslant 0,$$

i.e.

$$\frac{\mathrm{d}}{\mathrm{d}t} \|\boldsymbol{\sigma}_1 - \boldsymbol{\sigma}_2\|^2 \leqslant 0,$$

which, through integration, gives

$$\|\boldsymbol{\sigma}_1(t) - \boldsymbol{\sigma}_2(t)\|^2 \leqslant \|\boldsymbol{\sigma}_1(0) - \boldsymbol{\sigma}_2(0)\|^2, \qquad \forall t \geqslant 0.$$

This inequality shows that the distance (measured by the energy norm) between two possible solutions can only decrease. The uniqueness is then the result of the initial conditions $\boldsymbol{\sigma}_1(0) = \boldsymbol{\sigma}_2(0) = \boldsymbol{\sigma}_0$. Q.E.D.

Let us recall that there is nothing we may conclude concerning the eventual nonunique displacements.

4.2.6 Evolution of plastic strains

Strain rates are not necessarily regular. Since stress is regular, this irregularity is due to the plastic strain rate. More specifically, plastic strain may be concentrated in relatively small regions, eventually producing surfaces of discontinuity. This is the phenomenon of *localized strain.* As long as the plastic strain remains regular (e.g., when components of $\boldsymbol{\varepsilon}^{\mathrm{p}}$ are $L^2(\Omega)$)† we may formulate its evolution on the basis of the expression

$$\boldsymbol{\sigma}^{\mathrm{r}} = -\mathbf{Z}\boldsymbol{\varepsilon}^{\mathrm{p}}, \tag{4.40}$$

where $\mathbf{Z}$ is the linear operator giving the stress due to a residual strain $\boldsymbol{\varepsilon}^{\mathrm{p}}$. With $\boldsymbol{\sigma} = \boldsymbol{\sigma}^{\mathrm{E}} + \boldsymbol{\sigma}^{\mathrm{r}}$ we can then re-write the normality law $\dot{\boldsymbol{\varepsilon}}^{\mathrm{p}} \in N_{\mathrm{p}}(\boldsymbol{\sigma})$ in its general form

$$\langle \dot{\boldsymbol{\varepsilon}}^{\mathrm{p}}, \boldsymbol{\sigma}^{\mathrm{E}} - \mathbf{Z}\boldsymbol{\varepsilon}^{\mathrm{p}} - \boldsymbol{\sigma}^* \rangle \geqslant 0, \qquad \forall \boldsymbol{\sigma}^* \in P, \tag{4.41}$$

† $L^2(\Omega)$ denotes the space of square-integrable functions (Hilbert space); $H^1(\Omega)$ denotes the Sobolev space $W^{1,2}(\Omega)$ (see Duvaut and Lions (1972)).

or, also, as

$$\dot{\boldsymbol{\varepsilon}}^p \in N_p(\boldsymbol{\sigma}^E - \mathbf{Z}\boldsymbol{\varepsilon}^p). \tag{4.42}$$

This kind of equation is useful in studies of plastic stability.

Note: In *discretized* system such as

$$\boldsymbol{\varepsilon} = \mathbf{Bu}, \qquad \boldsymbol{\sigma} = \mathbf{D}(\boldsymbol{\varepsilon} - \boldsymbol{\varepsilon}^p), \qquad \tilde{\mathbf{B}}\boldsymbol{\sigma} = \mathbf{F},$$

we work directly on the discretized form, with

purely elastic response $\boldsymbol{\varepsilon}^E = \mathbf{Bu}^E$, $\boldsymbol{\sigma}^E = \mathbf{D}\boldsymbol{\varepsilon}^E$, $\tilde{\mathbf{B}}\boldsymbol{\sigma}^E = \mathbf{F}$,
residual response $\boldsymbol{\varepsilon}^r = \mathbf{Bu}^r$, $\boldsymbol{\sigma}^r = \mathbf{D}(\boldsymbol{\varepsilon}^r - \boldsymbol{\varepsilon}^p)$, $\tilde{\mathbf{B}}\boldsymbol{\sigma}^r = 0$,

and

$$\boldsymbol{\sigma}^r = -\mathbf{Z}\boldsymbol{\varepsilon}^p, \qquad \boldsymbol{\sigma} = \boldsymbol{\sigma}^E + \boldsymbol{\sigma}^r, \qquad \mathbf{u} = \mathbf{u}^E + \mathbf{u}^r. \tag{4.43}$$

We show that

$$\mathbf{Z} = -\mathbf{DBK}^{-1}\tilde{\mathbf{B}}\mathbf{D} + \mathbf{D} \tag{4.44}$$

where $\mathbf{K} = \tilde{\mathbf{B}}\mathbf{DB}$ is the rigidity matrix

4.3 Asymptotic behaviour: shakedown

4.3.1 Practical motivation

We observe that an alternating plastic strain quickly causes a fracture (example, the string we just cut); an *accumulated* plastic strain also causes a fracture. We easily understand that we would like to avoid this and that we should wish to be able to control these two phenomena. We shall pay particular attention to the phenomenon of *shakedown* (see Fig. 1.8) where the plastic strain is stabilized (in certain cases the response may even become purely elastic). The initial studies in this field were carried out by the Austrian engineer Melan (1938) with some highly interesting results by Symonds (1951) and Koiter (1960). The problem is to study the *asymptotic* behaviour of the solutions $(\mathbf{u}(t), \boldsymbol{\sigma}(t))$ when $t \to \infty$, while the load path is described by the given purely elastic response $(\mathbf{u}^E(t), \boldsymbol{\sigma}^E(t))$. The exact definition of shakedown is as follows:

Definition We say that a structure adapts to the loads in question if, for $t \to \infty$, the system's response tends to become purely elastic, i.e.

$$\mathbf{u}^r(t) \to \mathbf{u}^r_\infty, \quad \boldsymbol{\sigma}^r(t) \to \boldsymbol{\sigma}^r_\infty, \quad \boldsymbol{\varepsilon}^p(t) \to \boldsymbol{\varepsilon}^p_\infty. \tag{4.45}$$

We can then state the following essential theorem.

4.3.2 The Melan–Koiter Theorem

If there are a safety coefficient $m > 1$ and a constant field of residual stresses σ^r_* such that

$$m(\boldsymbol{\sigma}^r_* + \boldsymbol{\sigma}^E(t)) \in P, \quad \forall t, \tag{4.46}$$

then there is shakedown for any initial data.

Proof This is given in three steps:

(i) we prove first that the plastic power is bounded;

(ii) we show that, when $t \to \infty$, the stress $\boldsymbol{\sigma}(t)$ converges toward $\boldsymbol{\sigma}^r_* + \boldsymbol{\sigma}^E$; then,

(iii) if $\boldsymbol{\sigma} - \boldsymbol{\sigma}^E(t)$ converges when $t \to \infty$, then $\boldsymbol{\varepsilon}^p(t)$ converges when $t \to \infty$ (the strain hardening must be positive or zero.)

Let us examine the first two steps.

(i) We have a dissipated plastic power (positive or zero) expressed by

$$\Phi_p = \int_\Omega \boldsymbol{\sigma} : \dot{\boldsymbol{\varepsilon}}^p \, d\Omega. \tag{4.47}$$

The normality law integrated over Ω gives (eqn(4.29))

$$\int_\Omega (\boldsymbol{\sigma} - \boldsymbol{\sigma}^*) : \dot{\boldsymbol{\varepsilon}}^p \, d\Omega \geqslant 0. \tag{4.48}$$

Let us take successively

$$\boldsymbol{\sigma}^* = \boldsymbol{\sigma}^*_1 = \boldsymbol{\sigma}^r_* + \boldsymbol{\sigma}^E(t) \in P$$

and

$$\boldsymbol{\sigma}^* = \boldsymbol{\sigma}^*_2 = m(\boldsymbol{\sigma}^r_* + \boldsymbol{\sigma}^E(t)) \in P,$$

when m represents a homothety of the convex. Starting from (4.48) we obtain then

$$\left.\begin{aligned} &\int_\Omega (\boldsymbol{\sigma} - \boldsymbol{\sigma}^r_* - \boldsymbol{\sigma}^E) : \dot{\boldsymbol{\varepsilon}}^p \, d\Omega \geqslant 0, \\ &\int_\Omega (\boldsymbol{\sigma} - m\boldsymbol{\sigma}^r_* - m\boldsymbol{\sigma}^E) : \dot{\boldsymbol{\varepsilon}}^p \, d\Omega \geqslant 0, \end{aligned}\right\} \tag{4.49}$$

or else, by combining $[(4.49)_2 + (1/m - 1)\,(4.49)_2]$,

$$\int_\Omega \dot{\boldsymbol{\varepsilon}}^p : \left[\boldsymbol{\sigma} - \frac{m-1}{m}\boldsymbol{\sigma} - (\boldsymbol{\sigma}^r_* + \boldsymbol{\sigma}^E) \right] d\Omega \geqslant 0,$$

that is, the left hand side of $(4.49)_1$ is such that

$$\int_\Omega \dot{\boldsymbol{\varepsilon}}^{\mathrm{p}} : [\boldsymbol{\sigma} - (\boldsymbol{\sigma}^{\mathrm{r}}_* + \boldsymbol{\sigma}^{\mathrm{E}})]\,\mathrm{d}\Omega \geqslant \frac{m-1}{m}\Phi_{\mathrm{p}}, \tag{4.50}$$

and by integration over time we get

$$\int_0^t \mathrm{d}\tau \int_\Omega \dot{\boldsymbol{\varepsilon}}^{\mathrm{p}} : (\boldsymbol{\sigma} - \boldsymbol{\sigma}^{\mathrm{r}}_* - \boldsymbol{\sigma}^{\mathrm{E}})\,\mathrm{d}\Omega \geqslant \frac{m-1}{m}\int_0^t \Phi_{\mathrm{p}}(\tau)\,\mathrm{d}\tau, \tag{4.51}$$

But

$$\int_\Omega \dot{\boldsymbol{\varepsilon}}^{\mathrm{p}} : (\boldsymbol{\sigma} - \boldsymbol{\sigma}^{\mathrm{r}}_* - \boldsymbol{\sigma}^{\mathrm{E}})\,\mathrm{d}\Omega = -\int_\Omega (\boldsymbol{\sigma} - \boldsymbol{\sigma}^{\mathrm{r}}_* - \boldsymbol{\sigma}^{\mathrm{E}}) : (\dot{\boldsymbol{\varepsilon}}^{\mathrm{e}} - \dot{\boldsymbol{\varepsilon}}^{\mathrm{E}})\,\mathrm{d}\Omega \tag{4.52}$$

because

$$\int_\Omega (\dot{\boldsymbol{\varepsilon}} - \dot{\boldsymbol{\varepsilon}}^{\mathrm{E}}) : (\boldsymbol{\sigma} - \boldsymbol{\sigma}^{\mathrm{r}}_* - \boldsymbol{\sigma}^{\mathrm{E}})\,\mathrm{d}\Omega = 0 \tag{4.53}$$

(this last result arises from the following fact: if the stress $\hat{\boldsymbol{\sigma}}$ satisfies $\operatorname{div}\hat{\boldsymbol{\sigma}} = \mathbf{0}$ in Ω, $\hat{\boldsymbol{\sigma}}\cdot\mathbf{n} = \mathbf{0}$ on $\partial\Omega_T$, and $\mathbf{v}$ satisfies $\mathbf{v} = \mathbf{0}$ on $\partial\Omega_u$, then $\int_\Omega \boldsymbol{\varepsilon}(\mathbf{v}) : \hat{\boldsymbol{\sigma}}\,\mathrm{d}\Omega = 0$ (PVP); we may say that $\hat{\boldsymbol{\sigma}}$ and $\boldsymbol{\varepsilon}(\mathbf{v})$ are orthogonal, which is the case with the values present in (4.53) because $\hat{\boldsymbol{\sigma}} = \boldsymbol{\sigma} - \boldsymbol{\sigma}^{\mathrm{r}}_* - \boldsymbol{\sigma}^{\mathrm{E}} \in S_0$ and $\mathbf{v} = \dot{\mathbf{u}} - \dot{\mathbf{u}}^{\mathrm{E}} \in \mathscr{U}_0$ according to eqns (4.27)). Moreover, $\dot{\boldsymbol{\varepsilon}}^{\mathrm{p}} = \dot{\boldsymbol{\varepsilon}} - \dot{\boldsymbol{\varepsilon}}^{\mathrm{e}} = \dot{\boldsymbol{\varepsilon}} - \dot{\boldsymbol{\varepsilon}}^{\mathrm{E}} - (\dot{\boldsymbol{\varepsilon}}^{\mathrm{e}} - \dot{\boldsymbol{\varepsilon}}^{\mathrm{E}})$, QED. We may then rewrite (4.52) in the form

$$\int_\Omega \dot{\boldsymbol{\varepsilon}}^{\mathrm{p}} : (\boldsymbol{\sigma} - \boldsymbol{\sigma}^{\mathrm{r}}_* - \boldsymbol{\sigma}^{\mathrm{E}})\,\mathrm{d}\Omega = -\langle \boldsymbol{\sigma} - \boldsymbol{\sigma}^{\mathrm{r}}_* - \boldsymbol{\sigma}^{\mathrm{E}}, \dot{\boldsymbol{\sigma}} - \dot{\boldsymbol{\sigma}}^{\mathrm{E}} \rangle, \tag{4.54}$$

with $\langle\cdot,\cdot\rangle$ denoting the energy scalar-product

$$\langle \mathbf{A}, \mathbf{B} \rangle \int_\Omega \mathbf{A} : \mathbf{S} : \mathbf{B}\,\mathrm{d}\Omega$$

and

$$\dot{\boldsymbol{\varepsilon}}^{\mathrm{e}} - \dot{\boldsymbol{\varepsilon}}^{\mathrm{E}} = \mathbf{S} : \dot{\boldsymbol{\sigma}} - \mathbf{S} : \dot{\boldsymbol{\sigma}}^{\mathrm{E}} = \mathbf{S} : (\dot{\boldsymbol{\sigma}} - \dot{\boldsymbol{\sigma}}^{\mathrm{E}}) = \mathbf{S}\,(\dot{\boldsymbol{\sigma}} - \dot{\boldsymbol{\sigma}}^{\mathrm{r}}_* - \dot{\boldsymbol{\sigma}}^{\mathrm{E}}), \tag{4.55}$$

since $\dot{\boldsymbol{\sigma}}^{\mathrm{r}}_* = \mathbf{0}$. Finally, (4.50) is written as

$$-\langle \boldsymbol{\sigma} - \boldsymbol{\sigma}^{\mathrm{r}}_* - \boldsymbol{\sigma}^{\mathrm{E}}, \dot{\boldsymbol{\sigma}} - \dot{\boldsymbol{\sigma}}^{\mathrm{r}}_* - \dot{\boldsymbol{\sigma}}^{\mathrm{E}} \rangle \geqslant \frac{m-1}{m}\Phi_{\mathrm{p}}, \tag{4.56}$$

and by integration over time

$$-\int_0^t \frac{\mathrm{d}\tau}{2}\left[\frac{\mathrm{d}}{\mathrm{d}\tau}\|\boldsymbol{\sigma} - \boldsymbol{\sigma}^{\mathrm{r}}_* - \boldsymbol{\sigma}^{\mathrm{E}}\|^2\right] \geqslant \frac{m-1}{m}\int_0^t \Phi_{\mathrm{p}}(\tau)\,\mathrm{d}\tau, \tag{4.57}$$

so that

$$\tfrac{1}{2}\|\boldsymbol{\sigma}(0)-\boldsymbol{\sigma}^{\mathrm{r}}_{*}-\boldsymbol{\sigma}^{\mathrm{E}}(0)\|^2 \geqslant \int_0^t \mathrm{d}\tau \int_\Omega \dot{\boldsymbol{\varepsilon}}^{\mathrm{p}} : (\boldsymbol{\sigma}-\boldsymbol{\sigma}^{\mathrm{r}}_{*}-\boldsymbol{\sigma}^{\mathrm{E}})\,\mathrm{d}\Omega \geqslant \frac{m-1}{m} W_{\mathrm{p}}(t), \tag{4.58}$$

where

$$W_{\mathrm{p}}(t) = \int_0^t \Phi_{\mathrm{p}}(\tau)\,\mathrm{d}\tau \tag{4.59}$$

is the *dissipated plastic work*. As a consequence of (4.55), we see that, independently of the initial datum $\boldsymbol{\sigma}(0)$, the dissipated plastic work *remains bounded.*

(ii) By contraction, we show that the residual stress $\boldsymbol{\sigma}^{\mathrm{r}}$ has a limit $\boldsymbol{\sigma}^{\mathrm{r}}_{\infty}$. Let $t_1 < t_2$ and $\boldsymbol{\sigma}^{\mathrm{r}}(t_1) = \boldsymbol{\sigma}^{\mathrm{r}}_1$ and $\boldsymbol{\sigma}^{\mathrm{r}}(t_2) = \boldsymbol{\sigma}^{\mathrm{r}}_2$. We try to estimate $\|\boldsymbol{\sigma}^{\mathrm{r}}_1 - \boldsymbol{\sigma}^{\mathrm{r}}_2\|$. We verify that

$$\begin{aligned}\tfrac{1}{2}\|\boldsymbol{\sigma}^{\mathrm{r}}_1-\boldsymbol{\sigma}^{\mathrm{r}}_2\|^2 &= \int_{t_1}^{t_2} \langle \dot{\boldsymbol{\sigma}}^{\mathrm{r}}, \boldsymbol{\sigma}^{\mathrm{r}}-\boldsymbol{\sigma}^{\mathrm{r}}_1\rangle\,\mathrm{d}t \\ &= -\int_{t_1}^{t_2}\mathrm{d}t \int_\Omega \dot{\boldsymbol{\varepsilon}}^{\mathrm{p}} : (\boldsymbol{\sigma}^{\mathrm{r}}-\boldsymbol{\sigma}^{\mathrm{r}}_1)\,\mathrm{d}\Omega.\end{aligned} \tag{4.60}$$

But

$$\begin{aligned}-\dot{\boldsymbol{\varepsilon}}^{\mathrm{p}} : (\boldsymbol{\sigma}^{\mathrm{r}}-\boldsymbol{\sigma}^{\mathrm{r}}_1) &= -\dot{\boldsymbol{\varepsilon}}^{\mathrm{p}} : (\boldsymbol{\sigma}-\boldsymbol{\sigma}^{\mathrm{E}}-\boldsymbol{\sigma}^{\mathrm{r}}_{*}+(\boldsymbol{\sigma}_1-\boldsymbol{\sigma}^{\mathrm{E}}_1-\boldsymbol{\sigma}^{\mathrm{r}}_{*})) \\ &= -\dot{\boldsymbol{\varepsilon}}^{\mathrm{p}} : (\boldsymbol{\sigma}-\boldsymbol{\sigma}^{\mathrm{r}}_{*}-\boldsymbol{\sigma}^{\mathrm{E}}) + \dot{\boldsymbol{\varepsilon}}^{\mathrm{p}} : \boldsymbol{\sigma}_1 - \dot{\boldsymbol{\varepsilon}}^{\mathrm{p}} : (\boldsymbol{\sigma}^{\mathrm{E}}_1 + \sigma^{\mathrm{r}}_{*})\end{aligned} \tag{4.61}$$

The first contribution is nonpositive. The second is bounded from above by $\boldsymbol{\sigma} : \dot{\boldsymbol{\varepsilon}}^{\mathrm{p}}$ since $\boldsymbol{\sigma}_1 \in C$ and $(\boldsymbol{\sigma}-\boldsymbol{\sigma}_1) : \dot{\boldsymbol{\varepsilon}}^{\mathrm{p}} \geqslant 0$, and $-(\boldsymbol{\sigma}^{\mathrm{E}}_1 + \boldsymbol{\sigma}^{\mathrm{r}}_{*}) \in C$. Thus

$$\tfrac{1}{2}\|\boldsymbol{\sigma}^{\mathrm{r}}_1-\boldsymbol{\sigma}^{\mathrm{r}}_2\|^2 \leqslant 2[W_{\mathrm{p}}(t_2) - W_{\mathrm{p}}(t_1)], \tag{4.62}$$

which shows that the sequence $\boldsymbol{\sigma}^{\mathrm{r}}(t_i)$, $t_i \to \infty$, is a Cauchy sequence, and so convergent, given that $W(t)$ is bounded, according to point (i). So with (iii) we shall have proven the theorem (we must still prove (iii)).

Remarks (a) The manipulation (4.61) requires that the elasticity domain C be symmetrical with respect to the origin (which is the case for the Tresca and Mises criteria). If this is not the case, but it remains bounded, the method is still valid, provided that we bring to it some minor modifications.

(b) The Melan–Koiter Theorem corresponds to a *static* approach of shakedown. Koiter (1960) has also given a dual kinematic approach – see pp. 361–4 in Kachanov's book (1974).

(c) When the material is liable to be strain-hardened, we must apply the generalized notation of Chapter 5.

4.4 Remark on discontinuities

If a velocity discontinuity appears at a surface Σ in the volume Ω the strain rate $\dot{\boldsymbol{\varepsilon}}$ is a *distribution* (i.e., a generalized function in the sense of Schwartz and Sobolev) defined on Σ by

$$\dot{\varepsilon}_{ij} = \tfrac{1}{2}([\![v_i]\!]n_j + [\![v_j]\!]n_i) \tag{4.63}$$

where $[\![\ ]\!]$ is the discontinuity in the direction of unit normal $\mathbf{n}$ at Σ. Since the stress rate is regular (i.e., $\dot{\sigma}_{ij} \in L^2(\Omega)$) the decomposition $\dot{\boldsymbol{\varepsilon}} = \dot{\boldsymbol{\varepsilon}}^{\mathrm{e}} + \dot{\boldsymbol{\varepsilon}}^{\mathrm{p}}$ and $\dot{\boldsymbol{\varepsilon}}^{\mathrm{e}} = \mathbf{S} : \dot{\boldsymbol{\sigma}}$ show that the *plastic* strain rate at Σ is to be written

$$\dot{\varepsilon}^{\mathrm{p}}_{ij} = \tfrac{1}{2}([\![v_i]\!]n_j + [\![v_j]\!]n_i). \tag{4.64}$$

The constitutive equation (of normality) $(\boldsymbol{\sigma} - \boldsymbol{\sigma}^*) : \dot{\boldsymbol{\varepsilon}}^{\mathrm{p}} \geqslant 0, \forall \boldsymbol{\sigma}^* \in C$, entails certain restrictions on the discontinuity $[\![v]\!]$ because $\dot{\boldsymbol{\varepsilon}}^{\mathrm{p}}$ must be an outward normal to the elasticity domain. For example, for the Mises criterion we show that $[\![\mathbf{v}]\!] \cdot \mathbf{n} = 0$ (e.g., Salençon, 1979). If we accept the condition that velocity be regular outside the discontinuity surfaces, then the total dissipated plastic power Φ_{p} is

$$\Phi_{\mathrm{p}} = \int_{\Omega} \boldsymbol{\sigma} : \dot{\boldsymbol{\varepsilon}}^{\mathrm{p}} \,\mathrm{d}\Omega = \int_{\Omega-\Sigma} \boldsymbol{\sigma} : \dot{\boldsymbol{\varepsilon}}^{\mathrm{p}} \,\mathrm{d}\Omega + \int_{\Sigma} [\![\mathbf{v}]\!] \cdot (\boldsymbol{\sigma} \cdot \mathbf{n}) \,\mathrm{d}s. \tag{4.65}$$

The matter is developed in Kachanov (1974, pp. 308–318) and Halphen and Salençon (1987).

With this we conclude this chapter. The following appendix recalls the minimum principles in elasticity. The accompanying problems should give more insight into elastoplasticity problems for structures in the case of *perfect* elastoplasticity. More numerous cases of shakedown may be found in Sawczuk (1989), König (1987), Halphen and Salençon (1987), and Desbordes (1977, for mathematical properties).

Appendix to Chapter 4: Minimum principles in elasticity

4.A.1 Minimum principles in linear elasticity

We consider problems with *mixed boundary condition*, of the type

$$\left.\begin{aligned} \operatorname{div} \boldsymbol{\sigma} + \mathbf{g} = 0 &\quad \text{in } \Omega \\ \boldsymbol{\sigma} \cdot \mathbf{n} = \mathbf{T}^{\mathrm{d}} &\quad \text{on } \partial\Omega_T, \\ \mathbf{u} = \mathbf{u}^{\mathrm{d}} &\quad \text{on } \partial\Omega_u, \end{aligned}\right\} \tag{A.1}$$

with

$$\boldsymbol{\sigma} = \partial W/\partial \boldsymbol{\varepsilon} = \mathbf{E} : \boldsymbol{\varepsilon}, \qquad \boldsymbol{\varepsilon} = (\nabla \mathbf{u})_S, \tag{A.2}$$

where W is the energy per unit volume such that

$$W = W(\boldsymbol{\varepsilon}) = \tfrac{1}{2}\boldsymbol{\varepsilon} : \mathbf{E}\,\boldsymbol{\varepsilon} \geqslant m\varepsilon_{ij}\varepsilon_{ij} \geqslant 0, \qquad m > 0, \forall \varepsilon_{ij}. \tag{A.3}$$

We define the *complementary energy* per unit volume by the Legendre transform

$$W^*(\boldsymbol{\sigma}) = \boldsymbol{\sigma} : \boldsymbol{\varepsilon} - W(\boldsymbol{\varepsilon}) = \tfrac{1}{2}\boldsymbol{\sigma} : \boldsymbol{\varepsilon}(\boldsymbol{\sigma}) = \tfrac{1}{2}\boldsymbol{\sigma} : \mathbf{S} : \boldsymbol{\sigma} \tag{A.4}$$

with

$$\boldsymbol{\varepsilon} = \frac{\partial W^*}{\partial \boldsymbol{\sigma}} = \mathbf{S} : \boldsymbol{\sigma}. \tag{A.5}$$

We define the *total potential energy* $I(\xi)$ for the displacement field ξ and the total complementary *energy* $J(\mathbf{s})$ for the stress field $\mathbf{s}$ by

$$I(\xi) = \int_\Omega W(\boldsymbol{\varepsilon}(\xi))\,\mathrm{d}\Omega - \int_\Omega \mathbf{g}\cdot\xi\,\mathrm{d}\Omega - \int_{\partial\Omega} \mathbf{T}^{\mathrm{d}}\cdot\xi\,\mathrm{d}s \tag{A.6}$$

and

$$J(\mathbf{s}) = \int_\Omega W^*(\mathbf{s})\,\mathrm{d}\Omega - \int_{\partial\Omega_u} \mathbf{u}^{\mathrm{d}}\cdot(\mathbf{s}\cdot\mathbf{n})\,\mathrm{d}s, \tag{A.7}$$

respectively.

We have then the following minimum principles.

Minimum of potential energy *Among all the kinematically admissible fields $\xi \in \mathscr{U}$ for a given problem, the actual displacement field is the one that minimizes the total potential energy:*

$$I(\mathbf{u}) = \min_{\xi\in\mathscr{U}} I(\xi). \tag{A.8}$$

Minimum of total complementary energy *Among all the statically admissible fields $\mathbf{s} \in S$ for a given problem the field of actual stresses is the one that minimizes the total complementary energy:*

$$J(\boldsymbol{\sigma}) = \min_{\mathbf{s}\in S} J(\mathbf{s}). \tag{A.9}$$

Proof of (A.8) To the quadratic form $W(\boldsymbol{\varepsilon})$ defined in (A.8) we associate a symmetrical bilinear form $W(\boldsymbol{\varepsilon}_1, \boldsymbol{\varepsilon}_2)$ such that

$$W(\boldsymbol{\varepsilon}_1, \boldsymbol{\varepsilon}_2) = \tfrac{1}{2}\boldsymbol{\varepsilon}_1 : \mathbf{E} : \boldsymbol{\varepsilon}_2 = W(\boldsymbol{\varepsilon}_2, \boldsymbol{\varepsilon}_1) = \tfrac{1}{2}\boldsymbol{\sigma}_1 : \boldsymbol{\varepsilon}_2 = \tfrac{1}{2}\boldsymbol{\varepsilon}_1 : \boldsymbol{\sigma}_2, \tag{A.10}$$

We verify that

$$W^*(\boldsymbol{\sigma}_1, \boldsymbol{\sigma}_2) = \tfrac{1}{2}\boldsymbol{\sigma}_1 : \mathbf{S} : \boldsymbol{\sigma}_2 = W^*(\boldsymbol{\sigma}_2, \boldsymbol{\sigma}_1) \tag{A.11}$$

is a symmetric bilinear form with $\boldsymbol{\sigma}_1 = \mathbf{E} : \boldsymbol{\varepsilon}_1$ and $\boldsymbol{\sigma}_2 = \mathbf{E} : \boldsymbol{\varepsilon}_2$.

Let

$$\mathscr{W}(\boldsymbol{\varepsilon}_1, \boldsymbol{\varepsilon}_2) = \int_\Omega W(\boldsymbol{\varepsilon}_1, \boldsymbol{\varepsilon}_2)\,\mathrm{d}\Omega = \mathscr{W}^*(\boldsymbol{\sigma}_1, \boldsymbol{\sigma}_2). \tag{A.12}$$

The principle of virtual work (PVW; eqn (2.27))

$$\int_\Omega \boldsymbol{\sigma} : \boldsymbol{\varepsilon}^*\,\mathrm{d}\Omega = \int_\Omega \mathbf{g} \cdot \mathbf{u}^*\,\mathrm{d}\Omega + \int_{\partial\Omega} \mathbf{T}^\mathrm{d} \cdot \mathbf{u}^*\,\mathrm{d}s \tag{A.13}$$

with

$$\boldsymbol{\varepsilon}^* = \boldsymbol{\varepsilon}(\mathbf{u}^*) = (\nabla\mathbf{u}^*)_\mathrm{S}$$

given the definition (A.12) can also be written

$$2\mathscr{W}(\boldsymbol{\varepsilon}, \boldsymbol{\varepsilon}^*) = \int_\Omega \mathbf{g} \cdot \mathbf{u}^*\,\mathrm{d}\Omega + \int_{\partial\Omega} \mathbf{T}^\mathrm{d} \cdot \mathbf{u}^*\,\mathrm{d}s. \tag{A.14}$$

This formula leads directly to

the *work theorem* (if $\mathbf{u}^* = \mathbf{u}$)

$$2\mathscr{W} = \int_\Omega \mathbf{g} \cdot \mathbf{u}\,\mathrm{d}\Omega + \int_{\partial\Omega} \mathbf{T} \cdot \mathbf{u}\,\mathrm{d}s, \tag{A.15}$$

the *uniqueness theorem* in stresses,

the *reciprocity theorem* (resulting from the *symmetry* of $W(\boldsymbol{\varepsilon}_1, \boldsymbol{\varepsilon}_2)$)

$$\int_\Omega \mathbf{g}_1 \cdot \mathbf{u}_2\,\mathrm{d}\Omega + \int_{\partial\Omega} \mathbf{T}_1^\mathrm{d} \cdot \mathbf{u}_2\,\mathrm{d}s = \int_\Omega \mathbf{g}_2 \cdot \mathbf{u}_1\,\mathrm{d}\Omega + \int_{\partial\Omega} \mathbf{T}_2^\mathrm{d} \cdot \mathbf{u}_1\,\mathrm{d}s, \tag{A.16}$$

the expression (A.18).

In fact, we may write

$$2W(\boldsymbol{\varepsilon}_1, \boldsymbol{\varepsilon}_2) = W(\boldsymbol{\varepsilon}_1 + \boldsymbol{\varepsilon}_2) - W(\boldsymbol{\varepsilon}_1) - W(\boldsymbol{\varepsilon}_2) \tag{A.17}$$

and thus show that

$$\mathscr{W}(\boldsymbol{\varepsilon}_1 + \boldsymbol{\varepsilon}_2) - \mathscr{W}(\boldsymbol{\varepsilon}_1) = 2\mathscr{W}(\boldsymbol{\varepsilon}_1, \boldsymbol{\varepsilon}_2) + \mathscr{W}(\boldsymbol{\varepsilon}_2), \tag{A.18}$$

$$\mathscr{W}^*(\boldsymbol{\sigma}_1 + \boldsymbol{\sigma}_2) - \mathscr{W}^*(\boldsymbol{\sigma}_1) = 2\mathscr{W}^*(\boldsymbol{\sigma}_1, \boldsymbol{\sigma}_2) + \mathscr{W}^*(\boldsymbol{\sigma}_2). \tag{A.19}$$

Let it be noticed that we must take

$$\int_{\partial\Omega} \mathbf{T}^{\mathrm{d}} \cdot \mathbf{u}\, \mathrm{d}s = \int_{\partial\Omega_T} \mathbf{T}^{\mathrm{d}} \cdot \mathbf{u}\, \mathrm{d}s + \int_{\partial\Omega_u} \mathbf{u}^{\mathrm{d}} \cdot (\boldsymbol{\sigma} \cdot \mathbf{n})\, \mathrm{d}s. \tag{A.20}$$

We have

$$I(\boldsymbol{\varepsilon}') \equiv \mathscr{W}(\boldsymbol{\varepsilon}') - \int_{\Omega} \mathbf{g} \cdot \mathbf{u}'\, \mathrm{d}\Omega - \int_{\partial\Omega_T} \mathbf{T}^{\mathrm{d}} \cdot \mathbf{u}'\, \mathrm{d}s, \tag{A.21}$$

$$J(\mathbf{s}') \equiv \mathscr{W}^*(\boldsymbol{\sigma}') - \int_{\partial\Omega_u} \mathbf{u}^{\mathrm{d}} \cdot (\boldsymbol{\sigma}' \cdot \mathbf{n})\, \mathrm{d}s, \tag{A.22}$$

where

$$\mathscr{W}(\boldsymbol{\varepsilon}') = \int_{\Omega} W(\boldsymbol{\varepsilon}')\, \mathrm{d}\Omega, \qquad \mathscr{W}^*(\boldsymbol{\sigma}') = \int_{\Omega} W^*(\boldsymbol{\sigma}')\, \mathrm{d}\Omega. \tag{A.23}$$

Let us take $\mathbf{u}' = \mathbf{u} + \mathbf{u}^*$, where $\mathbf{u}$ is the real displacement field. According to (A.14) we have

$$2\mathscr{W}(\boldsymbol{\varepsilon}, \boldsymbol{\varepsilon}^*) = \int_{\Omega} \mathbf{g} \cdot \mathbf{u}^*\, \mathrm{d}\Omega + \int_{\partial\Omega_T} \mathbf{T}^{\mathrm{d}} \cdot \mathbf{u}^*\, \mathrm{d}s, \tag{A.24}$$

Let us calculate $I(\boldsymbol{\varepsilon}') - I(\boldsymbol{\varepsilon})$ with the help of the definition (A.21). We have

$$\begin{aligned} I(\boldsymbol{\varepsilon}') - I(\boldsymbol{\varepsilon}) &= \mathscr{W}(\boldsymbol{\varepsilon} + \boldsymbol{\varepsilon}^*) - \mathscr{W}(\boldsymbol{\varepsilon}) - \int_{\Omega} \mathbf{g} \cdot \mathbf{u}^*\, \mathrm{d}s - \int_{\partial\Omega_T} \mathbf{T}^{\mathrm{d}} \cdot \mathbf{u}\, \mathrm{d}s \\ &= \mathscr{W}(\boldsymbol{\varepsilon} + \boldsymbol{\varepsilon}^*) - \mathscr{W}(\boldsymbol{\varepsilon}) - 2\mathscr{W}(\boldsymbol{\varepsilon}, \boldsymbol{\varepsilon}^*) \\ &= \mathscr{W}(\boldsymbol{\varepsilon}^*) \end{aligned} \tag{A.25}$$

owing to the identity (A.18). But $\mathscr{W}(\boldsymbol{\varepsilon}^*)$ is nonnegative because it is the integral of a positive-definite quadratic form $W(\boldsymbol{\varepsilon}^*)$ which cannot be zero unless $\boldsymbol{\varepsilon}' = \boldsymbol{\varepsilon}$ (since $\boldsymbol{\varepsilon}' = \boldsymbol{\varepsilon} + \boldsymbol{\varepsilon}^*$), that is if $\mathbf{u}' = \mathbf{u}$ defined up to a solid displacement, whence (A.8).

Proof of A.9 This is analogous to that of (A.8) except that we set $\boldsymbol{\sigma}' = \boldsymbol{\sigma} + \boldsymbol{\sigma}^*$ and that we have

$$2\mathscr{W}^*(\boldsymbol{\sigma}, \boldsymbol{\sigma}') = \int_{\partial\Omega_u} \mathbf{u}^{\mathrm{d}} \cdot (\boldsymbol{\sigma}' \cdot \mathbf{n})\, \mathrm{d}s. \tag{A.26}$$

If we use this relation, the definition (A.22) and the identity (A.19), we can show that

$$J(\boldsymbol{\sigma}) - J(\boldsymbol{\sigma}') = \mathscr{W}^*(\boldsymbol{\sigma} + \boldsymbol{\sigma}^*) - \mathscr{W}^*(\boldsymbol{\sigma}) - 2\mathscr{W}^*(\boldsymbol{\sigma}, \boldsymbol{\sigma}^*)$$
$$= \mathscr{W}^*(\boldsymbol{\sigma}^*), \tag{A.27}$$

which shows that $J(\boldsymbol{\sigma}) - J(\boldsymbol{\sigma}')$ is a nonnegative quantity, which is zero at every point of Ω only if $\boldsymbol{\sigma}' = \boldsymbol{\sigma}$.

4.A.2 Minimum principles in nonlinear elasticity

It was H.J. Greenberg who, in 1949 (also Budiansky and Pearson, 1956/57), extended the statements (A.8) and (A.9) to the case of nonlinear constitutive laws, still keeping the hypothesis of small perturbations. It is evident that since the quadratic character of energy is then lost, what must be introduced in the proof of (A.8) and (A.9) is the *convexity* argument. On this subject we would recommend the reading of the general theorems (variational principles in finite elastic strains) in Ogden (1984) and Bufler (1988) and the studies on convexity in nonlinear elasticity in Ball (1977), Ciarlet (1988) and Sewell (1987).

Problems for Chapter 4

4.1. Huber–Mises plates – Examples (Problem 1.10 continued)
For a Huber–Mises material $f(\sigma)$ reads

$$f(\sigma) = 3\sigma_{\alpha\beta}\sigma_{\alpha\beta} - \sigma_{\alpha\alpha}\sigma_{\beta\beta} - 2\sigma_0^2 = 0, \tag{1}$$

while the flow rule reads

$$\dot{\varepsilon}^{\mathrm{P}}_{\alpha\beta} = \nu(3\sigma_{\alpha\beta} - \sigma_{\gamma\gamma}\delta_{\alpha\beta}), \qquad \nu \geqslant 0, \tag{2}$$

(a) By inversion when $\nu > 0$ show that

$$\nu = \frac{1}{\sigma\sqrt{6}}(\dot{\varepsilon}^{\mathrm{P}}_{\alpha\beta}\dot{\varepsilon}^{\mathrm{P}}_{\alpha\beta} + \dot{\varepsilon}^{\mathrm{P}}_{\alpha\alpha}\dot{\varepsilon}^{\mathrm{P}}_{\beta\beta})^{1/2}. \tag{3}$$

(b) In *pure bending* $n_{\alpha\beta} = 0$, $\lambda_{\alpha\beta} = 0$. Then show that the interaction surface has the expression

$$\varphi(\mathbf{m}) = 3m_{\alpha\beta}m_{\alpha\beta} - m_{\alpha\alpha}m_{\beta\beta} - 2 = 0. \tag{4}$$

What is the shape of this surface in **m**-space?

4.2. Plate in bending
Let w be the nondimensional deflection *rate*. The generalized strain rate is given by (see courses on the strength of materials)

$$\chi_x = -\frac{\partial^2 W}{\partial x^2}, \qquad \chi_y = -\frac{\partial^2 W}{\partial y^2}, \qquad \chi_{xy} = -2\frac{\partial^2 W}{\partial x \partial y}. \tag{1}$$

(a) Show that the power dissipated per unit area of the middle surface of the plate is given by

$$d = \frac{M_0}{h}(m_x\chi_x + m_y\chi_y + m_{xy}\chi_{xy}). \tag{2}$$

(b) Show that the equilibrium equations read

$$\frac{\partial m_x}{\partial x} + \frac{\partial m_{xy}}{\partial y} = s_x, \qquad \frac{\partial m_{xy}}{\partial x} + \frac{\partial m_y}{\partial y} = s_y, \qquad \frac{\partial s_x}{\partial x} + \frac{\partial s_y}{\partial y} + p = 0 \tag{3}$$

if the ms are the moments, s_x and s_y are transverse shear forces and p is a distributed load.

(c) Show that the interaction surface and the flow rule can be written as

$$\varphi(m_x, m_y, m_{xy}) - 1 = 0 \tag{4}$$

and

$$(\chi_x, \chi_y, \chi_{xy}) = \nu\left(\frac{\partial\varphi}{\partial m_x}, \frac{\partial\varphi}{\partial m_y}, \frac{\partial\varphi}{\partial m_{xy}}\right), \qquad \nu \geqslant 0, \tag{5}$$

4.3. Axisymmetric plate *b*

For a circular plate of radius R, r being a dimensionless measure of radius, eqns (2), (3), (4) and (5) of 4.2 are replaced respectively by

$$d = \frac{M_0}{h}(m_r\chi_r + m_\theta\chi_\theta), \tag{1}$$

$$(rs)' + rp = 0, \qquad (rm_r)' - m_\theta - rs = 0,$$
$$r = \frac{R}{A}, \qquad s = \frac{s_r A}{M_0}, \qquad (\cdot)' = \frac{d}{dr}, \tag{2}$$

$$\varphi(m_r, m_\theta) - 1 = 0, \tag{3}$$

$$(\chi_r, \chi_\theta) = \nu\left(\frac{\partial\varphi}{\partial m_r}, \frac{\partial\varphi}{\partial m_\theta}\right), \qquad \nu \geqslant 0. \tag{4}$$

For a Huber–Mises material, writing (3) in the form $m = m_\theta = g(m_r)$ and using the equilibrium equations, show that m_r is a solution of the following *nonlinear* differential equation:

$$\frac{\mathrm{d}m_r}{\mathrm{d}r} = \frac{1}{2r}[-m_r \pm \sqrt{(4 - 3m_r^2)}] - \int p\rho\,\mathrm{d}p + C. \tag{5}$$

What happens for $\mathrm{d}m_r/\mathrm{d}m = 0$, i.e., $m_r = 2m_\theta$ in the present case?

4.4. Existence of elastoplastic solutions (for mathematically oriented readers). Make a literature search and write a short report on the *functional framework* (functions of bounded variation) and the existence of solutions in elastoplasticity [*Hint*: see works by Strang, Suquet and Temam].

4.5. Discontinuity of velocity in a Mises material

In the absence of hardening, we often observe narrow zones (layers) of material in which a rapid change in velocity takes place. These can be interpreted as velocity-discontinuity surfaces Σ across which the discontinuity $[\![\mathbf{v}]\!] = \mathbf{v}_+ - \mathbf{v}_-$ occurs. We have then the expressions (4.64) and (4.65). For a material obeying the Mises criterion show that (or explain why) the following hold.

(a) A velocity discontinuity $[\![\mathbf{v}]\!]$ can occur through Σ only if one principal stress, say σ_2, is equal to half the sum of the other two principal stresses.

(b) The discontinuity surface is directed toward one of the bisector planes of the extreme principal directions (corresponding to σ_1 and σ_3).

(c) The velocity discontinuity is tangential, i.e., $[\![\mathbf{v}]\!] \cdot \mathbf{n} = 0$, so that Σ is a so-called *slip-surface*.

(d) The plastic dissipation per unit surface Σ is given by

$$\mathscr{D}([\![\mathbf{v}]\!]) = \frac{K_0}{\sqrt{2}} |[\![\mathbf{v}]\!]|.$$

(e) As a matter of fact,

$$\begin{aligned} \mathscr{D}(\mathbf{n}, [\![\mathbf{v}]\!]) &= +\infty \text{ if } [\![\mathbf{v}]\!] \cdot \mathbf{n} \neq 0 \\ &= \frac{K_0}{\sqrt{2}} |[\![\mathbf{v}]\!]| \text{ if } [\![\mathbf{v}]\!] \cdot \mathbf{n} = 0 \end{aligned}$$

and thus (by an argument of convex analysis)

$$\mathscr{D}(\mathbf{n}, [\![\mathbf{v}]\!]) = \frac{K_0}{\sqrt{2}} |[\![\mathbf{v}]\!] + \operatorname*{Sup}_{p \in \mathbb{B}} (p [\![\mathbf{v}]\!] \cdot \mathbf{n})$$

4.6. Discontinuity of velocity in a Tresca material

In this case there are no conditions on stresses for the existence of a discontinuity in velocity. Show that the results (b) and (c) of 4.5 are also true for a Tresca material but the plastic dissipation *per unit* surface Σ is given by (criterion given by (1.21))

$$\mathscr{D}([\![\mathbf{v}]\!]) = k |[\![\mathbf{v}]\!]|.$$

Also,

$$\mathscr{D}(\mathbf{n}, [\![\mathbf{v}]\!]) = k |[\![\mathbf{v}]\!]| + \operatorname*{Sup}_{p \in \mathbb{B}} (p [\![\mathbf{v}]\!] \cdot \mathbf{n}).$$

4.7. Minimum principles in the presence of velocity discontinuities

(i) Prove that the statement (4.16) will hold true in the presence of a surface Σ of velocity discontinuity in Ω.

(ii) In the same situation show that the statement (4.14) in displacement velocity must be replaced by

$$A(\dot{\mathbf{u}}) = \int_{\Omega-\Sigma} \Psi^{\sigma}(\dot{\boldsymbol{\varepsilon}})\,\mathrm{d}\Omega + \operatorname*{Sup}_{\dot{\boldsymbol{\sigma}}\in P} \int_{\Sigma(\dot{\mathbf{u}})} [\![\dot{\mathbf{u}}]\!]\cdot(\dot{\boldsymbol{\sigma}}\cdot\mathbf{n})\,\mathrm{d}s - \int_{\Omega-\Sigma} \dot{\mathbf{g}}\cdot\dot{\mathbf{u}}\,\mathrm{d}\Omega - \int_{\partial\Omega_T} \dot{\mathbf{T}}^{\mathrm{d}}\cdot\dot{\mathbf{u}}\,\mathrm{d}s.$$

4.8. Shakedown of a rectangular prism (static approach)

A prism of rectangular cross-section $2b \times 2h$ (Fig. 4.3(a)), made of a perfectly-plastic – elastic solid (Young's modulus E, elasticity limit in traction and simple compression σ_0) is in contact without friction with a fixed wall at the end $x = L$ while a load is applied to the end $x = 0$ through a rigid plate in contact without friction (Fig. 4.3(b)). This provides a three-parameter loading system (normal force N at $x = 0$, moment M_y about the y-axis and rotation θ about the z-axis). The loadings envisaged in the (θ, N)-plane in Fig. 4.3(d) correspond to a constant N, vanishing M, and θ_y varying from zero to a value θ.

Determine the set of pairs (N_1, θ_1) such that the corresponding domain $\Omega(N_1, \theta_1)$ is a domain of shakedown. [Result: Fig. 4.3(d)]. Use the static method by examining fields of self-stress. The shakedown domain is much wider than the elasticity domain from the natural state (triangle with vertices with coordinates $(-\sigma_0, 0)$, $(+\sigma_0, 0)$, $(0, \sigma_0)$).

Fig. 4.3. Figures for problem 4.8

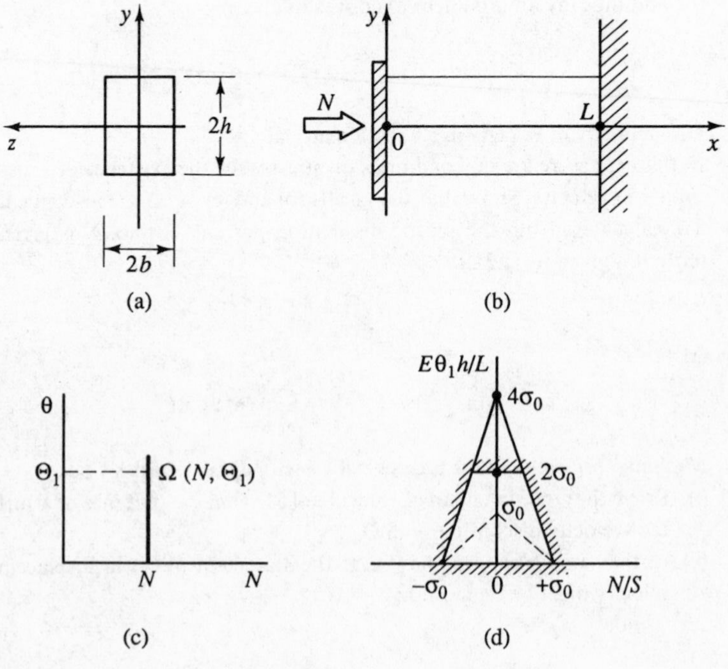

4.9. Shakedown of a rectangular prism (kinematic method)

Consider the same problem as in 4.8 but use the kinematic method, by considering special admissible histories of strain rates, to show that the *same* shakedown domain (d) as in 4.8 can be obtained (Halphen and Salençon, 1987)

4.10. Shakedown of a circular plate

A circular plate of Tresca material is subjected to uniform pressure p such that $p^- \leqslant p \leqslant p^+$. Find the values of p^+ and p^- for which no cyclic plastic strain will appear. Is p^+ equal to the collapse load (see Sawczuk, 1989)?

4.11. Shakedown of a cylindrical shell

Apply Koiter's theorem to estimate the incremental collapse of a cylindrical shell of radius R under uniform internal pressure P and loaded by a ring force Q at the end $x = L$ $(0 \leqslant p \leqslant p^+; 0 \leqslant Q \leqslant Q^+$; see Sawczuk, 1989).

5

Elastoplasticity with strain-hardening

The object of the chapter Our purpose here is to extend a number of results of Chapter 4 to the – often much more realistic – case of strain-hardened elastoplastic materials, i.e., to materials where the plasticity threshold depends upon the load's history. In the same context, we specify more clearly the notion of *generalized standard material* and we introduce certain theorems – such as Ilyushin's – concerning stability. What we must point out is the *regularizing* nature of positive hardening. The unified formulation presented here has been elaborated by Nguyen Quoc Son (1973), even though several scattered elements already existed before. A great number of examples are given.

5.1 Generalized standard media

5.1.1 The basic idea

We refer to Fig. 5.1 which presents an 'experimental' behaviour with hardening, and the schematization of the same through the use of segments of straight lines.

The schematization in Fig. 5.1 can be reproduced by a *rheological model* (Fig. 5.2) containing two springs of constants E and h and a friction element which introduces the plastic behaviour. According to this scheme, we have a Hookean element of spring constant E placed in series with a system made, in parallel, of an SV friction element of threshold k and another Hookean element (hardening) of spring constant h. We obviously have

$$\left.\begin{aligned} \varepsilon &= \varepsilon^{e} + \varepsilon^{p}, \\ \varepsilon^{e} &= \sigma/E, \\ \sigma &= h\varepsilon^{e} + \sigma_{f}, \end{aligned}\right\} \tag{5.1}$$

where σ_f is the force acting on the friction element. The '*normality law*' reads

$$\left.\begin{aligned}|\sigma_{\mathrm{f}}| = |\sigma - h\varepsilon^{\mathrm{p}}| \leqslant k,\\ \dot{\varepsilon}^{\mathrm{p}} = |\dot{\varepsilon}^{\mathrm{p}}|\,\mathrm{sign}(\sigma - h\varepsilon^{\mathrm{p}}),\end{aligned}\right\} \tag{5.2}$$

because the friction element slides in the direction of the force.

In this model, the *reversible energy*, stored through an elastic – ε^{e} – as well as plastic – ε^{p} – strain, is obviously written

$$\begin{aligned} W = W(\varepsilon^{\mathrm{e}}, \varepsilon^{\mathrm{p}}) &= \tfrac{1}{2}\varepsilon^{\mathrm{e}} E \varepsilon^{\mathrm{e}} + \tfrac{1}{2}\varepsilon^{\mathrm{p}} h \varepsilon^{\mathrm{p}} \\ &= \tfrac{1}{2}(\varepsilon - \varepsilon^{\mathrm{p}}) E (\varepsilon - \varepsilon^{\mathrm{p}}) + \tfrac{1}{2}\varepsilon^{\mathrm{p}} h \varepsilon^{\mathrm{p}}. \end{aligned} \tag{5.3}$$

The stress is given by

$$\sigma = \frac{\partial W(\varepsilon^{\mathrm{e}}, \varepsilon^{\mathrm{p}})}{\partial \varepsilon^{\mathrm{e}}} = \frac{\partial W(\varepsilon, \varepsilon^{\mathrm{p}})}{\partial \varepsilon}. \tag{5.4}$$

Fig. 5.1. One-dimensional elastoplasticity with hardening

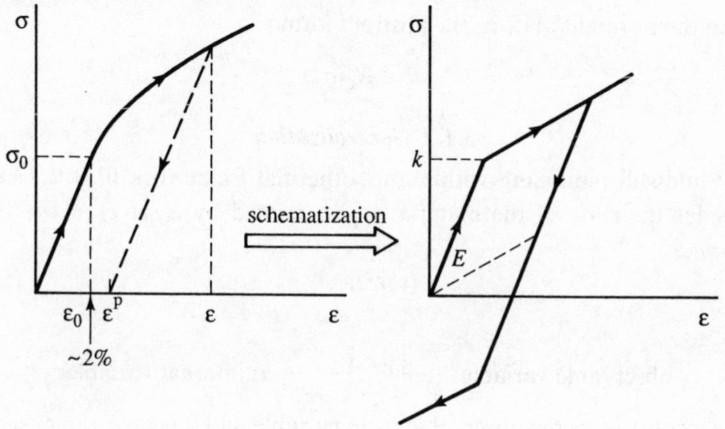

Fig. 5.2. Rheological model for schematized elastoplasticity with hardening

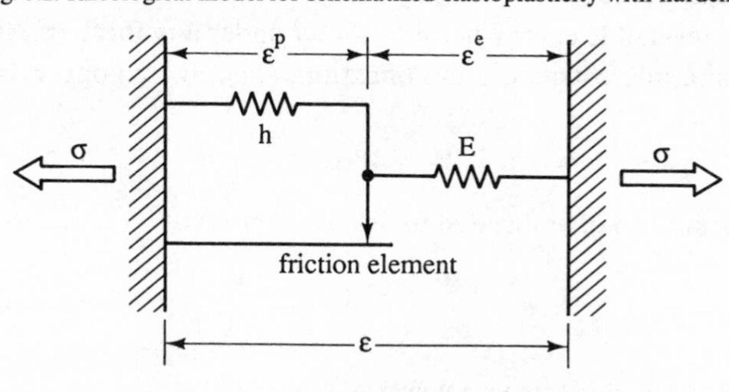

In the space of stresses, the *threshold* depends upon the plastic strain ε^{p}. Yet we notice that

$$\sigma_{\mathrm{f}} \equiv \sigma - h\varepsilon^{\mathrm{p}} = -\frac{\partial W(\varepsilon, \varepsilon^{\mathrm{p}})}{\partial \varepsilon^{\mathrm{p}}}. \tag{5.5}$$

We set then (defining a generalized thermodynamic force)

$$A = -\frac{\partial W(\varepsilon, \varepsilon^{\mathrm{p}})}{\partial \varepsilon^{\mathrm{p}}} \tag{5.6}$$

with

$$A \in C = \{A \mid |A| - k \leqslant 0\}, \tag{5.7}$$

where C is a convex (straight line segment). The generalized force A is nothing but the force acting on the friction element (eqns (5.5)–(5.6)). The evolution law of plastic strain (eqn (5.2)) is written

$$\dot{\varepsilon}^{\mathrm{p}}(A - A^*) \geqslant 0, \qquad \forall A^* \in C, \tag{5.8}$$

hence the normality law in the abstract form

$$\dot{\varepsilon}^{\mathrm{p}} \in N_C(\sigma). \tag{5.9}$$

5.1.2 Generalization

(i) While still remaining within an isothermal framework ($\theta = \theta_0$) let us consider the state of the material to be defined by *a set of* $6 + n$ *state variables*:

$$(\boldsymbol{\varepsilon}, \underbrace{\boldsymbol{\varepsilon}^{\mathrm{p}}, \ldots}), \tag{5.10}$$

observable variable ($\boldsymbol{\varepsilon}$) — $\boldsymbol{\alpha}$: internal variables

Here $\boldsymbol{\varepsilon}$ constitutes *the* observable state variable and $\boldsymbol{\alpha}$ is an n-dimensional vector accounting for irreversibility parameters. The plastic strain itself, $\boldsymbol{\varepsilon}^{\mathrm{p}}$, is included in the set $\boldsymbol{\alpha}$.

(ii) The reversible energy density, *stored* under any form, (macroscopic elastic strain, microstructure modification, etc.), by analogy with (5.3) is written

$$W = W(\boldsymbol{\varepsilon}, \boldsymbol{\alpha}) \tag{5.11}$$

with the definition of associated forces (state laws) by

$$\boldsymbol{\sigma} = \frac{\partial W}{\partial \boldsymbol{\varepsilon}}, \qquad \mathbf{A} = -\frac{\partial W}{\partial \boldsymbol{\alpha}}. \tag{5.12}$$

(iii) The *dissipated intrinsic power* is

$$\phi = A \cdot \dot{\boldsymbol{\alpha}} \geqslant 0 \tag{5.13}$$

and the internal variables $\boldsymbol{\alpha}$ evolve according to the *normality law*

$$\dot{\boldsymbol{\alpha}} \in N_C(\mathbf{A}) \Leftrightarrow \dot{\boldsymbol{\alpha}} \cdot (\mathbf{A} - \mathbf{A}^*) \geqslant 0, \quad \forall \mathbf{A}^* \in C, \tag{5.14}$$

where the *elasticity domain* C is a convex domain containing the origin in the space of generalized forces.

Following B. Halphen and Nguyen Quoc Son (1975) we say that a deformable material satisfying (5.10)–(5.14) is a *generalized standard material.* This notion, which offers a strict thermodynamic (and mathematical) framework, covers a number of mechanical behaviours. Here are some examples.

5.1.3 Examples

Example 1 *Perfect elastoplasticity.* This is the case where we consider $\alpha = \{\boldsymbol{\varepsilon}^{\mathrm{p}}\}$; consequently we have the equations (we consider the elastic part to be linear)

$$\left.\begin{aligned} \boldsymbol{\alpha} &= \boldsymbol{\varepsilon}^{\mathrm{p}}, \\ W &= \tfrac{1}{2}(\boldsymbol{\varepsilon} - \boldsymbol{\alpha}) : \mathbf{E} : (\boldsymbol{\varepsilon} - \boldsymbol{\alpha}), \\ \mathbf{A} &= -\frac{\partial W}{\partial \boldsymbol{\alpha}} = \boldsymbol{\sigma}, \\ f(\mathbf{A}) &\leqslant 0. \end{aligned}\right\} \tag{5.15}$$

Example 2 *Elastoplastic medium with kinematic hardening of the Prager–Ziegler type* (Prager, 1949). The model is the following:

$$\left.\begin{aligned} &\alpha = \{\boldsymbol{\varepsilon}^{\mathrm{p}}\}, \\ &W = \tfrac{1}{2}(\boldsymbol{\varepsilon} - \boldsymbol{\varepsilon}^{\mathrm{p}}) : \mathbf{E} : (\boldsymbol{\varepsilon} - \boldsymbol{\varepsilon}^{\mathrm{p}}) + \tfrac{1}{2}\boldsymbol{\varepsilon}^{\mathrm{p}} : \mathbf{h} : \boldsymbol{\varepsilon}^{\mathrm{p}}, \\ &\mathbf{A} = -\frac{\partial W}{\partial \boldsymbol{\varepsilon}^{\mathrm{p}}} = \boldsymbol{\sigma} - \mathbf{h} : \boldsymbol{\varepsilon}^{\mathrm{p}}, \qquad \boldsymbol{\sigma} = \frac{\partial W}{\partial \boldsymbol{\varepsilon}}, \\ &C = \{\mathbf{A} \mid |\mathbf{A}^{\mathrm{d}}| - K_0 \leqslant 0\}, \end{aligned}\right\} \tag{5.16}$$

where

$$|\mathbf{A}^{\mathrm{d}}|^2 = A^{\mathrm{d}}_{ij} A^{\mathrm{d}}_{ij}, \qquad A^{\mathrm{d}}_{ij} = A_{ij} - \tfrac{1}{3} A_{kk} \delta_{ij},$$

and $\mathbf{h}$ is the *kinematic-hardening modulus* tensor. This is the fully three-dimensional version of the rheological model of Fig. 5.2.

Remark In one dimension, the *tangent modulus* is defined as follows (all quantities being then scalar; $K_0 = k$ in Fig. 5.1(b)). On the threshold we have $A \equiv \sigma_f = \sigma - h\varepsilon^p = k$, and, consequently, by infinitesimal variation $\delta\sigma = E\delta\varepsilon^e = h\delta\varepsilon^p$. But, in general, $\delta\varepsilon = \delta\varepsilon^e + \delta\varepsilon^p$, so that we may write $\delta\varepsilon = \delta\sigma/E + \delta\sigma/h = \delta\sigma/E_T$, where the tangent modulus E_T is defined by

$$E_T = \frac{hE}{E+h} \qquad \text{or} \qquad \frac{1}{E_T} = \frac{1}{E} + \frac{1}{h}. \tag{5.17}$$

Example 3 *Isotropic hardening – identification of an experimental curve* (as with strongly hardened steel). We try to construct a generalized standard model that reproduces in a one-dimensional test the experimental curve of Fig. 5.3(a).

Fig. 5.3. Isotropic hardening. (a) σ vs. ε. (b) σ vs. ε^p.

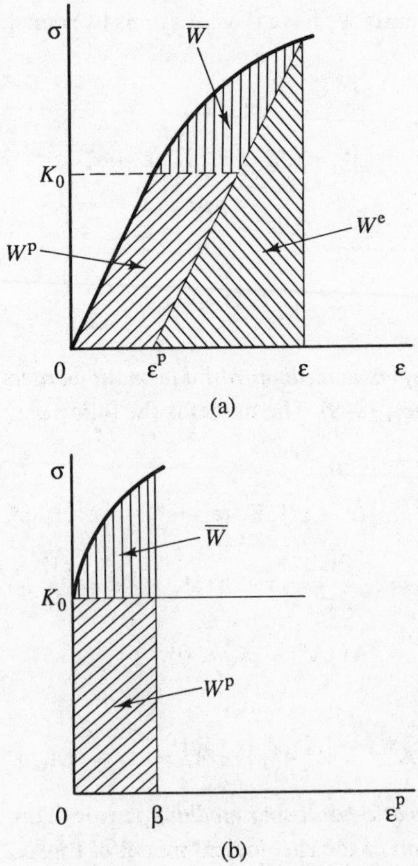

The curve of Fig. 5.3(a) defines in the plane $(\varepsilon^{p}, \sigma)$ – Fig. 5.3(b) – an area $\overline{W}(\beta)$, a function of the cumulative plastic strain (equivalent strain)

$$\beta = \int_0^t |\dot{\varepsilon}^{p}| \mathrm{d}t. \tag{5.18}$$

The standard generalized model is constructed with the following:

the internal variables $(\boldsymbol{\varepsilon}^{p}, \beta)$, where β is a scalar to be more exactly determined;

an energy of the type

$$W = \tfrac{1}{2}(\boldsymbol{\varepsilon} - \boldsymbol{\varepsilon}^{p}) : \mathbf{E} : (\boldsymbol{\varepsilon} - \boldsymbol{\varepsilon}^{p}) + \overline{W}(\beta), \tag{5.19}$$

where $\overline{W}(\beta)$ is the *reversible energy*, stored by modification of the internal structure;

the state laws

$$\boldsymbol{\sigma} = \frac{\partial W}{\partial \boldsymbol{\varepsilon}}, \qquad A = \left\{ \mathbf{A}_1 = -\frac{\partial W}{\partial \boldsymbol{\varepsilon}^{p}} \equiv \boldsymbol{\sigma}, A_2 = -\frac{\partial \overline{W}}{\partial \beta} = -\overline{W}' \right\}; \tag{5.20}$$

the plasticity convex (of the Mises criterion type with hardening with yield function

$$f = |\boldsymbol{\sigma}^{d}| + B - K_0. \tag{5.21}$$

In other words

$$C = \{ |\boldsymbol{\sigma}^{d}| + B - K_0 \leqslant 0 \} \tag{5.22}$$

is a convex in the space of As; more explicitly

$$0 = f(A_1, A_2) \Rightarrow C = \{ |\mathbf{A}_1^{d}| + A_2 - K_0 \leqslant 0 \}. \tag{5.23}$$

With this model the normality law yields

$$\dot{\varepsilon}^{p}_{ij}(\sigma_{ij} - \sigma^{*}_{ij}) + \dot{\beta}(B - B^{*}) \geqslant 0$$

or equivalently

$$\dot{\varepsilon}^{p} = \dot{\lambda} \frac{\partial f}{\partial \sigma}, \qquad \dot{\beta} = \dot{\lambda} \frac{\partial f}{\partial B} = \dot{\lambda} \qquad \left(\text{since } \frac{\partial f}{\partial B} = 1 \right). \tag{5.24}$$

From these two equations we deduce $\dot{\lambda}$ by

$$\dot{\lambda}^2 = \dot{\varepsilon}^{p}_{ij} \dot{\varepsilon}^{p}_{ij} = \dot{\beta}^2$$

in such a manner that

$$\beta(t) = \int_0^t (\dot{\varepsilon}^{\mathrm{p}}_{ij}(\tau)\dot{\varepsilon}^{\mathrm{p}}_{ij}(\tau))^{1/2}\,\mathrm{d}\tau \equiv \overline{\varepsilon^{\mathrm{p}}}, \tag{5.25}$$

which is the so-called *cumulative plastic strain or Odqvist's parameter*.

In applications it is specified that $|\boldsymbol{\sigma}^{\mathrm{d}}| = \sigma_Y \equiv (\frac{3}{2}\sigma^{\mathrm{d}}_{ij}\sigma^{\mathrm{d}}_{ij})^{1/2}$ (which is Von Mises' *equivalent stress*) so that (5.25) is replaced by $\beta(t) = \int_0^t(\frac{2}{3}\dot{\varepsilon}^{\mathrm{p}}_{ij}(\tau)\dot{\varepsilon}^{\mathrm{p}}_{ij}(\tau))^{1/2}\,\mathrm{d}\tau$.

Let us compute the corresponding *dissipated power*. We have $\phi = A\dot{\alpha}$, that is

$$\begin{aligned}\phi &= \boldsymbol{\sigma} : \dot{\boldsymbol{\varepsilon}}^{\mathrm{p}} + B\dot{\beta} \\ &= \dot{\lambda}|\mathbf{A}^{\mathrm{d}}_1| + \dot{\lambda}A_2 = \dot{\lambda}K_0 = K_0|\dot{\boldsymbol{\varepsilon}}^{\mathrm{p}}|,\end{aligned} \tag{5.26}$$

which is very much homogeneous of order 1 in $\dot{\boldsymbol{\varepsilon}}^{\mathrm{p}}$, but it is *not* the same thing as the plastic strain power.

Example 4 *Unified presentation* This elegant presentation is due to Nguyen Quoc Son (1973). We take ($\alpha = \{\boldsymbol{\varepsilon}^{\mathrm{p}}\}$).

$$W = W^{\mathrm{e}}(\boldsymbol{\varepsilon} - \boldsymbol{\varepsilon}^{\mathrm{p}}) + W_1(\boldsymbol{\varepsilon}^{\mathrm{p}}), \tag{5.27}$$

where W is convex with respect to the pair $(\boldsymbol{\varepsilon}, \boldsymbol{\alpha})$. We have the following *state laws*:

$$\boldsymbol{\sigma} = \frac{\partial W}{\partial \boldsymbol{\varepsilon}}, \qquad \mathbf{A} = \boldsymbol{\sigma} - \frac{\partial W_1}{\partial \boldsymbol{\varepsilon}^{\mathrm{p}}} = \boldsymbol{\sigma} + \mathbf{B}, \qquad \mathbf{B} = -\frac{\partial W_1}{\partial \boldsymbol{\varepsilon}^{\mathrm{p}}}. \tag{5.28}$$

We introduce the (obvious) notations

$$\left.\begin{aligned}\boldsymbol{\Sigma} &= (\boldsymbol{\sigma}, \mathbf{B}), \\ \mathscr{E} &= (\boldsymbol{\varepsilon}, 0), \\ \mathscr{E}^{\mathrm{p}} &= (\boldsymbol{\varepsilon}^{\mathrm{p}}, \boldsymbol{\varepsilon}^{\mathrm{p}}), \\ \mathscr{E}^{\mathrm{e}} &= (\boldsymbol{\varepsilon}^{\mathrm{e}}, -\boldsymbol{\varepsilon}^{\mathrm{p}}),\end{aligned}\right\} \tag{5.29}$$

so that we may write (in 12-dimensional spaces of generalized strains $\mathscr{E}$ and stresses $\boldsymbol{\Sigma}$) the following:

(i) the total *generalized* strain

$$\mathscr{E} = \mathscr{E}^{\mathrm{e}} + \mathscr{E}^{\mathrm{p}}; \tag{5.30}$$

(ii) the *normality law*

$$\dot{\mathscr{E}}^{\mathrm{p}} : (\boldsymbol{\Sigma} - \boldsymbol{\Sigma}^*) \geqslant 0, \qquad \text{i.e., } \dot{\mathscr{E}}^{\mathrm{p}} \in N_C(\boldsymbol{\Sigma}); \tag{5.31}$$

(iii) the *state law* (the case of *any* energy whatsoever) in *incremental* form

$$\dot{\boldsymbol{\Sigma}} = \mathscr{L} : \dot{\mathscr{E}}^{\mathrm{e}} \tag{5.32}$$

i.e.,

$$\begin{pmatrix} \dot{\boldsymbol{\sigma}} \\ \dot{\mathbf{B}} \end{pmatrix} = (\mathscr{L}) \begin{pmatrix} \dot{\boldsymbol{\varepsilon}}^{\mathrm{e}} \\ -\dot{\boldsymbol{\varepsilon}}^{\mathrm{p}} \end{pmatrix} \tag{5.33}$$

with

$$(\mathscr{L}) = \begin{bmatrix} \{E_{ijkl}\} = \dfrac{\partial^2 W^{\mathrm{e}}}{\partial \varepsilon_{ij} \partial \varepsilon_{kl}} & 0 \\ 0 & \{H_{ijkl}\} = \dfrac{\partial^2 W_1}{\partial \varepsilon^{\mathrm{p}}_{ij} \partial \varepsilon^{\mathrm{p}}_{kl}} \end{bmatrix} \tag{5.34}$$

We verify that (5.31) is equivalent to

$$\dot{\boldsymbol{\varepsilon}}^{\mathrm{p}} : (\boldsymbol{\sigma} - \boldsymbol{\sigma}^*) + \dot{\boldsymbol{\varepsilon}}^{\mathrm{p}} : (\mathbf{B} - \mathbf{B}^*) \geqslant 0$$

or

$$\dot{\boldsymbol{\varepsilon}}^{\mathrm{p}} : [(\boldsymbol{\sigma} + \mathbf{B}) - (\boldsymbol{\sigma}^* + \mathbf{B}^*)] \geqslant 0$$

or even

$$\dot{\boldsymbol{\varepsilon}}^{\mathrm{p}} : (\mathbf{A} - \mathbf{A}^*) \geqslant 0 \tag{5.35}$$

since $\{\boldsymbol{\varepsilon}^{\mathrm{p}}\} = \alpha$.

If W^{e} is convex in $\boldsymbol{\varepsilon}^{\mathrm{e}}$ and W_1 is convex in $\boldsymbol{\varepsilon}^{\mathrm{p}}$, then $\mathscr{L}$ is *positive definite.* In the formalism (5.29)–(5.32) we can apply the results of the *perfect-plasticity* type by direct transposition. The orthogonality relation (3.45), in particular, is generalized in $\mathbb{R}^{12}$ by

$$\dot{\boldsymbol{\Sigma}} : \dot{\mathscr{E}}^{\mathrm{p}} = 0 \tag{5.36}$$

with

$$C = \{\boldsymbol{\sigma} + \mathbf{B} | f(\boldsymbol{\sigma} + \mathbf{B}) \leqslant 0\}. \tag{5.37}$$

More explicitly (5.36) is written

$$\dot{\boldsymbol{\sigma}} : \dot{\boldsymbol{\varepsilon}}^{\mathrm{p}} + \dot{\mathbf{B}} : \dot{\boldsymbol{\varepsilon}}^{\mathrm{p}} = 0 \tag{5.38}$$

or

$$\dot{\lambda} \dot{\boldsymbol{\sigma}} : \frac{\partial f}{\partial \boldsymbol{\sigma}} - \dot{\lambda} \frac{\partial f}{\partial \boldsymbol{\sigma}} : \mathbf{H} : \frac{\partial f}{\partial \boldsymbol{\sigma}} \dot{\lambda} = 0,$$

so that, if $\dot{\lambda} \neq 0$,

$$\dot{\lambda} = \frac{1}{h}\left\langle \frac{\partial f}{\partial \boldsymbol{\sigma}} : \dot{\boldsymbol{\sigma}} \right\rangle, \tag{5.39}$$

where

$$h = \frac{\partial f}{\partial \boldsymbol{\sigma}} : \mathbf{H} : \frac{\partial f}{\partial \boldsymbol{\sigma}} \tag{5.40}$$

is the *hardening modulus*, a scalar quantity.

So we have

$$\dot{\boldsymbol{\varepsilon}}^{\mathrm{p}} = \dot{\lambda}\frac{\partial f}{\partial \boldsymbol{\sigma}} = \frac{1}{h}\left\langle \frac{\partial f}{\partial \boldsymbol{\sigma}} : \dot{\boldsymbol{\sigma}} \right\rangle \frac{\partial f}{\partial \boldsymbol{\sigma}}. \tag{5.41}$$

If we know $\dot{\boldsymbol{\sigma}}$ we can then calculate $\dot{\boldsymbol{\varepsilon}}^{\mathrm{p}}(\dot{\boldsymbol{\sigma}})$.

Another expression for the plastic multiplier $\dot{\lambda}$. The equation after (5.38) may also be written ($\dot{\lambda} \neq 0$)

$$\frac{\partial f}{\partial \boldsymbol{\sigma}} : [\mathbf{E} : (\dot{\boldsymbol{\varepsilon}} - \dot{\boldsymbol{\varepsilon}}^{\mathrm{p}})] - h\dot{\lambda} = 0, \tag{5.42}$$

or

$$\frac{\partial f}{\partial \boldsymbol{\sigma}} : \mathbf{E} : \dot{\boldsymbol{\varepsilon}} - \dot{\lambda}\left(h + \frac{\partial f}{\partial \boldsymbol{\sigma}} : \mathbf{E} : \frac{\partial f}{\partial \boldsymbol{\sigma}} \right) = 0.$$

As a consequence

$$\dot{\lambda} = \frac{\left\langle \dfrac{\partial f}{\partial \boldsymbol{\sigma}} : \mathbf{E} : \dot{\boldsymbol{\varepsilon}} \right\rangle}{h + \left(\dfrac{\partial f}{\partial \boldsymbol{\sigma}} : \mathbf{E} : \dfrac{\partial f}{\partial \boldsymbol{\sigma}} \right)}. \tag{5.43}$$

Perfect plasticity corresponds to the case $h = 0$.

Moreover, (5.32) shows that $\dot{\boldsymbol{\varepsilon}}^{\mathrm{p}}$ is determined by $\dot{\boldsymbol{\varepsilon}}$ and that

$$\dot{\boldsymbol{\sigma}}(\dot{\boldsymbol{\varepsilon}}) = \frac{\partial \Psi^{\sigma}(\dot{\boldsymbol{\varepsilon}})}{\partial \dot{\boldsymbol{\varepsilon}}} \tag{5.44}$$

with

$$\Psi^{\sigma}(\dot{\boldsymbol{\varepsilon}}) = \tfrac{1}{2}\dot{\boldsymbol{\varepsilon}} : \mathbf{E} : \dot{\boldsymbol{\varepsilon}} - \tfrac{1}{2}\frac{\left\langle \dfrac{\partial f}{\partial \boldsymbol{\sigma}} : \mathbf{E} : \dot{\boldsymbol{\varepsilon}} \right\rangle^{2}}{h + \left(\dfrac{\partial f}{\partial \boldsymbol{\sigma}} : \mathbf{E} : \dfrac{\partial f}{\partial \boldsymbol{\sigma}} \right)}. \tag{5.45}$$

Contrary to the case of perfect plasticity, if we consider the remark following (5.41) and if we consider (5.44), we have the *one-to-one* relation

$$\dot{\boldsymbol{\sigma}} \Leftrightarrow \dot{\boldsymbol{\varepsilon}}. \tag{5.46}$$

Since Ψ^σ is convex, by the Legendre – Fenchel transform, we have (exercise)

$$\Psi^*(\boldsymbol{\sigma}) = \tfrac{1}{2}\boldsymbol{\sigma} : \mathbf{S} : \dot{\boldsymbol{\sigma}} + \frac{1}{2h}\left\langle \frac{\partial f}{\partial \boldsymbol{\sigma}} : \dot{\boldsymbol{\sigma}} \right\rangle^2 \tag{5.47}$$

and

$$\dot{\boldsymbol{\varepsilon}}(\dot{\boldsymbol{\sigma}}) = \frac{\partial \Psi^*(\dot{\boldsymbol{\sigma}})}{\partial \dot{\boldsymbol{\sigma}}}. \tag{5.48}$$

The one-to-one correspondence between increments in strain and stress is obvious in both elastic and plastic regimes and both loading *and* unloading in a (σ, ε) plot such as in Fig. 5.4 in the presence of hardening.

Note in passing that nothing distinguishes the *loading* behaviour in Fig. 5.4(a) from that of a *nonlinear-elastic* solid having the same $\sigma(\varepsilon)$ curve. It is only on unloading (Fig. 5.4(b)) that *plasticity* appears typically with its unloading path differing from the loading one. Still the correspondence (5.46) holds true for the difference is taken care of in the same expression (5.48).

Example 5 *Rheological models with n constituents*

(a) Friction elements and springs in series This is the so-called Prager–Mroz (and others) model (after Prager, 1955, and Mroz, 1967) in which

Fig. 5.4. One-to-one correspondence between strain and stress increments. (a) Loading. (b) Unloading.

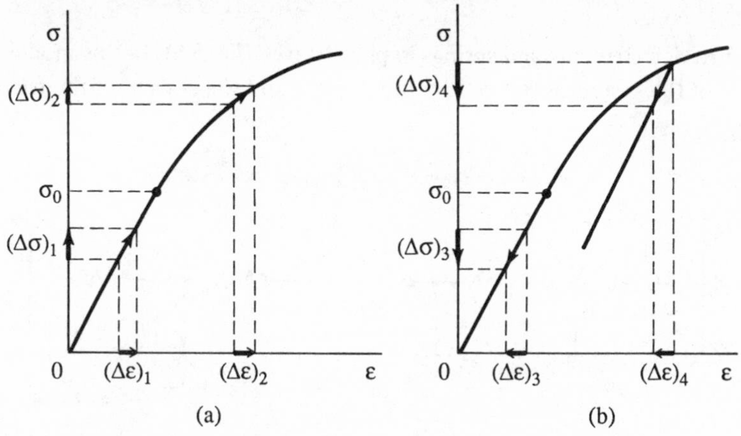

from our unified viewpoint the internal variables α are the plastic strain ε^p (the first friction element slip) and the spring stretches $\beta_1, \ldots, \beta_n$ as shown in Fig. 5.5 on a rheological model. Clearly, this is an extension of the model in Fig. 5.2.

In this model the free energy is the energy stored in all springs, or

$$W = \tfrac{1}{2}(\varepsilon - \varepsilon^p)E(\varepsilon - \varepsilon^p) + \tfrac{1}{2}E_1\beta_1^2 + \cdots + \tfrac{1}{2}E_n\beta_n^2. \tag{5.49}$$

The *state laws* are written

$$\sigma = E(\varepsilon - \varepsilon^p), \qquad \mathbf{A} = (\sigma, B_1, \ldots, B_n), \qquad B_i = -E_i\beta_i. \tag{5.50}$$

The B_i are nothing but the *spring tensions.* If k_i designates the threshold of the friction element i, then the *elasticity domain* is defined by the inequalities (compare to eqn $(5.2)_1$)

$$\left.\begin{aligned} |\sigma + B_1| - k_1 &\leqslant 0, \\ |B_2 - B_1| - k_2 &\leqslant 0, \\ &\vdots \\ |B_n - B_{n-1}| - k_n &\leqslant 0, \end{aligned}\right\} \tag{5.51}$$

which is a convex in the product space $\sigma \times B_1 \times \cdots \times B_n$.

The *normality law* is written (compare to eqn $(5.2)_2$)

$$\left.\begin{aligned} \dot{\varepsilon}^p &= \dot{\lambda}_1 \operatorname{sign}(\sigma + B_1), \\ \dot{\beta} &= \dot{\lambda}_1 \operatorname{sign}(\sigma + B_1) - \dot{\lambda}_2 \operatorname{sign}(B_2 - B_1), \\ &\vdots \\ \dot{\beta}_n &= \dot{\lambda}_n \operatorname{sign}(B_n - B_{n-1}). \end{aligned}\right\} \tag{5.52}$$

(b) Friction elements and springs in parallel (see Fig. 5.6) In this model the state of the system is defined by the strain ε and the slips α_i of the friction

Fig. 5.5. Friction elements and springs in series

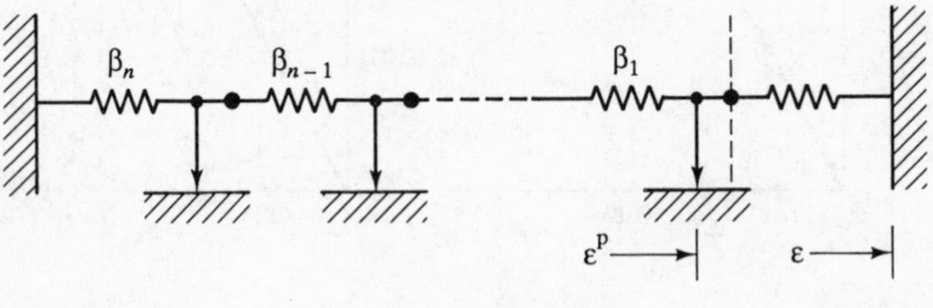

elements. The plastic strain ε^p is obtained by elastic unloading. It is defined, beginning from the actual state of the system, through an imaginary unloading (thought experiment), while all friction elements remain blocked.

The *free energy* is the energy stored in all the springs, that is,

$$W = \tfrac{1}{2}E_1(\varepsilon - \alpha_1)^2 + \cdots + \tfrac{1}{2}E_n(\varepsilon - \alpha_n)^2. \tag{5.53}$$

The *stress* is the sum of the tensions in the springs, that is,

$$\begin{aligned}\sigma &= E_1(\varepsilon - \alpha_1) + \cdots + E_n(\varepsilon - \alpha_n)\\ &= \left(\sum_i E_i\right)\varepsilon - E_i\alpha_i = E(\varepsilon - \varepsilon^p)\end{aligned} \tag{5.54}$$

if we set

$$E = \sum_i E_i,$$

and the last of eqn (5.54) *defines* ε^p by

$$\varepsilon^p = \frac{1}{E}\sum_i E_i\alpha_i. \tag{5.55}$$

The energy split in W^e and $W(\boldsymbol{\alpha})$ results from (5.53) and (5.55), which allows us to write

Fig. 5.6. Friction elements and springs in parallel

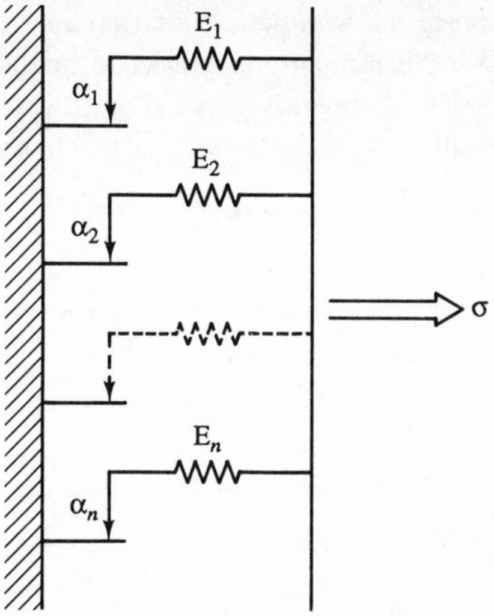

$$W = \tfrac{1}{2}E(\varepsilon - \varepsilon^{\mathrm{p}})^2 - \tfrac{1}{2}E(\varepsilon^{\mathrm{p}})^2 + \tfrac{1}{2}\sum_i E_i\alpha_i^2. \tag{5.56}$$

The *associated forces* are the spring tensions, that is,

$$\mathbf{A} = \{A_i\}, \qquad A_i = E_i(\varepsilon - \alpha_i) \Rightarrow \sigma = \sum_i A_i. \tag{5.57}$$

If k_i is the slip threshold of spring i, the elasticity domain C is defined by the inequalities

$$|A_i| - k_i \leqslant 0, \qquad i = 1, \ldots, n, \tag{5.58}$$

and the evolution laws

$$\dot{\alpha}_i = \dot{\lambda}_i \operatorname{sign} A_i, \qquad \dot{\lambda}_i \geqslant 0 \quad \text{if the } i\text{-friction-element is active,} \tag{5.59}$$

represent the *normality law*. Finally, after (a) and (b), we may consider the case of any rheological model whatsoever.

General rheological model In the light of the two preceding examples we have the

Proposition *Any instantaneous-elasticity rheological model, composed of any assembly of friction elements and springs, is of the standard generalized type.*

This can be proven in a number of ways, but the simplest is to isolate all the friction elements. Let then α_i be the slip of friction element i (see Fig. 5.7). After the friction elements have been isolated, the rest of the model is then an elastic structure, *but with* perfect internal links. Compared to the stressless initial state, this structure is subject to imposed displacements $(\varepsilon, \boldsymbol{\alpha})$ and it stores in all its springs an energy $W(\varepsilon, \boldsymbol{\alpha})$, which we can write (as we can easily verify, this is a particular form of (5.53) or (5.56))

$$W = \tfrac{1}{2}\varepsilon \cdot E \cdot \varepsilon - \varepsilon \cdot G_i\alpha_i + \tfrac{1}{2}\boldsymbol{\alpha} \cdot \mathbf{H} \cdot \boldsymbol{\alpha}, \tag{5.60}$$

The *state laws* are written

$$\sigma = \frac{\partial W}{\partial \varepsilon}, \qquad A_i = -\frac{\partial W}{\partial \alpha_i}, \tag{5.61}$$

With (5.60), they give

$$\sigma = E\varepsilon - G_i\alpha_i, \qquad A_i = G_i\varepsilon - H_{ij}\alpha_j \tag{5.62}$$

with a plastic strain

$$\varepsilon^{\mathrm{p}} = SG_i\alpha_i \quad (S = E^{-1}). \tag{5.63}$$

The elasticity domain C and the normality law will be written

$$\left.\begin{aligned} |A_i| - k_i \leqslant 0, \\ \dot{\alpha}_i = \dot{\lambda}_i \operatorname{sign} A_i. \end{aligned}\right\} \tag{5.64}$$

The model is then *generalized standard.*

5.2 Relations between velocities; incremental constitutive equations

Although we have put together, inside the same m-vector $\boldsymbol{\alpha}$, all the internal variables, including $\boldsymbol{\varepsilon}^{\mathrm{p}}$, it is often useful to distinguish $\boldsymbol{\varepsilon}^{\mathrm{p}}$ from the other internal variables, which we shall now denote by the $(m-6)$-vector $\boldsymbol{\beta}$, as we already did in (5.49)–(5.50). We then write the free energy in the form

$$W = W^{\mathrm{e}}(\boldsymbol{\varepsilon} - \boldsymbol{\varepsilon}^{\mathrm{p}}) + W^{\alpha}(\boldsymbol{\varepsilon}^{\mathrm{p}}, \boldsymbol{\beta}). \tag{5.65}$$

The *state laws* are given by

$$\left.\begin{aligned} \boldsymbol{\sigma} &= \frac{\partial W^{\mathrm{e}}}{\partial \boldsymbol{\varepsilon}}, & \mathbf{A} &= (\boldsymbol{\sigma} + \mathbf{B}^{\mathrm{p}}, \mathbf{B}^{\beta}), \\ \mathbf{B}^{\mathrm{p}} &= -\frac{\partial W^{\alpha}}{\partial \boldsymbol{\varepsilon}^{\mathrm{p}}}, & \mathbf{B}^{\beta} &= -\frac{\partial W^{\alpha}}{\partial \boldsymbol{\beta}}. \end{aligned}\right\} \tag{5.66}$$

The elasticity domain in a convex domain in the space of forces $(\boldsymbol{\sigma} + \mathbf{B}^{\mathrm{p}}) \times (\mathbf{B}^{\beta})$ and the normality law is written

Fig. 5.7. Rheological model with instantaneous elasticity

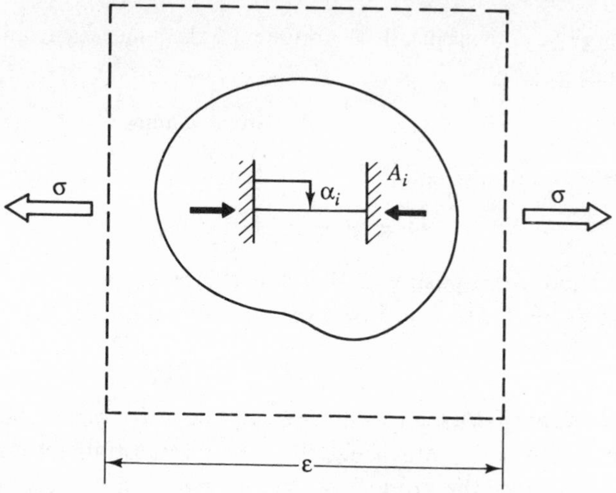

$$\dot{\boldsymbol{\alpha}} \in N_C(\mathbf{A}) \qquad \text{or} \qquad (\dot{\boldsymbol{\varepsilon}}^{\mathrm{p}}, \dot{\boldsymbol{\beta}}) \in N_C(\boldsymbol{\sigma} + \mathbf{B}^{\mathrm{p}}, \mathbf{B}^{\beta}), \tag{5.67}$$

or, with a threshold

$$f(\boldsymbol{\sigma} + \mathbf{B}^{\mathrm{p}}, \mathbf{B}^{\mathrm{p}}) = 0, \tag{5.68}$$

$$\left.\begin{aligned} \dot{\boldsymbol{\varepsilon}}^{\mathrm{p}} &= \dot{\lambda}\frac{\partial f}{\partial \boldsymbol{\sigma}}, & \dot{\boldsymbol{\beta}} &= \dot{\lambda}\frac{\partial f}{\partial \mathbf{B}^{\beta}}, \\ \dot{\lambda} &\geqslant 0 \quad \text{if } f = 0, & \dot{\lambda} &= 0 \quad \text{if } f \leqslant 0. \end{aligned}\right\} \tag{5.69}$$

Just as previously – eqn (3.45) – the normality law implies the condition of orthogonality

$$\dot{\mathbf{A}} \cdot \dot{\boldsymbol{\alpha}} = 0 \tag{5.70}$$

or more explicitly with the notations (5.66)

$$\dot{\boldsymbol{\sigma}} : \dot{\boldsymbol{\varepsilon}}^{\mathrm{p}} + \dot{\mathbf{B}}^{\mathrm{p}} : \dot{\boldsymbol{\varepsilon}}^{\mathrm{p}} + \dot{\mathbf{B}}^{\beta} \cdot \dot{\boldsymbol{\beta}} = 0. \tag{5.71}$$

Let us set (T = transpose)

$$\mathbb{E} = \frac{\partial^2 W^{\mathrm{e}}}{\partial \boldsymbol{\varepsilon}^{\mathrm{e}} \partial \boldsymbol{\varepsilon}^{\mathrm{e}}}, \qquad \mathbb{H} = \begin{pmatrix} \mathbf{L} & \mathbf{N} \\ \mathbf{N}^{\mathrm{T}} & \mathbf{M} \end{pmatrix}, \tag{5.72}$$

where

$$\left.\begin{aligned} \mathbf{L} &= \frac{\partial^2 W^{\alpha}}{\partial \boldsymbol{\varepsilon}^{\mathrm{p}} \partial \boldsymbol{\varepsilon}^{\mathrm{p}}} = \mathbf{L}^{\mathrm{T}}, \\ \mathbf{N} &= \frac{\partial^2 W^{\alpha}}{\partial \boldsymbol{\varepsilon}^{\mathrm{p}} \partial \boldsymbol{\beta}}, \qquad \mathbf{M} = \frac{\partial^2 W^{\alpha}}{\partial \boldsymbol{\beta} \partial \boldsymbol{\beta}} = \mathbf{M}^{\mathrm{T}}. \end{aligned}\right\} \tag{5.73}$$

If the energy is convex in all its arguments, the matrices **L** and **M** are positive, i.e..

$$W \textit{ convex} \Leftrightarrow \mathbb{E}, \mathbb{H} \textit{ positive definite.} \tag{5.74}$$

Under these conditions, eqn (5.71) is written

$$\dot{\boldsymbol{\sigma}} : \dot{\boldsymbol{\varepsilon}}^{\mathrm{p}} = \dot{\boldsymbol{\alpha}} \cdot \mathbb{H} \cdot \dot{\boldsymbol{\alpha}}$$

The right-hand side is positive if W is convex; then, in the presence of any *hardening*, we get the so-called *Drucker inequality* (Drucker, 1951):

$$\dot{\boldsymbol{\sigma}} : \dot{\boldsymbol{\varepsilon}}^{\mathrm{p}} \geqslant 0. \tag{5.75}$$

Through this approach, then, the free-energy convexity *implies* the Drucker condition, which represents the fact that the plastic strain–stress curve is *nondecreasing* (since the product of the increments in $\boldsymbol{\sigma}$ and $\boldsymbol{\varepsilon}^{\mathrm{p}}$ is always *nonnegative*, according to (5.75)); see Fig. 5.8. We say then that the *hardening is positive or zero.*

Relation $\dot{\boldsymbol{\sigma}}(\dot{\boldsymbol{\varepsilon}})$ Just as in perfect plasticity, then, if the actual state is considered as known, the stress rate $\dot{\boldsymbol{\sigma}}$ is expressed in terms of the strain rate $\dot{\boldsymbol{\varepsilon}}$. The calculation of $\dot{\boldsymbol{\sigma}}(\dot{\boldsymbol{\varepsilon}})$ is very simple when the convex C is expressed in the form of a single inequality:

$$C = \{\mathbf{A} | f(\boldsymbol{\sigma} + \mathbf{B}^{\mathrm{p}}, \mathbf{B}^{\beta}) \leqslant 0\} \tag{5.76}$$

or

$$\dot{\boldsymbol{\alpha}} \in N_C(\mathbf{A}) \Leftrightarrow \left\{ \begin{array}{ll} \dot{\boldsymbol{\varepsilon}}^{\mathrm{p}} = \dot{\lambda} \dfrac{\partial f}{\partial \boldsymbol{\sigma}}, & \dot{\lambda} \geqslant 0 \quad \text{if } f = 0, \\ \dot{\boldsymbol{\beta}} = \dot{\lambda} \dfrac{\partial f}{\partial \mathbf{B}^{\beta}}, & \dot{\lambda} = 0 \quad \text{if } f < 0. \end{array} \right\} \tag{5.77}$$

We have then the equivalence

$$\dot{\mathbf{A}} \cdot \dot{\boldsymbol{\alpha}} = 0 \Leftrightarrow \dot{\lambda} \dot{f} = 0. \tag{5.78}$$

So that, with $\dot{\lambda} \neq 0$,

$$\dot{f} = \frac{\partial f}{\partial \boldsymbol{\sigma}} : (\dot{\boldsymbol{\sigma}} + \dot{\mathbf{B}}^{\mathrm{p}}) + \frac{\partial f}{\partial \mathbf{B}^{\beta}} \cdot \dot{\mathbf{B}} = 0 \tag{5.79}$$

or

$$\frac{\partial f}{\partial \boldsymbol{\sigma}} : (\mathbf{E}\dot{\boldsymbol{\varepsilon}}) - \left(\mathbf{N} : \mathbf{H} : \mathbf{N} + \frac{\partial f}{\partial \boldsymbol{\sigma}} : \mathbf{E} : \frac{\partial f}{\partial \boldsymbol{\sigma}} \right) \dot{\lambda} = 0 \tag{5.79)α}$$

Fig. 5.8. Positive hardening (Drucker's inequality)

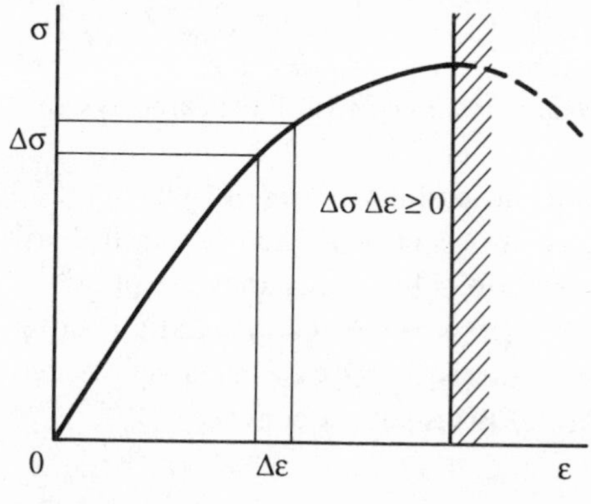

with

$$\mathbf{N} = \left(\frac{\partial f}{\partial \boldsymbol{\sigma}}, \frac{\partial f}{\partial \mathbf{B}^{\mathrm{p}}} \right)$$

so that

$$\dot{\lambda} = \frac{\left\langle \dfrac{\partial f}{\partial \boldsymbol{\sigma}} : \mathbf{E} : \dot{\boldsymbol{\varepsilon}} \right\rangle}{\left(\mathbf{N} : \mathbf{H} : \mathbf{N} + \dfrac{\partial f}{\partial \boldsymbol{\sigma}} : \mathbf{E} : \dfrac{\partial f}{\partial \boldsymbol{\sigma}} \right)}. \tag{5.80}$$

According to (5.77), this yields

$$\dot{\boldsymbol{\varepsilon}}^{\mathrm{p}} = \frac{1}{h} \left\langle \frac{\partial f}{\partial \boldsymbol{\sigma}} : \dot{\boldsymbol{\sigma}} \right\rangle \frac{\partial f}{\partial \boldsymbol{\sigma}} \qquad \text{if } h \equiv \mathbf{N} : \mathbf{H} : \mathbf{N} > 0. \tag{5.81}$$

It follows that

$$\dot{\boldsymbol{\sigma}}(\dot{\boldsymbol{\varepsilon}}) = \mathbf{E} : \dot{\boldsymbol{\varepsilon}} - \frac{\left\langle \dfrac{\partial f}{\partial \boldsymbol{\sigma}} : \mathbf{E} : \dot{\boldsymbol{\varepsilon}} \right\rangle}{h + \left(\dfrac{\partial f}{\partial \boldsymbol{\sigma}} : \mathbf{E} : \dfrac{\partial f}{\partial \boldsymbol{\sigma}} \right)} \mathbf{E} : \frac{\partial f}{\partial \boldsymbol{\sigma}}, \tag{5.82}$$

which we can re-write as

$$\dot{\boldsymbol{\sigma}}(\dot{\boldsymbol{\varepsilon}}) = \frac{\partial \Psi^{\alpha}}{\partial \dot{\boldsymbol{\varepsilon}}}(\dot{\boldsymbol{\varepsilon}}) \tag{5.83}$$

with a potential Ψ^{α} defined by

$$\Psi^{\alpha}(\dot{\boldsymbol{\varepsilon}}) = \tfrac{1}{2} \dot{\boldsymbol{\varepsilon}} : \mathbf{E} : \dot{\boldsymbol{\varepsilon}} - \tfrac{1}{2} \frac{\left\langle \dfrac{\partial f}{\partial \boldsymbol{\sigma}} : \mathbf{E} : \dot{\boldsymbol{\varepsilon}} \right\rangle^2}{h + \left(\dfrac{\partial f}{\partial \boldsymbol{\sigma}} : \mathbf{E} : \dfrac{\partial f}{\partial \boldsymbol{\sigma}} \right)}, \tag{5.84}$$

where h is the *hardening modulus*. This is the same as eqn (5.45).

Remarks on the hardening positivity We saw that after (5.75) we demanded that the hardening be always positive. Actually, if the hardening is *strictly* positive, then the relation (5.81) shows that the plastic strain rate is expressed in terms of the stress rate $\dot{\boldsymbol{\sigma}}$ and the relation between rates $\dot{\boldsymbol{\sigma}}$ and $\dot{\boldsymbol{\varepsilon}}$ is one-to-one, whereas this is *not* the case in perfect plasticity! h, in particular, is strictly positive if W^{α} is *strictly* convex.

Generalized strains and stresses We set the following notations (see eqns (5.29)):

$$\left.\begin{aligned} \mathscr{E}^{\mathrm{p}} &= (\boldsymbol{\varepsilon}^{\mathrm{p}}, \boldsymbol{\varepsilon}^{\mathrm{p}}, \boldsymbol{\beta}), \\ \mathscr{E}^{\mathrm{e}} &= (\boldsymbol{\varepsilon}^{\mathrm{e}}, -\boldsymbol{\varepsilon}^{\mathrm{p}}, -\boldsymbol{\beta}), \\ \boldsymbol{\Sigma} &= (\boldsymbol{\sigma}, \mathbf{B}^{\mathrm{p}}, \mathbf{B}^{\beta}), \\ \mathscr{E} &= (\boldsymbol{\varepsilon}, \mathbf{0}, \mathbf{0}). \end{aligned}\right\} \tag{5.85}$$

With these notations we find once more a formulation of the perfect plasticity type:

$$\left.\begin{aligned} \mathscr{E} &= \mathscr{E}^{\mathrm{e}} + \mathscr{E}^{\mathrm{p}} && \text{(split of strains)}, \\ \dot{\boldsymbol{\Sigma}} &= \mathscr{L} : \dot{\mathscr{E}}^{\mathrm{e}} && \text{(state laws)}, \\ \dot{\mathscr{E}}^{\mathrm{p}} &\in N_C(\boldsymbol{\Sigma}) && \text{(normality laws)}, \end{aligned}\right\} \tag{5.86}$$

with a *generalized elastic coefficient* matrix $\mathscr{L}$, of the form

$$\mathscr{L} = \begin{bmatrix} \mathbb{E} & 0 & 0 \\ 0 & \mathbf{L} & \mathbf{N} \\ 0 & \mathbf{N}^{\mathrm{T}} & \mathbf{M} \end{bmatrix}. \tag{5.87}$$

This matrix represents the *second* derivatives of the energy W with respect to the state variables $(\boldsymbol{\varepsilon}^{\mathrm{e}}, -\boldsymbol{\varepsilon}^{\mathrm{p}}, -\boldsymbol{\beta})$ since $(5.86)_2$ is written in *incremental* form. If W is *quadratic*, the coefficients of $\mathscr{L}$ are constants.

5.3 Stability in Ilyushin's sense

The generalized standard materials satisfy the *Ilyushin 'postulate'* which we can state as follows (Ilyushin, 1948).

For any strain cycle $\boldsymbol{\varepsilon}(t)$, $t \in [0, 1]$, $\boldsymbol{\varepsilon}(0) = \boldsymbol{\varepsilon}(1)$, *the strain power is positive or zero*:

$$\int_0^1 \boldsymbol{\sigma} : \dot{\boldsymbol{\varepsilon}}(t)\,\mathrm{d}t \geqslant 0. \tag{5.88}$$

Let us show now that this holds true for generalized standard materials such that

$$W = W(\boldsymbol{\varepsilon}, \alpha), \qquad \boldsymbol{\sigma} = \frac{\partial W}{\partial \boldsymbol{\varepsilon}}, \qquad A = -\frac{\partial W}{\partial \alpha}, \qquad A \in C. \tag{5.89}$$

Proof We have

$$\dot{W} = \frac{\partial W}{\partial \boldsymbol{\varepsilon}} : \dot{\boldsymbol{\varepsilon}} + \frac{\partial W}{\partial \alpha}\dot{\alpha} = \boldsymbol{\sigma} : \dot{\boldsymbol{\varepsilon}} - A\dot{\alpha}. \tag{5.90}$$

Integrating this on the time interval [0, 1] yields

$$\int_0^1 \dot{W}(t)\,\mathrm{d}t = \int_0^1 \boldsymbol{\sigma} : \dot{\boldsymbol{\varepsilon}}\,\mathrm{d}t - \int_0^1 A\dot{\alpha}\,\mathrm{d}t$$

or

$$W(\boldsymbol{\varepsilon}(1), \alpha(1)) - W(\boldsymbol{\varepsilon}(0), \alpha(0)) = \int_0^1 \boldsymbol{\sigma} : \dot{\boldsymbol{\varepsilon}}\,\mathrm{d}t - \int_0^1 A\dot{\alpha}\,\mathrm{d}t. \tag{5.91}$$

By adding and subtracting the quantity $A_0\,(\alpha(1) - \alpha(0))$ where A_0 is constant and, with the normality law,

$$(A(t) - A_0)\dot{\alpha}(t) \geqslant 0, \tag{5.92}$$

we get the following syllogism. According to (5.91),

$$\begin{aligned}\int_0^1 \boldsymbol{\sigma} : \dot{\boldsymbol{\varepsilon}}(t)\,\mathrm{d}t &= W(\boldsymbol{\varepsilon}(1), \alpha(1)) - W(\boldsymbol{\varepsilon}(0), \alpha(0)) + \int_0^1 A\dot{\alpha}\,\mathrm{d}t \\ &= W(\boldsymbol{\varepsilon}(0), \alpha(1)) - W(\boldsymbol{\varepsilon}(0), \alpha(0)) + \int_0^1 A\dot{\alpha}\,\mathrm{d}t, \end{aligned} \tag{5.93}$$

since $\boldsymbol{\varepsilon}(0) = \boldsymbol{\varepsilon}(1)$. According to the convexity of W with respect to α we have

$$W(\boldsymbol{\varepsilon}(0), \alpha(1)) - W(\boldsymbol{\varepsilon}(0), \alpha(0)) \geqslant \left(\frac{\partial W}{\partial \alpha}\right)_0 [\alpha(1) - \alpha(0)], \tag{5.94}$$

and so, with $(\partial W/\partial \alpha)_0 = -A(0)$, (5.93) gives

$$\begin{aligned}\int_0^1 \boldsymbol{\sigma} : \dot{\boldsymbol{\varepsilon}}(t)\,\mathrm{d}t &\geqslant -A(0)[\alpha(1) - \alpha(0)] + \int_0^1 A\dot{\alpha}\,\mathrm{d}t \\ &\geqslant \int_0^1 (A - A(0))\dot{\alpha}\,\mathrm{d}t \geqslant 0,\end{aligned}$$

where the last inequality is the result of the integration of (5.92). So we have proven (5.88), which means that the *loops* of response $\boldsymbol{\sigma}(\boldsymbol{\varepsilon})$ always follow a clockwise direction (Fig. 5.9). Compared to (5.75) which is a local stability condition, (5.89) is a *global* stability condition.

Only cycle (b) in Fig. 5.9. to which (5.88) applies, satisfies (5.75) at all of its points, even along the unloading part.

5.4 Elastoplastic evolution in the presence of hardening

We consider the quasi-static evolution (the inertia forces being left out), of an elastoplastic material with hardening, represented by a scheme of generalized standard materials. The results (5.82), (5.85) and (5.86) show that all the results established for perfect plasticity in Chapter 4 can be transposed to the case of generalized standard materials (B. Halphen and Nguyen Quoc Son, 1975). We may, in particular, state the variational principles of the Greenberg and Hodge–Prager type and note that the hardening has a regularizing effect.

(i) Variational principle in terms of displacement rates $\dot{\mathbf{u}}$ (Greenberg) *The velocity $\dot{\mathbf{u}}$ minimizes the Greenberg functional*

$$A(\dot{\mathbf{u}}) = \int_{\Omega} \Psi^{\alpha}(\dot{\boldsymbol{\varepsilon}})\,\mathrm{d}\Omega - \int_{\Omega} \dot{\mathbf{g}}\cdot\dot{\mathbf{u}}\,\mathrm{d}\Omega - \int_{\partial\Omega_{\mathrm{T}}} \dot{\mathbf{T}}^{\mathrm{d}}\cdot\dot{\mathbf{u}}\,\mathrm{d}s \tag{5.95}$$

among the velocities $\dot{\mathbf{u}}^*$ that are compatible with the kinematic conditions, i.e., $\dot{\mathbf{u}}|\partial\Omega_u = \dot{\mathbf{u}}^{\mathrm{d}}$.

(ii) Variational principle in terms of stress rates $\dot{\boldsymbol{\sigma}}$ (Hodge–Prager) *The rate $\dot{\boldsymbol{\sigma}}$ minimizes the Hodge–Prager functional*

$$B(\dot{\boldsymbol{\sigma}}) = \int_{\Omega} \tfrac{1}{2}\dot{\boldsymbol{\sigma}}\cdot\mathbf{S}\cdot\dot{\boldsymbol{\sigma}}\,\mathrm{d}\Omega + \int_{\Omega_{\mathrm{p}}} \frac{1}{2h}\left\langle\frac{\partial f}{\partial\boldsymbol{\sigma}}:\dot{\boldsymbol{\sigma}}\right\rangle^{2}\mathrm{d}\Omega - \int_{\partial\Omega_u} \dot{\mathbf{u}}^{\mathrm{d}}\cdot(\dot{\boldsymbol{\sigma}}\cdot\mathbf{n})\,\mathrm{d}s \tag{5.96}$$

Fig. 5.9. Ilyushin's 'postulate'

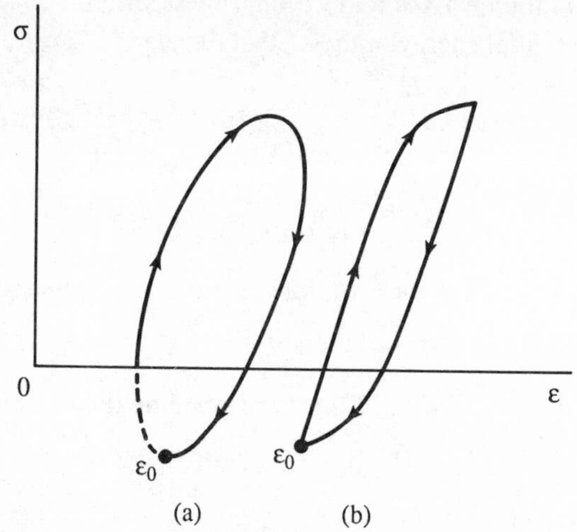

among the statically admissible stress rates $\dot{\boldsymbol{\sigma}}^*$; Ω_p *is the plastic volume at time t, that is, the set of points where the plasticity criterion* $f(\boldsymbol{\sigma} + \mathbf{B}^p, \mathbf{B}^\beta) = 0$ *is attained.*

(iii) Regularization by positive hardening The introduction of positive hardening regularizes the solution of the evolution problem in displacement, because the plastic strain $\boldsymbol{\varepsilon}^p$ becomes a regular function (for example L^2) – compare with the remark in section 4.4 – and the *discontinuity surfaces need be no more considered* in positive hardening.

We can show that $\Psi^\alpha(\dot{\boldsymbol{\varepsilon}})$ is *strictly* convex if the hardening is strictly positive. In fact, then, $\Psi^\alpha(\dot{\boldsymbol{\varepsilon}})$ satisfies the inequality (problem)

$$\left(\frac{\partial \Psi^\alpha}{\partial \dot{\boldsymbol{\varepsilon}}} - \frac{\partial \Psi^\alpha}{\partial \dot{\boldsymbol{\varepsilon}}^*}\right) : (\dot{\boldsymbol{\varepsilon}} - \dot{\boldsymbol{\varepsilon}}^*) \geqslant h_0(\dot{\boldsymbol{\varepsilon}} - \dot{\boldsymbol{\varepsilon}}^*) : \mathbf{E} : (\dot{\boldsymbol{\varepsilon}} - \dot{\boldsymbol{\varepsilon}}^*) \geqslant 0 \tag{5.97}$$

This inequality can then be explored to prove *uniqueness* in $\dot{\boldsymbol{\sigma}}$ and $\dot{\mathbf{u}}$. We also have regularity and uniqueness in $\dot{\boldsymbol{\alpha}}$. We can, more particularly, show the inequality

$$\Psi^\alpha(\dot{\boldsymbol{\varepsilon}}) > \tfrac{1}{2} h \dot{\boldsymbol{\varepsilon}}(\dot{\mathbf{u}}) : \mathbf{E} : \dot{\boldsymbol{\varepsilon}}(\dot{\mathbf{u}}). \tag{5.98}$$

5.5 Simplified abstract formulation

The standard generalized models suggest that the plasticity equations can be summed up under an abstract general form that we shall explain. In order to simplify we shall assume that the data are controlled by load parameters λ_i, $i = 1, \ldots, m$, i.e., $\mathbf{g} = \lambda_i \bar{\mathbf{g}}_i$, $\mathbf{T}^d = \lambda_i \bar{\mathbf{T}}_i$, $\mathbf{u}^d = \lambda_i \bar{\mathbf{u}}_i$, where the functions $(\bar{\mathbf{g}}_i, \bar{\mathbf{T}}_i, \bar{\mathbf{u}}_i)$ are *fixed* and $\lambda_i = \lambda_i(t)$ describes the load path. Let I be the total potential energy of the system (see eqn (4.12)),

$$I = \int_\Omega W(\boldsymbol{\varepsilon}(\mathbf{u}), \boldsymbol{\alpha})\, d\Omega - \int_\Omega \lambda_i \bar{\mathbf{g}}_i \cdot \mathbf{u}\, d\Omega - \int_{\partial\Omega_T} \lambda_i \bar{\mathbf{T}} \cdot \mathbf{u}\, ds, \tag{5.99}$$

i.e.,

$$I = I(\mathbf{u}, \boldsymbol{\alpha}, \boldsymbol{\lambda}) \tag{5.100}$$

with $\mathbf{u} \in \mathscr{U}(\boldsymbol{\lambda})$, where $\mathscr{U}$ is the set of the kinematically admissible fields,

$$\mathscr{U}(\boldsymbol{\lambda}) = \{\mathbf{u} \mid \mathbf{u} = \lambda_i \mathbf{u}_i \text{ on } \partial\Omega_u\}. \tag{5.101}$$

Evidently, the equilibrium conditions are given by the first variation:

$$I_{,\mathbf{u}}(\mathbf{u}, \boldsymbol{\alpha}, \boldsymbol{\lambda}) \cdot \delta\mathbf{u} = 0, \tag{5.102}$$

that is, the principle of virtual work (2.27):

$$\int_{\Omega} W_{,\varepsilon}(\boldsymbol{\varepsilon}(\mathbf{u}), \boldsymbol{\alpha}) : \delta\boldsymbol{\varepsilon}\, d\Omega - \int_{\Omega} \lambda_i \bar{\mathbf{g}}_i \cdot \delta\mathbf{u}\, d\Omega - \int_{\partial\Omega_T} \lambda_i \bar{\mathbf{T}}_i \cdot \delta\mathbf{u}\, ds = 0$$

$$\forall \delta u = \mathbf{u}^* - \mathbf{u},\ \mathbf{u}^* \in \mu(\boldsymbol{\lambda}), \tag{5.103}$$

The evolution of the internal parameters $\boldsymbol{\alpha}$, tensorial functions on Ω, is given by the *normality law*

$$\dot{\boldsymbol{\alpha}} \in N_C(A) \quad \text{or} \quad \dot{\boldsymbol{\alpha}}(A - A^*) \geqslant 0, \quad \forall A^* \in C. \tag{5.104}$$

In *summary* we have the following problem:

Let $I(\mathbf{u}, \boldsymbol{\alpha}, \boldsymbol{\lambda})$ be the global potential energy, $\mathbf{u} \in \mathscr{U}$, $\alpha \in V$ and $\boldsymbol{\lambda}(t)$ given; determine $\mathbf{u}(t)$, $\boldsymbol{\alpha}(t)$, satisfying for every t the following conditions:

$I_{,\mathbf{u}}(\mathbf{u}, \boldsymbol{\alpha}, \boldsymbol{\lambda}) \cdot \delta\mathbf{u} = 0, \forall \delta\mathbf{u}$ (equilibrium equation),

$A = -I_{,\boldsymbol{\alpha}}(\mathbf{u}, \boldsymbol{\alpha}, \boldsymbol{\lambda})$ (generalized force),

$\dot{\boldsymbol{\alpha}} \in N_C(A)$ (normality law),

$\boldsymbol{\alpha}(0) = \boldsymbol{\alpha}_0$ (initial conditions).

Let us point out that the nature of the problem remains unchanged if we take

$$\bar{\mathbf{u}} = \mathbf{0}, \qquad \bar{\mathbf{u}} \in \mathscr{U}_0 = \{\mathbf{u} | \mathbf{u} = \mathbf{0} \text{ on } \partial\Omega_u\}. \tag{5.105}$$

We may then ask the following question. *Is it possible to take* $\boldsymbol{\alpha}$ *as the main unknown*? (just as we can do in ordinary plasticity where $\alpha = \{\boldsymbol{\varepsilon}^p\}$).

The equilibrium conditions (5.102) provide a solution

$$\mathbf{u} = \mathbf{u}(\boldsymbol{\alpha}, \boldsymbol{\lambda}). \tag{5.106}$$

The potential energy *in equilibrium* is then written

$$\mathscr{W} = \mathscr{W}(\boldsymbol{\alpha}, \boldsymbol{\lambda}) = I(\mathbf{u}(\boldsymbol{\alpha}, \boldsymbol{\lambda}), \boldsymbol{\alpha}, \boldsymbol{\lambda}). \tag{5.107}$$

Yet, is it true that $\mathbf{A} = -\mathscr{W}_{,\boldsymbol{\alpha}}(\boldsymbol{\alpha}, \boldsymbol{\lambda})$? Yes, since we can immediately verify it by the calculation

$$\begin{aligned}\mathscr{W}_{,\boldsymbol{\alpha}}(\boldsymbol{\alpha}, \boldsymbol{\lambda})\delta\boldsymbol{\alpha} &= \frac{\partial I}{\partial \boldsymbol{\alpha}}(\mathbf{u}(\boldsymbol{\alpha}, \boldsymbol{\lambda}), \boldsymbol{\alpha}, \boldsymbol{\lambda})\delta\boldsymbol{\alpha} \\ &= I_{,\mathbf{u}}(\mathbf{u}, \boldsymbol{\alpha}, \boldsymbol{\lambda}) \cdot \frac{\partial \mathbf{u}}{\partial \boldsymbol{\alpha}}\delta\boldsymbol{\alpha} + I_{,\boldsymbol{\alpha}}(\mathbf{u}, \boldsymbol{\alpha}, \boldsymbol{\lambda})\delta\boldsymbol{\alpha} \\ &= I_{,\mathbf{u}} \cdot \delta\mathbf{u} + I_{,\boldsymbol{\alpha}}\delta\boldsymbol{\alpha},\end{aligned} \tag{5.108}$$

and, according to (5.102), the first contribution is zero. So

$$\mathcal{W}_{,\boldsymbol{\alpha}}(\boldsymbol{\alpha}, \boldsymbol{\lambda})\delta\boldsymbol{\alpha} = I_{,\boldsymbol{\alpha}}(\mathbf{u}, \boldsymbol{\alpha}, \boldsymbol{\lambda})\delta\boldsymbol{\alpha} \tag{5.109}$$

and

$$\mathbf{A} = -I_{,\boldsymbol{\alpha}}(\mathbf{u}, \boldsymbol{\alpha}, \boldsymbol{\lambda}) = -\mathcal{W}_{,\boldsymbol{\alpha}}(\boldsymbol{\alpha}, \boldsymbol{\lambda}). \tag{5.110}$$

We may then state the following so-called *simplified abstract formulation*, in $\boldsymbol{\alpha}$ (Nguyen Quoc Son).

Problem Let $\mathcal{W}(\boldsymbol{\alpha}, \boldsymbol{\lambda})$ be the energy *in equilibrium*, $\boldsymbol{\alpha} \in V$, $\lambda(t)$ given; determine $\boldsymbol{\alpha}(t)$ satisfying the following equations:

$$\begin{aligned} \mathbf{A} &= -\mathcal{W}_{,\boldsymbol{\alpha}}(\boldsymbol{\alpha}, \boldsymbol{\lambda}) && \text{(state laws)},\\ \dot{\boldsymbol{\alpha}} &\in N_C(A) && \text{(normality laws)},\\ \boldsymbol{\alpha}(0) &= \boldsymbol{\alpha}_0 && \text{(initial conditions)}. \end{aligned}$$

This concludes the formal study of the equations of elastoplasticity whether without or with hardening. The following problems illustrate the notion of *strain hardening*, starting with the Hencky – Nadai (and Ilyushin) model valid only for very small strains, and considering the determination of the hardening modulus.

Problems for Chapter 5

5.1. Hencky–Ilyushin model (for loading only)

Let

$$\boldsymbol{\sigma} = \boldsymbol{\sigma}(\boldsymbol{\varepsilon}) = \varphi_0 \mathbf{I} + \varphi_1 \boldsymbol{\varepsilon}, \qquad \boldsymbol{\varepsilon} = \boldsymbol{\varepsilon}^e + \boldsymbol{\varepsilon}^p = (\nabla \mathbf{u})_s, \tag{1}$$

in small-strain theory $\boldsymbol{\varepsilon}^e$ satisfies Hooke's law. Thus, as $\operatorname{tr} \boldsymbol{\varepsilon}^p = 0$,

$$\operatorname{tr} \boldsymbol{\varepsilon} = \operatorname{tr} \boldsymbol{\varepsilon}^e = \frac{1-2\nu}{E} \operatorname{tr} \boldsymbol{\sigma} \quad (E \text{ is Young's modulus}). \tag{2}$$

Let $\mathbf{e}$ be the deviator of $\boldsymbol{\varepsilon}$. For small elastic–plastic strains, referring to shearing strains and considering *loading only*, we have

$$\mathbf{e} = \varphi \mathbf{s} = \left(\frac{1}{2\mu} + \psi\right)\mathbf{s}, \qquad \mathbf{s} \geqslant 0, \tag{3}$$

where μ is the elastic shear modulus. Show that

$$\psi = \varphi - \frac{1}{2\mu}, \qquad \varphi = \frac{\operatorname{tr}(\mathbf{se})}{\operatorname{tr} \mathbf{s}^2}. \tag{4}$$

5.2. **P5.1 continued**

If the material obeys the Huber–Mises criterion, show that

$$\varphi = \frac{1}{2k}(2e_{ij}e_{ij})^{1/2}. \tag{1}$$

As another example we may take $\varphi = \varphi(\sqrt{\operatorname{tr}\mathbf{e}^2})$. Can this be a law for material exhibiting hardening (in loading)?

5.3. **Stress-rate potential**

Show that in the presence of strain-hardening the stress-rate potential is given by expression (5.47) [*Hint*: evaluate the Legendre–Fenchel transform of $\Psi^\sigma(\dot{\boldsymbol{\varepsilon}})$ given by eqn (5.45)].

5.4. **Stable materials** (Drucker's postulate)

Drucker's stability postulate reads

$$W = \int_0^t (\boldsymbol{\sigma} - \boldsymbol{\sigma}^0) : \dot{\boldsymbol{\varepsilon}}^{\mathrm{p}}\, \mathrm{d}t \geqslant 0. \tag{1}$$

In plastic strains, where $\boldsymbol{\sigma}^0$ is the original state of stresses and $[0, t]$ is a closed time-cycle of loading and unloading.

(a) Show that the above expression reduces to

$$W^{\mathrm{p}} = \int_{t_1}^{t_2} (\boldsymbol{\sigma} - \boldsymbol{\sigma}^0) : \dot{\boldsymbol{\varepsilon}}^{\mathrm{p}}\, \mathrm{d}t \geqslant 0. \tag{2}$$

What are t_1 and t_2? [*Hint*: use reversibility of elastic strains]

(b) Then, expanding W^{p} into Taylor's series about $\boldsymbol{\sigma}(t_1)$, show that, if t_2 is in a neighourhood of t_1, (2) implies that

$$\left.\begin{aligned} (\boldsymbol{\sigma} - \boldsymbol{\sigma}^0) : \dot{\boldsymbol{\varepsilon}}^{\mathrm{p}} \geqslant 0 \quad & \text{if } \boldsymbol{\sigma} \neq \boldsymbol{\sigma}^0 \quad \text{(at first order in the expansion)}, \\ \dot{\boldsymbol{\sigma}} : \dot{\boldsymbol{\varepsilon}}^{\mathrm{p}} \geqslant 0 \quad & \text{if } \boldsymbol{\sigma} = \boldsymbol{\sigma}^0 \quad \text{(at second order in the expansion)}, \end{aligned}\right\} \tag{3}$$

5.5. **Associated flow rule**

Assume that the yield surface $f(\boldsymbol{\sigma}) = 0$ is *convex* in stress space. Such a convex surface has a unique normal at every point if and only if

$$(\boldsymbol{\sigma} - \boldsymbol{\sigma}^0) : \frac{\partial f}{\partial \boldsymbol{\sigma}} \geqslant 0, \qquad \boldsymbol{\sigma}^0 \in C. \tag{1}$$

Using the result of 5.4 deduce that

$$\dot{\boldsymbol{\varepsilon}}^{\mathrm{p}} = \dot{\lambda} \frac{\partial f}{\partial \boldsymbol{\sigma}}, \tag{2}$$

i.e., the *associated flow rule*. In this presentation, the associated flow rule follows from Drucker's postulate and the convexity of the yield surface!

5.6. **Hardening**

Assume that the yield surface has equations $f(\boldsymbol{\sigma}, W^{\mathrm{p}}) = 0$. Specify the form of the associated flow rule by determining the scalar multiplier $\dot{\lambda}$.

5.7. Strictly convex strain-rate potential
Prove the inequality (5.97) if the strain-rate potential $\Psi^{\alpha}(\dot{\boldsymbol{\varepsilon}})$ is *strictly* convex.

5.8. Locking materials and hysteresis
Make a literature search on locking materials (notion due to W. Prager: materials that exhibit a *saturation* in plastic strain) and draw an analogy between plastic locking of solid materials and magnetic hysteresis or ferroelectric hysteresis.

5.9. Uniqueness and regularity of solutions (for mathematically oriented readers). Make a literature search (and use the result of 5.6) to write a short report on the uniqueness of $\dot{\boldsymbol{\sigma}}$ and $\dot{\mathbf{u}}$ and the regularity and uniqueness of $\dot{\boldsymbol{\alpha}}$ for plasticity problems in the presence of hardening.

5.10. Elastoplastic small-strain constitutive equations Consider a hardening elastic–plastic material satisfying Hooke's law in the elastic regime ($\sigma_{ij} = E_{ijkl}\varepsilon_{kl}$) and the following *associated flow* rule for the plastic strain:

$$\dot{\varepsilon}^{\mathrm{p}}_{ij} = \frac{1}{h}(n_{kl}\dot{\sigma}_{kl})n_{ij}, \tag{1}$$

where h is the *hardening modulus* and n_{kl} is the unit normal to the yield surface in stress space, i.e., $n_{kl} = (\partial F/\partial\sigma_{kl})(\partial F/\partial\sigma_{ij}\partial F/\partial\sigma_{ij})^{-1/2}$. What is the expression of the plastic multiplier $\dot{\lambda}$? Show that the material obeys the following elastic–plastic small-strain constitutive equations:

$$\dot{\sigma}_{ij} = E^{\mathrm{ep}}_{ijkl}\dot{\varepsilon}_{kl}. \tag{2}$$

What is the general expression for E^{ep}_{ijkl}?

5.11. Determining the hardening modulus from uniaxial test data.
Suppose that the Huber–Mises criterion holds good: $F(\boldsymbol{\sigma}) = (\frac{3}{2}\sigma^{\mathrm{d}}_{ij}\sigma^{\mathrm{d}}_{ij})^{1/2} - \sigma_0 = 0$, where σ_0 is the initial yield stress taken from the uniaxial test. We have at hand experimental curves σ_{ef} versus $\varepsilon_{\mathrm{ef}}$ and σ_{ef} versus $\varepsilon^{\mathrm{p}}_{\mathrm{ef}}$ – with $\sigma_{\mathrm{ef}} = (\frac{3}{2}\sigma^{\mathrm{d}}_{ij}\sigma^{\mathrm{d}}_{ij})^{1/2}$ and $\varepsilon^{\mathrm{p}}_{\mathrm{ef}} = \int_0^t(\frac{2}{3}\dot{\varepsilon}^{\mathrm{p}}_{ij}\dot{\varepsilon}^{\mathrm{p}}_{ij})^{1/2}\,\mathrm{d}t$ – as shown in Fig. 5.10.
(a) Show, in the uniaxial case, $\varepsilon^{\mathrm{p}} = \varepsilon^{\mathrm{p}}_{11}$, $\varepsilon = \varepsilon_{11}$, $\sigma = \sigma_{11}$, that

Fig. 5.10. Figures for Problem 5.11

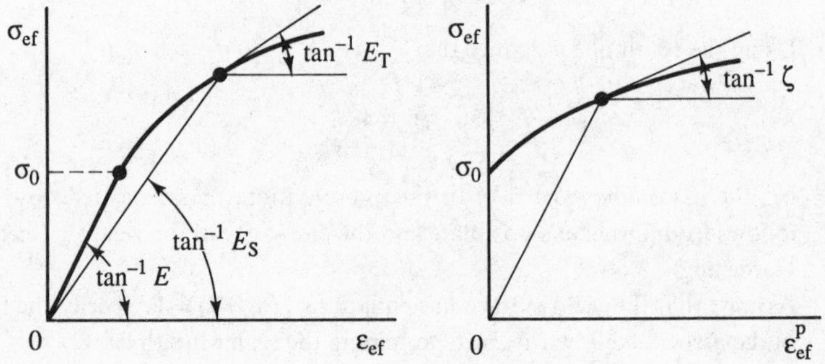

$$h = \tfrac{2}{3}\zeta, \qquad E = \frac{E\zeta}{E+\zeta}, \qquad \zeta = \frac{EE_T}{E - E_T}. \tag{1}$$

Here E, E_T and E_S (Fig. 5.10) are the Young's modulus, the tangent instantaneous modulus, and the so-called *secant* modulus, respectively.

(b) Then using the explicit form for E_{ijkl} for linear isotropic elasticity, show that E^{ep}_{ijkl} is given by

$$E^{ep}_{ijkl} = \frac{E}{1+\nu}\left[\frac{\nu}{1-2\nu}\delta_{ij}\delta_{kl} + \tfrac{1}{2}(\delta_{ik}\delta_{jl} + \delta_{il}\delta_{jk}) - \alpha\frac{\alpha^d_{ij}\sigma^d_{kl}}{\varpi}\right], \tag{2}$$

where

$$\varpi = \tfrac{2}{3}\zeta\sigma_0^2\left(\frac{1}{E_T} - \frac{1}{9B}\right), \qquad B = \frac{E}{3(1-2\nu)} \tag{3}$$

and

$$\alpha = \begin{cases} 1 & \text{for } \dot{\sigma} = \tfrac{3}{2}(\sigma^d_{ij}\dot{\sigma}^d_{ij}/\sigma_{ef}) \geqslant 0 \textit{ and } \sigma = \sigma_{max}, \\ 0 & \text{for } \dot{\sigma} < 0 \textit{ or } \qquad\qquad \sigma < \sigma_{max}, \end{cases} \tag{4}$$

where ν and B are Poisson's ratio and the bulk modulus, respectively.

(c) Invert the elastic–plastic constitutive relation to obtain

$$\dot{\varepsilon}_{ij} = S^{ep}_{ijkl}\dot{\sigma}_{kl} \tag{5}$$

with

$$S^{ep}_{ijkl} = \frac{1}{E}\left[(1+\nu)\delta_{ik}\delta_{jl} - \nu\delta_{ij}\delta_{kl} + \alpha\frac{\sigma^d_{ij}\sigma^d_{kl}}{\omega}\right], \qquad \omega \equiv \alpha h\sigma_0^2/3E. \tag{6}$$

5.12. Small-strain 'deformation' theory of elastoplasticity (no unloading)

Assume for isotropic bodies that, in the spirit of the Hencky–Ilyushin model, we have a behaviour described without rates by

$$\left.\begin{aligned} &\boldsymbol{\varepsilon} = \boldsymbol{\varepsilon}^e + \boldsymbol{\varepsilon}^p, \\ &\varepsilon^e_{ij} = \frac{1}{E}[(1+\nu)\sigma_{ij} - \nu\sigma_{kk}\delta_{ij}] \quad \text{(Hooke's law)}, \\ &\varepsilon^p_{ij} = \frac{1}{E}h_2(\sigma_{ef})\sigma^d_{ij}, \quad h_2 = \tfrac{3}{2}\left[\frac{E}{E_s(\sigma_{ef})} - 1\right], \end{aligned}\right\} \tag{1}$$

where the secant modulus E_S is shown in Fig. 5.10 ($\sigma_{ef} = E_S\varepsilon_{ef}$ at any point along the experimental curve). By time differentiation of the above formulas show that the rate equation $\dot{\sigma}_{ij} = E^{ep}_{ijkl}\dot{\varepsilon}_{kl}$ holds with

$$E^{ep}_{ijkl} = \frac{E}{1+\nu_S}\left[\frac{\nu_S}{1-2\nu_S}\delta_{ij}\delta_{kl} + \tfrac{1}{2}(\delta_{ik}\delta_{jl} + \delta_{il}\delta_{jk}) - \tfrac{3}{2}\frac{(E_S/E_T) - 1}{(E_S/E_T) - (1-2\nu_S)/3}\frac{\sigma^d_{ij}\sigma^d_{ij}}{\sigma_{ef}}\right], \tag{2}$$

where the 'secant' Poisson ratio ν_S is defined by

$$\frac{\nu_S}{E_S} = \frac{\nu}{E} + \frac{h_2}{3E}. \tag{3}$$

Remark: The present formulation is similar to the one given in 5.11 but for the fact that E_S and ν_S replace E and ν, respectively. However, as unloading is *not* allowed, the α coefficient does not appear in the present case.

5.13. Nonlinear waves in elastoplasticity (term paper)
Make a literature search and write a report on the propagation of one-dimensional plane *simple* and *shock* waves in an elastoplastic material. [*Hint*: see works by J. Mandel and W. K. Nowacki (especially the latter's 1978 book) from 1960 to 1980; study first the elements of propagation of singular surfaces in a continuum.]

6

Elements of limit analysis

The object of the chapter In this chapter we are interested in the ruin of perfectly-plastic–elastic structures and we introduce the notions of *limit load* and of *maximum admissible load*, the determination of which constitutes the essential object of every engineer's office computations. We shall only attempt an *introduction* to this type of calculation, which will be illustrated by two examples. Certain minimum principles apply to *velocities* and to *stresses*. The static and dynamic methods in the determination of the maximum admissible load are only given in a rough draft.

6.1 The notion of limit load

The object of our attention is the notion of *limit load* and the *collapse* of *perfectly* plastic structures under unrestrained plastic strains. What do we mean by that? As certain deformable structures evolve, we observe that the elastoplastic response is produced in *three* stages. The first phase is elastic, the material being elastic *everywhere*. This phase lasts until the appearance of the first yielding. But the fact that the criterion of plasticity $f(\boldsymbol{\sigma}) = 0$ is reached at one point does not necessarily mean that there is collapse. If the strain rate $\dot{\boldsymbol{\varepsilon}}$ is still *controlled*, the plastic strain rate is not unlimited, since it is expressed in terms of $\dot{\boldsymbol{\varepsilon}}$ (Section 4.2). We say then that the plastic strain is still *controlled*. The second phase of the response corresponds to the appearance and the extension of one or more regions of the structure, generally called *plastic zones*; in these, the plasticity criterion is satisfied at all the points. In this second phase, we can still increase the applied load, until the moment when the elastic regions cease to allow the control of the strain velocity. Then the third phase begins and this means the *destruction* of the structure through *uncontrolled* plastic strain. In what follows we are mostly interested in this last phase. In order to see this, let us consider a *progressive* loading with one parameter δ of any given structure Ω. This structure is subject, on a part $\partial\Omega_T$ of its boundary, to surface forces $\mathbf{T}^d = \delta\mathbf{R}^e$, while a *zero* displacement ($\mathbf{u}^d = \mathbf{0}$) is imposed on the complement $\partial\Omega_u$ in $\partial\Omega$. To simplify, we assume the absence of the volume force ($\mathbf{g} = \mathbf{0}$). The progressive increase of δ beginning from zero defines a *one*-parameter (δ) load path. We have of course the intuitive notion that no real structure may resist indefinitely. If the structure has an *elastic–perfectly-plastic* behaviour, and if the geometrical changes during evolution with increasing δ are negligible, then a complete study of the quasi-static evolution must be able to show the steps of the elastoplastic response. The last step, in particular, should provide *the maximum admissible load* δ_m.

Estimate of $\boldsymbol{\delta}_m$ This is a problem *in stresses*. We should then recall the results of Chapter 4 (eqn 4.39):

$$(\dot{\boldsymbol{\sigma}}^E - \dot{\boldsymbol{\sigma}}) \in N_{P\cap S(t)}(\boldsymbol{\sigma}), \qquad \boldsymbol{\sigma}(0) = \boldsymbol{\sigma}_0, \tag{6.1}$$

which we can also write

$$\left.\begin{aligned} \dot{\boldsymbol{\sigma}} + N_{P\cap S(t)}(\boldsymbol{\sigma}) = \dot{\boldsymbol{\sigma}}^E, \\ \boldsymbol{\sigma}(0) = \boldsymbol{\sigma}_0. \end{aligned}\right\} \tag{6.2}$$

As this point we should remember that P is the set of plastically admissible stress fields and $S(t)$ or $S(\delta)$ – in our case – is the set of statically admissible stress fields, i.e. (The spaces L^2 are indicated as an example of regularity),

$$P = \{\sigma_{ij} \in L^2(\Omega) | \boldsymbol{\sigma} \in C, \forall \mathbf{x} \in \Omega\} \tag{6.3}$$

and

$$S(\delta) = \{\sigma_{ij} \in L^2(\Omega) | \operatorname{div} \boldsymbol{\sigma} = \mathbf{0} \text{ in } \Omega, \boldsymbol{\sigma}\cdot\mathbf{n} = \delta\mathbf{R}^0 \text{ on } \partial\Omega_T\}. \tag{6.4}$$

At the moment of collapse we have $\delta = \delta_m$ and $\sigma \in P \cap S(\delta_m)$. Let us introduce the quantity δ_L by

$$\delta_L = \sup_{P\cap S(\delta)\neq\phi} \delta = \sup_{\substack{(\sigma,\delta)\\ \sigma\in P\\ \sigma\in S(\delta]}} \delta, \tag{6.5}$$

We necessarily have

$$\delta_m \leqslant \delta_L. \tag{6.6}$$

The quantity δ_L is such that $P \cap S(\delta)$ is the *empty* set for any $\delta > \delta_L$. Therefore, this is an upper bound of the *maximum admissible load* δ_m.

6.1.1 Characterization of the limit load

Definition. We call *limit load* the maximum admissible load δ_m.

At the moment of collapse the plastic strains are no longer controlled, so that the state of stresses at that moment is in a state *of limit equilibrium* defined by

$$\left.\begin{aligned} &\boldsymbol{\sigma} \in P \cap S(\delta_m), \\ &\exists \mathbf{v} \in \mathscr{U}_0 = \{\mathbf{v} | \mathbf{v}_+ = \mathbf{0} \text{ on } \partial\Omega_u\}, \\ &\dot{\boldsymbol{\varepsilon}} = (\nabla \mathbf{v})_S \neq \mathbf{0} \quad \text{such that} \\ &\dot{\boldsymbol{\varepsilon}}(\mathbf{v}) \in N_P(\boldsymbol{\sigma}), \end{aligned}\right\} \tag{6.7}$$

or, in other terms: we can associate to a statically and plastically admissible stress field $\boldsymbol{\sigma}$ a *compatible* plastic deformation rate field (i.e., derivable from velocity $\mathbf{v}$ by $\boldsymbol{\varepsilon}(\mathbf{v}) = (\nabla \mathbf{v})_S$ obeying the normality law).

6.1.2 Case of the rigid–plastic model

In this case the evolution law (6.7), or

$$\dot{\boldsymbol{\varepsilon}}(\mathbf{v}) \in N_C(\boldsymbol{\sigma}) \Leftrightarrow \dot{\boldsymbol{\varepsilon}} : (\boldsymbol{\sigma} - \boldsymbol{\sigma}^*) \geqslant 0, \qquad \forall \boldsymbol{\sigma}^* \in C, \tag{6.8}$$

defines a particular type of material that we may call rigid–plastic, *associated* to the elastoplastic model from which we started. It is an identical material as far as the plastic characteristics are concerned, but the strain rate is purely plastic (one-dimensional illustration in Fig. 6.1).

6.1.3 Example: spherical envelope under pressure

This is a classical elastoplasticity problem. We are dealing with a *thick* ($b - a$ thickness) spherical envelope, subject to an internal pressure p (Fig. 6.2). The pressure at the exterior is zero.

Fig. 6.1. Associated rigid – plastic material (rheological models)

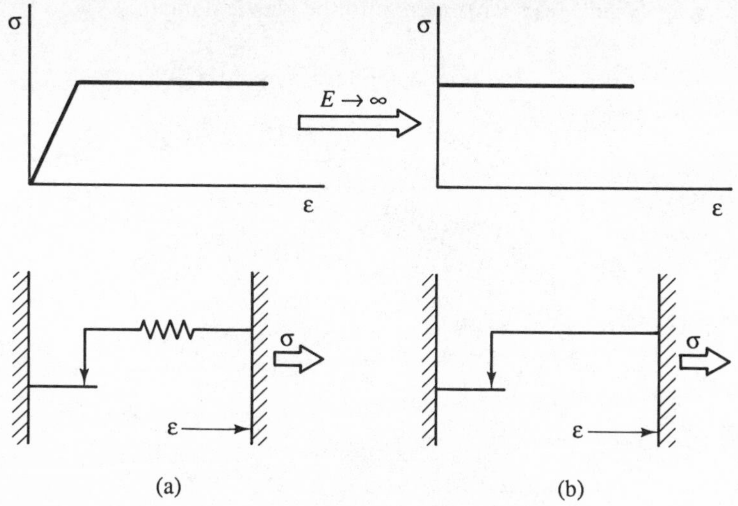

The problem admits the spherical symmetry with $\mathbf{T}^d = \mathbf{0}$ for $r = b$ and $\mathbf{T}^d = -p\mathbf{n}$ for $r = a$. The only nonzero displacement and stress components are the radial displacement $\xi(r)$ and the stresses $\sigma_{rr}(r)$, $\sigma_{\theta\theta}(r) = \sigma_{\varphi\varphi}(r)$. The isotropic linear elastic solution (see Germain, 1962) with spherical symmetry is written

$$\left.\begin{aligned} \xi(r) &= \alpha r + \frac{\beta}{r^2}, \\ \sigma_{rr} = \mathrm{A} - \frac{B}{r^3}, &\qquad \sigma_{\theta\theta} = \sigma_{\varphi\varphi} = A + \frac{B}{2r^3}, \end{aligned}\right\} \tag{6.9}$$

with the integration constants

$$A = \frac{p}{(b/a)^3 - 1}, \quad B = A(b/a)^3, \quad \alpha = \frac{A}{3\lambda + 2\mu}, \quad \beta = \frac{B}{2\mu}, \tag{6.10}$$

where λ and μ are the Lamé coefficients.

We admit that the material is elastic–plastic and obeys the Tresca criterion (1.10), i.e. (Actually, the Mises and Tresca criteria produce here the *same* results. The solution is the one given in Hill's book (1950), pp. 96–106),

$$\sup_{\alpha,\beta} |\sigma_\alpha - \sigma_\beta| \leqslant 2k, \qquad \alpha, \beta = 1, 2, 3. \tag{6.11}$$

If we calculate the left-hand side of this with the help of (6.9)–(6.10) we see that, when the interior pressure p reaches the critical value

$$p^* = \frac{2k}{a}\left[1 - \left(\frac{a}{b}\right)^3\right], \tag{6.12}$$

the *interior boundary* ($r = a$) enters into the plastic domain.

Fig. 6.2. Spherical envelope

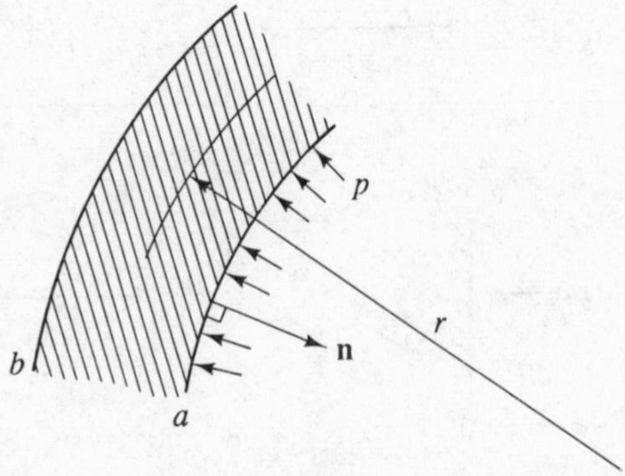

Therefore, if we still increase p (beyond p^*) we must distinguish two concentric regions (according to the problem's symmetry) in the spherical envelope, i.e.,

$$\Omega_{\mathrm{p}} : a < r < c \quad \text{plastic region (zone),}$$

$$\Omega_{\mathrm{el}} : c < r < b \quad \text{still elastic region.}$$

The problem is to determine c, but it is much more subtle to reason by fixing c and looking for the pressure p, which corresponds to this position, at the boundary between the domains Ω_{p} and Ω_{el}.

In the region Ω_{el} of thickness $b - c$ we now have an elastic solution (E is Young's modulus v is Poisson's coefficient)

$$\left.\begin{aligned} \xi(r) &= \frac{kc^3}{2E}\left[\frac{1+v}{r^2} + 2(1-2v)\frac{r}{b^3}\right], \\ \sigma_{rr} &= \frac{2k}{3}\left(\frac{c}{b}\right)^3\left[1 - \left(\frac{b}{r}\right)^3\right], \\ \sigma_{\theta\theta} = \sigma_{\varphi\varphi} &= \frac{2k}{3}\left(\frac{c}{b}\right)^3\left[1 + \frac{1}{2}\left(\frac{b}{r}\right)^3\right]. \end{aligned}\right\} \tag{6.13}$$

Actually, this is a solution in a spherical envelope of thickness $b - c$ with $\mathbf{T}^{\mathrm{d}} = \mathbf{0}$ in $r = b$, while the Tresca criterion (6.11) in $r = c$ is reached in such a way that

$$\sigma_{\theta\theta} - \sigma_{rr} = 2k \qquad \text{at } r = c. \tag{6.14}$$

Let us now look for the solution in Ω_{p}. This solution satisfies the Tresca criterion as well as the equation of radial equilibrium

$$\frac{\mathrm{d}\sigma_{rr}}{\mathrm{d}r} + \frac{2}{r}(\sigma_{rr} - \sigma_{\theta\theta}) = 0. \tag{6.15}$$

If we take into account $\sigma_{rr} - \sigma_{\theta\theta} = -2k$, the solution of (6.15) can be written

$$\sigma_{rr} = 2k \operatorname{Log} r + C. \tag{6.16}$$

The integration constant C is determined by the condition of the *radial stress continuity* at $r = c$, or, with (6.16) and $(6.13)_2$,

$$C = -2k \operatorname{Log} c = -\frac{2k}{3}\left[1 - \left(\frac{c}{b}\right)^3\right], \tag{6.17}$$

therefore, in Ω_p

$$\left.\begin{aligned}\sigma_{rr} &= 2k\left\{\mathrm{Log}\frac{r}{c} - \frac{1}{3}\left[1 - \left(\frac{c}{b}\right)^3\right]\right\},\\ \sigma_{\theta\theta} &= \sigma_{rr} + 2k = \sigma_{\varphi\varphi}.\end{aligned}\right\} \tag{6.18}$$

This solution depends upon c, but for the time being we will ignore the exact place of this boundary.

It is obvious that the maximum admissible pressure corresponds to the solution where the plastic zone extends all over the spherical envelope, consequently then $c = b$, and this, according to $(6.18)_1$, gives

$$p_{\max} = 2k\,\mathrm{Log}\left(\frac{b}{a}\right), \tag{6.19}$$

Let us now compute the *radial displacement* $\xi(r)$. We will notice that the *volume* dilatation is purely elastic, because the normality law and the Tresca criterion imply that $\mathrm{tr}\,\boldsymbol{\varepsilon}^{\mathrm{p}} = 0$. The dilatation per unit volume is written

$$\Theta = \mathrm{tr}\,\boldsymbol{\varepsilon}^{\mathrm{e}} = \frac{\mathrm{d}\xi}{\mathrm{d}r} + 2\frac{\xi}{r}, \tag{6.20}$$

and the inverse state law $\boldsymbol{\varepsilon}^{\mathrm{e}} = \mathbf{S} : \boldsymbol{\sigma}$, or

$$\Theta = \mathrm{tr}\,\boldsymbol{\varepsilon}^{\mathrm{e}} = \frac{(1-2\nu)}{E}(\sigma_{rr} + \sigma_{\theta\theta}), \tag{6.21}$$

provides ξ, considering (6.18), in the form

$$\xi(r) = 2\frac{(1-2\nu)}{E}k\left\{r\,\mathrm{Log}\frac{r}{c} - \frac{r}{3}\left[1 - \left(\frac{c}{b}\right)^3\right] + \frac{D}{r^2}\right\}, \tag{6.22}$$

where the integration constant D is determined by the *displacement continuity* at $r = c$. Given these two expressions, (6.13), and (6.22), for $\xi(r)$, $r \in \Omega_{\mathrm{p}}$, we get

$$\xi(r) = \frac{k}{E}\left\{2(1-2\nu)r\left[\mathrm{Log}\frac{r}{c} - \tfrac{1}{3}\left[1 - \left(\frac{c}{b}\right)^3\right]\right] + (1-\nu)\left(\frac{c}{r}\right)^2\right\}. \tag{6.23}$$

This means that the problem is *statically determined*, since the field of stress $\boldsymbol{\sigma}$ in $\Omega_{\mathrm{p}} = \{a < r < c\}$ can be determined on the basis of the equilibrium equations and the plasticity conditions (eqn (6.18)). This is certainly a particular case. But, what about the *limit load* problem? If we admit that there is no symmetry loss as a result of the stress distribution, then, according to the definition (6.5) of the limit load, this latter is the maximum of coefficients δ (that is, in our case, p) that satisfy the following equations:

$$\left.\begin{array}{ll}\dfrac{d\sigma_{rr}}{dr}+\dfrac{2}{r}(\sigma_{rr}-\sigma_{\theta\theta})=0 & \text{(equilibrium);}\\ |\sigma_{rr}-\sigma_{\theta\theta}|\leqslant 2k & \text{(Tresca);}\\ \sigma_{rr}(r=b)=0 & \text{(no traction applied in } r=b\text{);}\\ \sigma_{rr}(a)=\delta & \text{(internal presure in } r=a\text{).}\end{array}\right\} \tag{6.24}$$

We have immediately

$$\delta_{\mathrm{L}} = p_{\max} = 2k\,\mathrm{Log}\left(\frac{b}{a}\right), \tag{6.25}$$

which corresponds to the complete plasticization of the spherical envelope.

Let us observe that this simple evaluation of δ_{L} is exceptional. All the same, we observe that the structure resists for

$$p^* < p < p_{\max}. \tag{6.26}$$

6.2 Computation of the limit load

6.2.1 Generalities

The limit load is obtained on the basis of the determination of the equilibrium states (6.7) represented by the triplet $(\boldsymbol{\sigma}, \mathbf{u}, \delta_m)$ such that

$$\left.\begin{array}{ll}\boldsymbol{\sigma}\in C & \text{in }\Omega,\\ \operatorname{div}\boldsymbol{\sigma}=0 & \text{in }\Omega,\\ \boldsymbol{\sigma}\cdot\mathbf{n}=\delta_{\mathrm{m}}\mathbf{R}^0 & \text{on }\partial\Omega_T,\\ v_i=0 & \text{on }\partial\Omega_u,\ \forall v_i,\\ \dot{\boldsymbol{\varepsilon}}(v)\in N_C(\boldsymbol{\sigma}), & \end{array}\right\} \tag{6.27}$$

Here, the velocity $\mathbf{v}$ may eventually be discontinuous at the crossing of certain surfaces, but this discontinuity is not arbitrary; it must be compatible with the normality law, as we explained in Section 4.4.

Let us examine the system (6.27). It is a *generalized eigenvalue system.* As a matter of fact, supposing that the 'eigenvalue' δ_{m} were determined, then the 'eigenvector' $\mathbf{v}$ could be obtained by solving a *nonlinear boundary-value* problem. The field $\mathbf{v}$ determined in this manner is then called the ruin mode. The associated boundary-value problem corresponds to the following local equations:

$$\left.\begin{array}{ll} \boldsymbol{\sigma} \in C & \text{in } \Omega, \\ \operatorname{div} \boldsymbol{\sigma} = \mathbf{0} & \text{in } \Omega, \\ \boldsymbol{\sigma} \cdot \mathbf{n} = \mathbf{T}^{\mathrm{d}} & \text{on } \partial\Omega_T, \\ v_i = v_i^{\mathrm{d}} & \text{on } \partial\Omega_{\mathrm{u}}, \\ \dot{\boldsymbol{\varepsilon}}(\mathbf{v}) \in N_C(\boldsymbol{\sigma}). & \end{array}\right\} \tag{6.28}$$

We recall that the dissipated plastic power $\phi(\dot{\boldsymbol{\varepsilon}})$ is defined by

$$\phi(\dot{\boldsymbol{\varepsilon}}) = \boldsymbol{\sigma} : \dot{\boldsymbol{\varepsilon}} = \sup_{\sigma^* \in C} (\sigma^* : \dot{\boldsymbol{\varepsilon}}). \tag{6.29}$$

This only represents the dissipation potential, since

$$\boldsymbol{\sigma} = \frac{\partial \phi}{\partial \dot{\boldsymbol{\varepsilon}}}. \tag{6.30}$$

According to the nature of the strain rate $\dot{\boldsymbol{\varepsilon}}$, $\phi(\dot{\boldsymbol{\varepsilon}})$ is a density per unit volume or a surface distribution. So, for Mises' criterion, C is defined by $|\sigma^{\mathrm{d}}| - K_0 \leqslant 0$ and

$$\phi(\dot{\boldsymbol{\varepsilon}}) = \operatorname*{Sup}_{|\sigma^{\mathrm{d}}| \leqslant K_0} \boldsymbol{\sigma}^* : \dot{\boldsymbol{\varepsilon}} = K_0 |\dot{\boldsymbol{\varepsilon}}|. \tag{6.31}$$

If $\dot{\boldsymbol{\varepsilon}}$ is a surface distribution such as

$$\dot{\boldsymbol{\varepsilon}} = \tfrac{1}{2}([\![\mathbf{v}]\!] \otimes \mathbf{n} + \mathbf{n} \otimes [\![\mathbf{v}]\!]), \qquad [\![\mathbf{v}]\!] \cdot \mathbf{n} = 0, \tag{6.32}$$

then (since the discontinuity is purely tangential) (6.31) takes the form

$$\phi(\dot{\boldsymbol{\varepsilon}}) = \frac{K_0}{\sqrt{2}} |[\![\mathbf{v}]\!]|. \tag{6.33}$$

We should point out that the relation (6.30) is absolutely identical to an elastic law where W and $\boldsymbol{\varepsilon}$ were replaced by ϕ and $\dot{\boldsymbol{\varepsilon}}$. This similarity may be used in order to state directly the *minimum principles* for the strain rate $\dot{\boldsymbol{\varepsilon}}$. We have then (compare with the principles in Section 4.2) the following.

Minimum principle for velocities *The velocity* $\mathbf{v}$ *of the problem* (6.28) *minimizes the functional*

$$X(\mathbf{v}^*) = \int_{\Omega} \phi(\dot{\boldsymbol{\varepsilon}}(\mathbf{v}^*))\, \mathrm{d}\Omega - \int_{\partial\Omega_T} \mathbf{T}^{\mathrm{d}} \cdot \mathbf{v}\, \mathrm{d}s \tag{6.34}$$

among the velocity fields $\mathbf{v}^*$ *that are kinematically admissible, have* $v_i^* = v_i^{\mathrm{d}}$ *on* $\partial\Omega_u$ *and are compatible with the normality law.*

Through duality we get the following.

Minimum principle for stresses (elaborated by Hill) *The solution in terms of stress* $\boldsymbol{\sigma}$ *of the boundary problem (6.28) minimizes the functional*

$$\mathbf{Y}(\boldsymbol{\sigma}^*) = -\int_{\partial\Omega_u} \mathbf{v}^{\mathrm{d}} \cdot \boldsymbol{\sigma} \cdot \mathbf{n}\, \mathrm{d}s \tag{6.35}$$

among the statically ($\operatorname{div} \boldsymbol{\sigma}^* = \mathbf{0}$ *in* Ω, $\boldsymbol{\sigma}^* \cdot \mathbf{n} = \mathbf{T}^{\mathrm{d}}$ *on* $\partial\Omega_T$) *and plastically* ($\boldsymbol{\sigma}^* \in C$ *in* Ω) *admissible stress fields.*

6.2.2 Static method

Let us now consider the load problem (6.27). Which are the properties that are naturally satisfied by the eventual solution $(\boldsymbol{\sigma}, \mathbf{v}, \delta_{\mathrm{m}})$? We have in particular, the following.

Proposition *Let* δ *be a given load level. If* $P \cap S(\delta) \neq \{\varnothing\}$, *i.e., if there is at least* one stress field $\boldsymbol{\sigma}^* \in P \cap S(\delta)$, *then* δ *satisfies the inequality*

$$\delta \leqslant \delta_{\mathrm{m}}. \tag{6.36}$$

Proof. According to the virtual-power principle (2.26) we have

$$\int_\Omega \boldsymbol{\sigma}^* : \dot{\boldsymbol{\varepsilon}}(\mathbf{v})\, \mathrm{d}\Omega = \int_{\partial\Omega_T} \delta \mathbf{R}^0 \cdot \mathbf{v}\, \mathrm{d}s \tag*{(6.37)$_1$}$$

and

$$\int_\Omega \boldsymbol{\sigma} : \dot{\boldsymbol{\varepsilon}}(\mathbf{v})\, \mathrm{d}\Omega = \int_{\partial\Omega_T} \delta_{\mathrm{m}} \mathbf{R}^0 \cdot \mathbf{v}\, \mathrm{d}s \tag*{(6.37)$_2$}$$

since $\boldsymbol{\sigma}^* \in S(\delta)$ and $\boldsymbol{\sigma} \in S(\delta)$. If we subtract $(6.37)_1$ from $(6.37)_2$, and we take into account the normality law $(\boldsymbol{\sigma} - \boldsymbol{\sigma}^*) : \dot{\boldsymbol{\varepsilon}}(\mathbf{v}) \geqslant 0$, $\forall \boldsymbol{\sigma}^* \in C$, we obtain

$$\int_\Omega (\boldsymbol{\sigma} - \boldsymbol{\sigma}^*) : \dot{\boldsymbol{\varepsilon}}(\mathbf{v})\, \mathrm{d}\Omega = \int_{\partial\Omega_T} (\delta_{\mathrm{m}} - \delta)(\mathbf{R}^0 \cdot \mathbf{v})\, \mathrm{d}s \geqslant 0 \tag{6.38}$$

and, therefore, $\delta_{\mathrm{m}} \geqslant \delta$, QED.

The proposition (6.36) shows that the estimation δ_{L} is *exactly* the limit load.

According to (6.37) and (6.38) the limit load is defined by *Rayleigh's quotient*,

$$\delta = \frac{\int_\Omega \phi(\dot{\boldsymbol{\varepsilon}}(\mathbf{v}))\, \mathrm{d}\Omega}{\int_{\partial\Omega_T} \mathbf{R}^0 \cdot \mathbf{v}\, \mathrm{d}s} = \operatorname*{Inf}_{\mathbf{v}^*} \frac{\int_\Omega \phi(\dot{\boldsymbol{\varepsilon}}(\mathbf{v}^*))\, \mathrm{d}\Omega}{\int_{\partial\Omega_T} \mathbf{R}^0 \cdot \mathbf{v}^*\, \mathrm{d}s}, \tag{6.39}$$

where the velocity fields $\mathbf{v}^* \in \mathscr{U}_0$ are chosen in such a way that the quotient is positive. In fact, if $\mathbf{T}^d = \delta \mathbf{R}^0$, the velocity field $\mathbf{v}$ satisfies the minimum principles (6.34), which is written

$$\begin{aligned} 0 &= \int_\Omega \phi(\dot{\boldsymbol{\varepsilon}}(\mathbf{v}))\,\mathrm{d}\Omega - \int_{\partial\Omega_T} \delta_\mathrm{m}\mathbf{R}^0 \cdot \mathbf{v}\,\mathrm{d}s \\ &\leqslant \int_\Omega \phi(\dot{\boldsymbol{\varepsilon}}(\mathbf{v}^*))\,\mathrm{d}\Omega - \int_{\partial\Omega_T} \delta\mathbf{R}^0 \cdot \mathbf{v}^*\,\mathrm{d}s, \end{aligned} \tag{6.40}$$

where the first expression is nothing but the virtual-power principle, so it entails (6.39), which means that it is possible to approach δ_m from below, by applying the proposition (6.36). This is what is called the *static method* of computing the limit load.

6.2.3 Kinematic method

We may equally express (6.39) in the following form

Proposition *Let δ be a given load level. If there is a kinematically admissible velocity field $\mathbf{v}^*$, compatible with the normality law, such that*

$$\int_{\bar{\Omega}} \phi(\dot{\boldsymbol{\varepsilon}}(\mathbf{v}^*))\,\mathrm{d}\Omega \leqslant \int_{\partial\Omega_T} \delta\mathbf{R}^0 \cdot \mathbf{v}^*\,\mathrm{d}s, \tag{6.41}$$

then necessarily

$$\delta \geqslant \delta_\mathrm{m}. \tag{6.42}$$

This proposition allows the approach to δ_m from above. This is what is called the *kinematic method* of limit-load computation. We must notice that, in practice, given the extreme difficulty of the exact solution of the problem (6.5), (6.27) and (6.39), we are usually contented with an estimate of δ_m, by excess or defect (see example below).

6.3 Example of a foundation's limit load

This example is due to J. Mandel (Fig. 6.3) in soil mechanics. We consider a foundation of $AB = a$ width and infinite length, on a plastic, homogeneous soil, obeying the Tresca criterion (1.20).

We admit that the foundation is acting on the soil by a uniform pressure δ and we wish to figure out the maximum admissible pressure δ_m. The problem is expressed by means of plane (x, y) strains. The static method gives $\delta_m \geqslant 4k$ where k is the Tresca criterion coefficient. The kinematic method makes use of the Fellenius circles (Salençon, 1974) and leads to $\delta_m \leqslant 5.54k$, whence the bounds

$$4k \leqslant \delta \leqslant 5.54k.$$

The exact solution of the problem may be obtained as $\delta_m = (\pi + 2)k$, a result obviously within the above bounds.

The reader will find in J. Salençon (1983) a long discussion of limit analysis applied to structures and soil mechanics. Other recommended references dealing with this matter are Johnson and Mellor (1962), Calladine (1985), Chen and Baladi (1985), Lescouarch (1983), Massonet and Save (1963), Salençon (1974), Hodge (1959, 1963), Aker and Hayman (1969), Heyman (1971), Save and Massonet (1972), Sobotka (1989), Wood (1961), Neal (1977) and Kwiecinski (1989). Many of these are concerned with special cases of structures (beams, trusses, plates, etc.) and present a marked engineering touch related to the *plastic design* of structures. In this regard, it seems that the first attempt along this line goes back to Kazinczy (1914). However, the important breakthrough for engineering applications came with the notion of *elementary* and *combined collapse mechanisms* by Symonds and co-workers (Symonds and Neal, 1951, Neal and Symonds, 1952) and its application to framed structures, while general theorems were elaborated upon by Gvozdev (1938), Hill (1951), Drucker *et al.* (1952), Lee (1952), and others (French school in 1960s and 1970s: J. Salençon, M. Frémond and J. Mandel with a marked interest in soil mechanics).

The following problems deal more thoroughly with the hollow-sphere problem, simple evaluations of upper and lower bounds for collapse loads, and the collapse mechanisms of framed structures.

Fig. 6.3. Limit load of a foundation; ground = $\{y \leqslant 0\}$

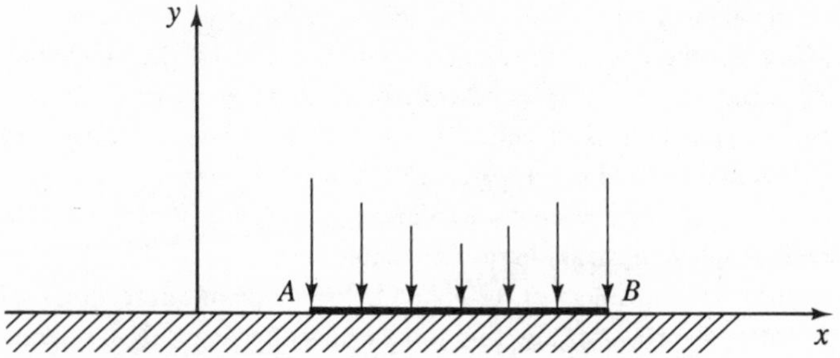

Problems for Chapter 6

6.1. Hollow sphere with residual stresses

Consider the problem of the spherical envelope of Subsection 6.1.3 but imagine now that internal pressure is removed from the sphere after plastic flow has occurred over part of the sphere (say, between $r = a$ and $r = c < b$). Residual stresses will occur. Determine this field of *residual* stresses $\{\sigma_{rr}, \sigma_{\theta\theta}\}$ in the two regions of the sphere.

[*Hint*: Superimpose on the stress system due to the internal pressure p a *completely elastic* stress system due to a pressure $-p$; this will be correct in so far as yielding in reverse does not occur, i.e., if the residual stresses are *not* large enough to produce yielding.]

Then imagine a scenario for alternating loadings and unloadings of the sphere.

6.2. Hollow sphere with thermal loading

Reconsider the problem of the spherical envelope of Subsection 6.1.3 in the absence of internal pressure but with thermal loading corresponding to a *temperature gradient* such as the steady-state temperature distribution given by (this is produced by an outer flow of heat due to an inner-surface temperature θ_0 and an outer-surface temperature of zero)

$$\theta = \frac{\theta_0 a}{b - a}\left(\frac{b}{r} - 1\right).$$

Consider *thermoelastic* linear isotropic stress – strain relations for the elastic behaviour and Mises yield criterion with elastic limit σ_0.

(i) Find the expression of the condition of yielding in terms of θ_0 and σ_0.

$$\frac{\beta\tau}{2(\beta^3 - 1)}\left|\frac{3\beta^2}{\rho^3} - \frac{\beta^2 + \beta + 1}{\rho}\right| = 1$$

[*Answer*:
when $\tau_0 = E\alpha\theta_0/(1 - \nu)\theta_0$, where α, E and ν are, respectively, thermal expansion, Young's modulus and Poisson's ratio, and $\rho = r/a$, $\beta = b/a$.]

(ii) Yielding starts at $r = a$. Then show that as temperature θ_0 is increased a *second* plastic zone unconnected to the first may start at another radius depending on the value of β.

[*Hint*: As there are no prescribed boundary forces, the tangential stress in the structure may vary from compression at $r = a$ to tension at the outer surface. The inner surface will begin to flow plastically *in compression* but, with sufficiently high temperature-gradient, the outer surface may start to flow plastically *in tension*; see Johnson and Mellor, 1962.]

6.3. Hollow sphere of strain-hardening material

Consider the problem of the spherical envelope of Subsection 6.1.3 subject to internal pressure, but this time with a linear strain-hardening law as sketched in Fig. 6.4.

Redo the work of Subsection 6.1.3, and 6.1 and 6.2 above. Check that, for $\beta = b/a = 2$, $m = 0.1$ and $\nu = 0.3$, the internal pressure required for yielding of the complete envelope is 1.83 compared to 1.39 for a perfectly plastic material (a gain of 32%); check all steps of computation by making $m = 0$.

6.4. Beam with distributed load

Consider the beam in Fig. 6.5 with distributed load of intensity Q, where one end ($x = 0$) is built in and the other $x = L$ is simply supported. Suppose that the beam collapses with a hinge at distance $x = \xi$ from the origin. Use the yield criterion $f(M) = M \pm M_0 = 0$. Call R the reaction at the support $x = L$

(i) Show that the moment at any x is given by

$$M(x) = R(L - x) - \tfrac{1}{2}Q(L - x)^2.$$

(ii) Find the *exact* solution of the problem, i.e., show that the collapse load is $Q = Q_m = 11.6568(M_0/L^2)$ with $\xi = 0.5857\,L$. Call $q = QL/M_0$.

(iii) By using work arguments find upper and lower bounds q_1 and q_2 such that

$$q_1 \leqslant q_m \leqslant q_2$$

with $q_1 \approx 11.656\,74$ and $q_2 \approx 11.657\,14$.

Fig. 6.4. Figure for Problem 6.2

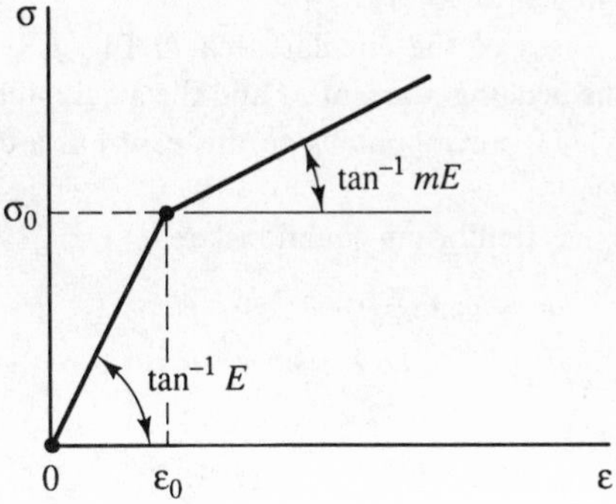

Fig. 6.5. Figure for Problem 6.4

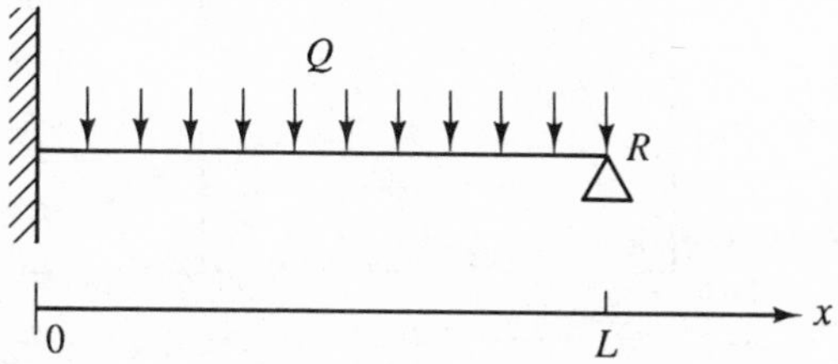

6.5 Limit load of a foundation

Give the detail of the proof of the results of Section 6.3, the problem being treated in plane strains for a Tresca material.

6.6. Limit analysis of rigid–plastic beams

Consider a simply supported beam of length L subject to uniformly distributed pressure $p = \delta p^0$ where $p^0 = \text{const.} = 1$ is the unit load. Find the general expression for the statically admissible load multiplier $\delta = \delta_S$ and that of the kinetical load multiplier δ_K. [*Hint*: use the principle of virtual work in bending.] What is the value of the *ultimate* multiplier δ_K when *one* single plastic hinge occurs or when *three* plastic hinges occur [*Note*: a plastic hinge, unlike a mechanical one, is capable of transmitting a definite value of bending moment, namely the moment M_0, if, in pure bending, the yield condition is of the form $f(M) = M \pm M_0 = 0$].

6.7. Collapse mechanisms of frame

Make a literature search and write a report on mechanisms of collapse of *portal frames* (see Fig. 6.6) via plastic hinges in pure bending (yield criterion; $f(M) = M \pm M_0 = 0$).

6.8. Complex loading of frames

Redo 6.7 but for the frame and load depicted in Fig. 6.7: find the safe load region in the plane of coordinates ($Th/4M_0$ v.s. $qL^2/16\, M_0$).

[*Hint*: consider a combined collapse mechanism with a hinge placed at unknown position ξL from one corner and use the principle of virtual work.]

6.9. Collapse of a circular arch

Consider the pieces of the circular arch in Fig. 6.8. Internal forces are described by the bending moment M and the axial force N. The interaction surface $f(M, N) = 0$ corresponding to this case has a curve which was described in 1.9.

(a) Show that the equilibrium equations are

$$\left.\begin{aligned} m &= p(\sin\varphi_0 - \sin\varphi) - r(\cos\varphi - \cos\varphi_0),\\ kn &= p\sin\varphi + r\cos\varphi, \end{aligned}\right\} \quad (1)$$

Fig. 6.6. Loads p_1, p_2 and p_3

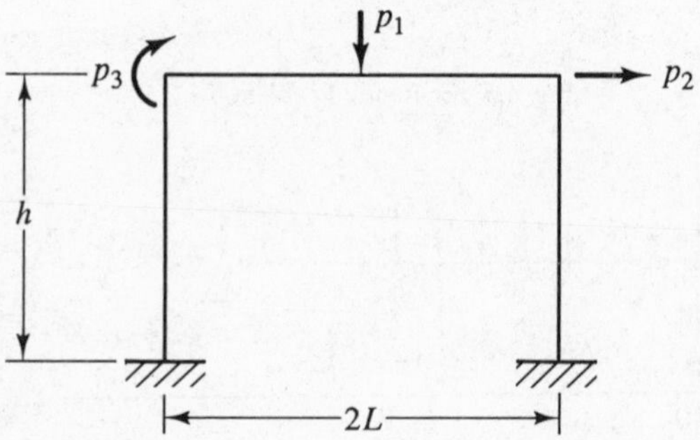

where $m = M/M_0, n = N/N_0, k = N_0R/M_0, p = PR/M_0, r =$ (horizontal reaction) $\times (L/M_0)$;

(b) show that m and n are related by

$$m + kn = p \sin \varphi_0 + r \cos \varphi_0; \qquad (2)$$

(c) show that the interaction conditions (interaction polygon) are

$$\left.\begin{aligned} f = m \pm 1 = 0, \qquad -1 \leqslant n \leqslant 1, \\ f = n \pm 1 = 0, \qquad -1 \leqslant m \leqslant 1; \end{aligned}\right\} \qquad (3)$$

(d) show that the collapse load is given by

$$p = \frac{2\cos(\varphi_0/2) + 1/\sqrt{2}}{\sin(\varphi_0/2)}.$$

Describe the collapse mechanism [*Hint*: at least *two* hinges must form in the structure which has one redundant support reaction; using symmetry, *three* hinges in fact develop.]

Fig. 6.7. Loads q and T

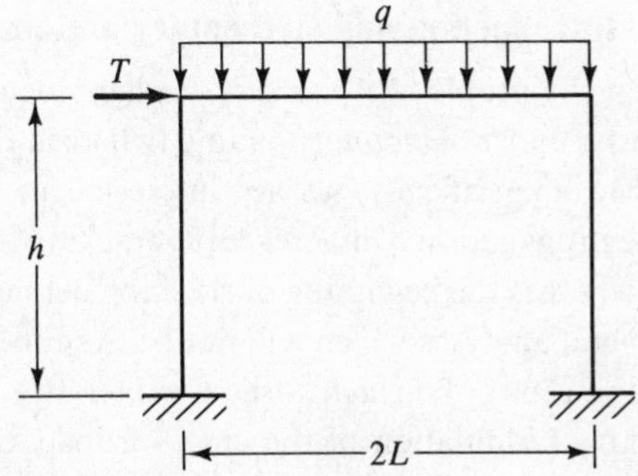

Fig. 6.8

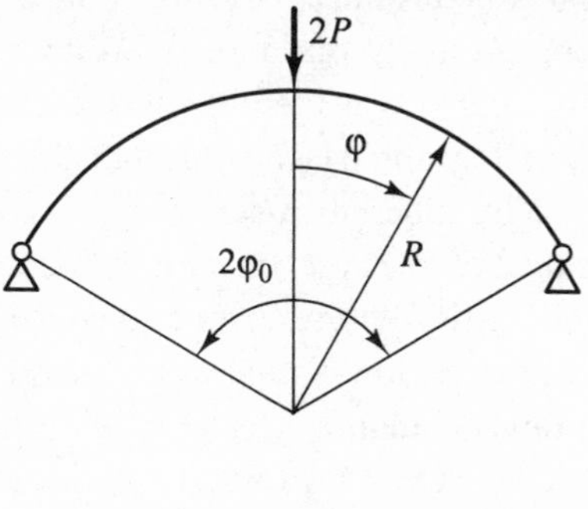

7

Crack propagation and fracture mechanics

The object of the chapter In the absence of plastic strain, the problem of *brittle fracture* by extension of cracks can be presented in a thermodynamic framework, analogous to that of elastoplasticity. This means that the *fracture criterion* (or the criterion of crack propagation) replaces the plasticity criterion. One important notion is the notion of mechanical field *singularity* (displacement, stresses).

7.1 Introduction and elementary notions

We are interested in the problem of *fracture*, a phenomenon that occurs, more or less violently, under *monotonic* loading (whereas *fatigue* concerns cyclical loading). More specifically, we are interested in the problem of *cracking*, that is, the progagation of macroscopic cracks (of size of the order of one millimetre), whereas the beginning of cracking belongs to the microscopic and to the metal analyses which will not be examined here. (Microscopic cracks are one cause of *damage* – see Chapter 10.) The aim of this study is to arrive at a formulation of the crack-propagation laws, based upon fracture criteria and the definition of the conditions that may insure resistance to this fracture. We are certainly aware of the interest that such a subject implies for industry; it suffices to think about aeronautical engines and nuclear installations. Actually, our main interest is *brittle fracture*, that is, the kind that occurs *without* considerable plastic strain (i.e. the separation mechanism of crystallographic facets through cleavage), whereas *ductile rupture* is produced by different mechanisms accompanied by great plastic strains). The model of *linear elasticity* is sufficiently known by now to allow for the study of brittle fracture so far as it concerns so-called *brittle* materials (e.g., glass, ceramics, low-temperature steels) (see Fig. 7.1). The particular case of ceramics demands great precision in calculation (as a result of adjustment imprecision, they tolerate no flexure).

As shown in Fig. 7.2, the *crack* problem is a field discontinuity (displacement) problem. Under the effect of traction the two lips Σ^+ and Σ^- are separated and, although under the effect of new forces they might later come back again into contact, we admit that this contact *cannot* solder back the lips of the crack, so that, along Σ, there is *absence of cohesion*, and, therefore, incapacity of the material to transmit across Σ a stress whose normal component is a traction: there is discontinuity of matter on the cracked surface. To simplify, we may suppose that this surface is plane and we shall call the *crack front* the neighbourhood of the endpoint of the crack to which we attach a local frame. This allows for the definition of *three* modes of cracking (Figs. 7.3 and 7.4) corresponding to the discontinuity of each one of the displacement vector's components:

Fig. 7.1. Brittle and ductile fractures

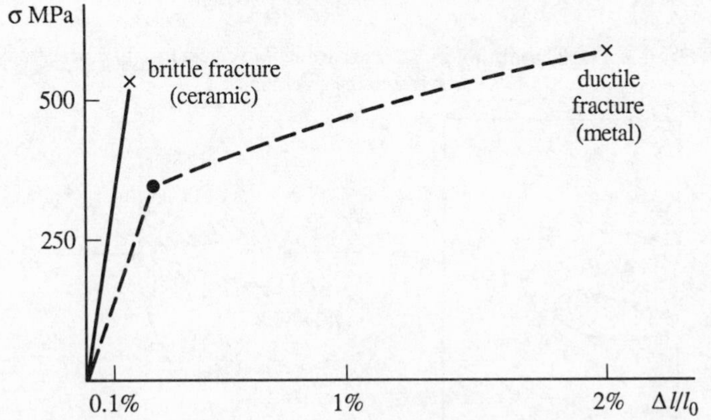

Fig. 7.2. Crack/discontinuity in displacement. (a) Unloaded body (the crack is already here). (b) Loaded body: $[\![\mathbf{u}]\!]_\Sigma = \mathbf{u}^+ - \mathbf{u}^-$.

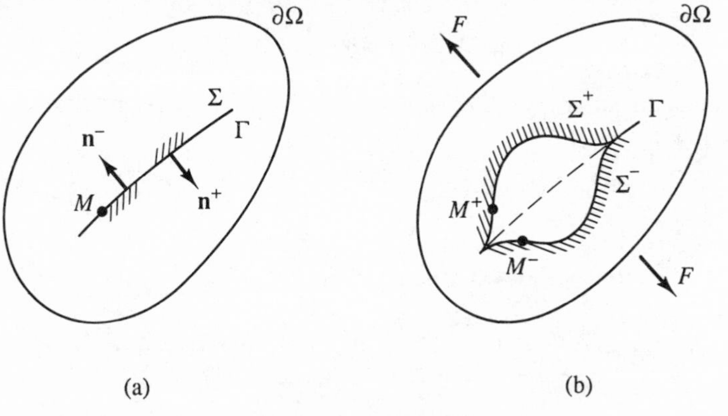

mode I, the opening mode $[\![u_3]\!] \neq 0$;

mode II, the shear plane mode $[\![u_1]\!] \neq 0$;

mode III, the shear antiplane mode $[\![u_2]\!] \neq 0$.

This defines three elementary modes, and a complex crack mode as a combination of two or three elementary modes.

Historical note. The essential dates in the history of the cracking theory are roughly as follows. 1920: introduction of the surface energy by A.A. Griffith. 1950: elasticity calculation by Irwin. 1960: importance of developments in aeronautics. 1970: importance of developments in nuclear power plants.

Fig. 7.3. Crack front/tip. (a) $\mathscr{C}$: crack front, M : crack tip. (b) Local frame attached to the crack tip M.

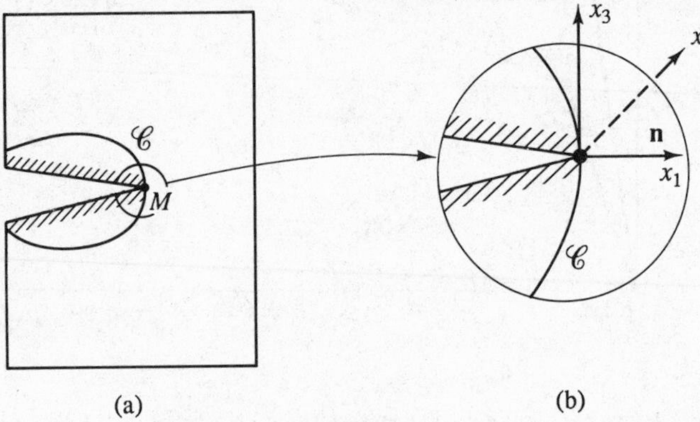

Fig. 7.4. Elementary modes of cracking

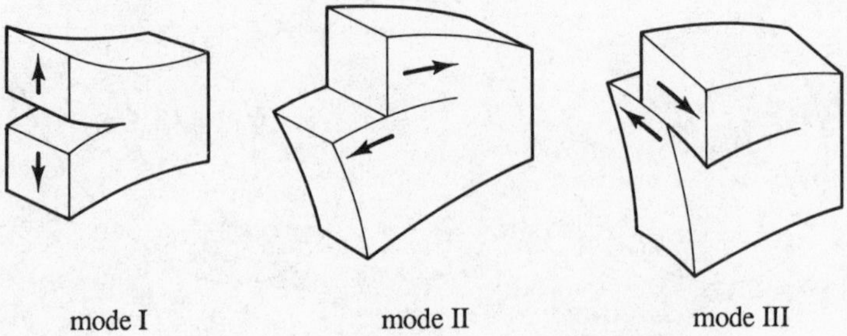

7.2 The notion of singularity

We suppose that the problem (Fig. 7.5) is *two-dimensional* in terms of plane strains or plane stresses (see Appendix 4 for these notions). We intend to determine the state of stresses and strains *in the vicinity* of the crack's endpoint, when the material is linear, elastic, homogeneous and isotropic, satisfying, that is, Hooke's law,

$$\boldsymbol{\sigma} = E : \boldsymbol{\varepsilon} \rightarrow \sigma_{ij} = \lambda \varepsilon_{kk} \delta_{ij} + 2\mu \varepsilon_{ij}, \tag{7.1}$$

The crack of length l at time t is represented open for the sake of illustration. The body Ω is fixed on one part $\partial\Omega_u$ of its boundary; a traction $\mathbf{T}^d$ is applied along $\partial\Omega_T$; no traction is applied along the lips of the crack. Body forces $\mathbf{g}$ are absent.

To illustrate the notion of singularity in mechanical fields in elasticity, let us consider the three 'real' situations of Fig. 7.6. In a homogeneous case (a), we have a strain tensor $\boldsymbol{\sigma} = \text{diag}(0, -p, 0)$. In the case (b) of a circular hole with a radius r we have (see Sokolnikoff (1956)) $\sigma_{yy} = CP$ where C is the stress concentration coefficient. Finally, in case (c), of a 'crack' with radius r at the bottom of the crack, the coefficient C introduced in (b) gradually becomes greater with $r \rightarrow 0$; therefore, C and σ_{yy} go to infinity with r going to zero. This illustration shows that the *singularity* must be analysed at one single point, the *crack tip*. In order to illustrate this we shall examine a typical problem, the problem of cracking in *mode III*, in antiplane shear (Fig. 7.7); other modes are examined in Appendix 4. In this case we have the following fields.

Fig. 7.5. Schematization of the problem

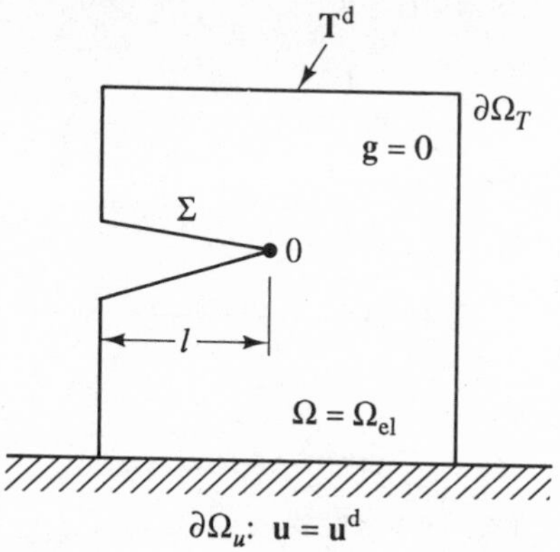

Displacement field

$$\mathbf{u} = \{-, -, w(x, y)\} \tag{7.2}$$

(the two nonspecified components do not enter into the considerations below).

Stress field: as a simplifying hypothesis we consider that the stress has the form of an *antiplane* shear in such a manner that only the components σ_{13} and σ_{23} are not zero.

Hooke's law is written

$$\sigma_{13} = 2\mu\varepsilon_{13}, \qquad \sigma_{23} = 2\mu\varepsilon_{23}, \tag{7.3}$$

Fig. 7.6. Some real situations. (a) Homogeneous case. (b) Circular hole (radius r). (c) crack (radius r).

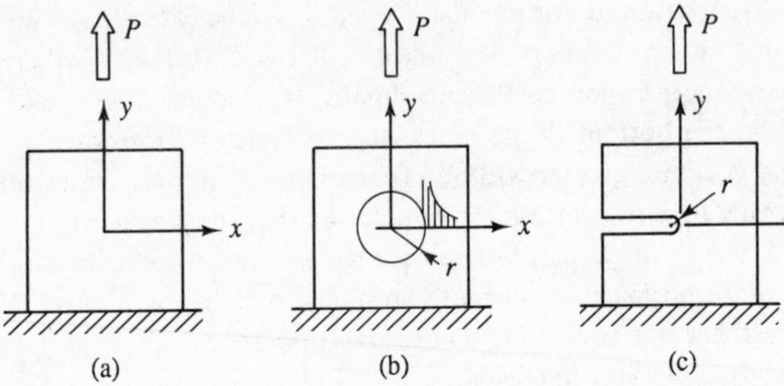

Fig. 7.7. Analysis of singularity in mode III

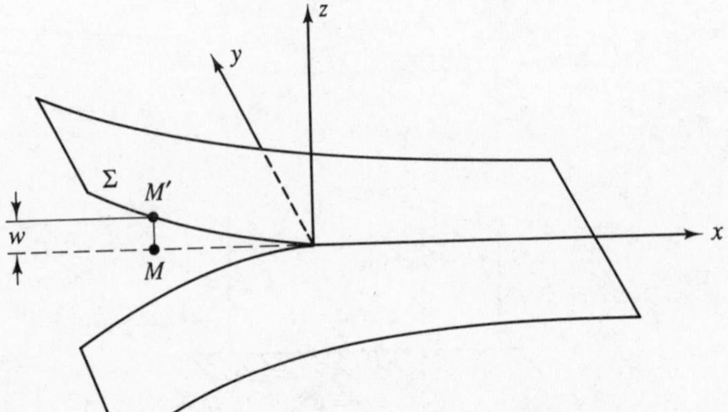

with

$$\varepsilon_{13} = \tfrac{1}{2} w_{,x}, \qquad \varepsilon_{23} = \tfrac{1}{2} w_{,y}. \tag{7.4}$$

The equilibrium equation $\operatorname{div} \boldsymbol{\sigma} = \mathbf{0}$ for the third component gives

$$\sigma_{31,1} + \sigma_{32,2} = 0 \tag{7.5}$$

or, with (7.8) and (7.4), $\Delta w = 0$, where Δ is the two-dimensional Laplacian.
The boundary condition $\boldsymbol{\sigma} \cdot \mathbf{n} = \mathbf{0}$ in $x \in \Sigma$ gives

$$\frac{\partial w}{\partial y} = 0 \qquad \text{at } x \in \Sigma^+ \text{ and } \Sigma^- \text{ when } x < 0. \tag{7.6}$$

Generally, then, we have the following problem:

$$(\mathrm{P}_1) \begin{cases} \Delta w = 0, \quad \forall \mathbf{x} \in \Omega, \\ \dfrac{\partial w}{\partial y} = 0, \quad \mathbf{x} \in \Sigma^+, \Sigma^- \text{ with } x < 0, \\ \text{boundary conditions at } \partial\Omega_u \text{ and } \partial\Omega_T. \end{cases} \tag{7.7}$$

The most interesting thing, though, is the *vicinity* of the crack tip, hence the problem extracted from (P_1):

$$(P_2) \begin{cases} \Delta w = 0, \quad \forall \mathbf{x} \in \Omega, \\ \dfrac{\partial w}{\partial y}(x, 0\pm) = 0, \quad x < 0, \end{cases} \tag{7.8}$$

where we seek all the possible solutions in terms of polar coordinates (r, θ) for *small r's under the form of a particular family* of solutions, such as

$$w(r, \theta) = r^{\alpha} \hat{w}(\theta), \tag{7.9}$$

where α is the degree of singularity and $\hat{w}(\theta)$ is a function of angular distribution.

In order to be somewhat less qualitative we should actually consider the following.

Proposition *In the vicinity of the point O (the crack's tip) the solution w presents an asymptotic expansion in $r^{\alpha_i}(\operatorname{Log} r)^{\beta_i} g_i(\theta)$ in the regularity order, i.e.*

$$w = C_1 r^{\alpha_1} (\operatorname{Log} r)^{\beta_1} g_1(\theta) + \text{'more } \textit{regular terms}\text{'}. \tag{7.10}$$

Let us point out that the continuity of w at the point O implies that

$$\alpha > 0. \tag{7.11}$$

Let us now consider the problem (7.8) in terms of polar coordinates; we have classically

$$\Delta f = \frac{\partial^2 f}{\partial r^2} + \frac{1}{r}\frac{\partial f}{\partial r} + \frac{1}{r^2}\frac{\partial^2 f}{\partial \theta^2},$$

hence, on account of (7.9), the problem:

$$(P_3)\begin{cases} \alpha^2 \hat{w} + \hat{w}'' = 0, & \forall\theta, \\ \hat{w}'(\theta) = 0 & \text{at } \theta = \pm\pi, \end{cases} \tag{7.12}$$

where the prime indicates the derivative with respect to θ. We get immediately

$$\hat{w}(\theta) = A\cos\alpha\theta + B\sin\alpha\theta \tag{7.13}$$

and, by applying the boundary conditions $(7.12)_2$, we obtain

$$\cos(\pm\alpha\pi) = 0, \text{ i.e., } \pm\alpha\pi = \frac{\pi}{2} + k\pi, \; k \text{ integer,} \tag{7.14}$$

in such a way that

$$\alpha = (\ldots, -\tfrac{3}{2}, -\tfrac{1}{2}, \tfrac{1}{2}, \tfrac{3}{2}, \ldots). \tag{7.15}$$

But given the restriction (7.11) there remains only the *positive* sequence

$$\alpha = (\tfrac{1}{2}, \tfrac{3}{2}, \ldots). \tag{7.16}$$

We must now, within this sequence, select α. Let us point out the following behaviours:

$$\left.\begin{aligned} \varepsilon_{13} &\sim \frac{\partial w}{\partial r} \sim r^{\alpha-1} \times (\ldots), \\ \sigma_{13} &\sim \varepsilon_{13} \sim r^{\alpha-1} \times (\ldots), \\ \sigma_{23} &\sim \varepsilon_{23} \sim r^{\alpha-1} \times (\ldots), \\ \textit{energy} &\sim \varepsilon^2 \sim r^{2(\alpha-1)} \times (\ldots). \end{aligned}\right\} \tag{7.17}$$

But, in order for us to define the total energy, the elastic energy must be *integrable*.

$$\int W \,\mathrm{d}\Omega \sim \int r^{2(\alpha-1)} \times (\ldots) \times r \,\mathrm{d}\theta \,\mathrm{d}r. \tag{7.18}$$

This allows for the selection of the smallest value in the sequence (7.16), i.e.,

$$\alpha = \tfrac{1}{2}. \tag{7.19}$$

In fact, we have the following proposition (see Appendix 4).

Proposition *We have*

$$w = w_{\text{regular}} + w_{\text{singular}},$$

where $w_{\text{singular}} = A\sqrt{r}\cdot\sin\frac{1}{2}\theta$ *and* w_{regular} *is, for example,* C^1*, where* A *depends upon* Ω*,* l*,* $\partial\Omega_u$ *and* $\partial\Omega_T$ *(therefore, upon the data* $\mathbf{T}^d$ *and* $\mathbf{u}^d$*). For a given structure with a given crack length* l *and for a loading* $(\mathbf{T}^d, \mathbf{u}^d) = \lambda(\mathbf{T}_0^d, \mathbf{u}_0^d)$ *then* $A \to \lambda A$ *because of the problem's linearity.*

Strain singularity Since $w = w_{\text{sing}} = A\sqrt{r}\cdot\sin\frac{1}{2}\theta$ for $r \ll l$, we have (eqn (7.17))

$$\varepsilon_{13} \sim 1/\sqrt{r}, \qquad \varepsilon_{23} \sim 1/\sqrt{r}, \qquad \sigma_{13} \sim 1/\sqrt{r} \qquad \text{and} \qquad \sigma_{23} \sim 1/\sqrt{r},$$

which we write alternatively (with $A \sim K_{\text{III}}$) as

$$\sigma_{23} = K_{\text{III}}\frac{1}{\sqrt{(2\pi r)}}\cos\frac{\theta}{2}, \qquad \sigma_{13} = -K_{\text{III}}\frac{1}{\sqrt{(2\pi r)}}\sin\frac{\theta}{2}, \tag{7.20}$$

where K_{III} is called the *stress intensity factor* for mode III and for the boundary-value elasticity problem just examined. Actually we have

$$K_{\text{III}} = \mu\sqrt{\frac{2}{\pi}}\cdot A, \tag{7.21}$$

but we *must* still compute A!

For a *mixed* (or complex) cracking mode we have the following.

Proposition

$$\mathbf{u} = \mathbf{u}_R + \mathbf{u}_S,$$

where $\mathbf{u}_R$ *is regular (i.e.* C^1*) and* $\mathbf{u}_S$ *is singular and such that*

$$\mathbf{u}_s = \sqrt{r}\cdot[K_{\text{I}}\hat{\mathbf{u}}_{\text{I}}(\theta) + K_{\text{II}}\hat{\mathbf{u}}_{\text{III}}(\theta) + K_{\text{III}}\hat{\mathbf{u}}_{\text{III}}(\theta)], \tag{7.22}$$

where the K_α*'s are the stress intensity factors of the elementary modes.*

What we should keep in mind as a *general idea* is that the strain singularity is defined by $K/\sqrt{r}$ *no matter what* the elasticity problem examined; and also, that the singularity (always in $1/\sqrt{r}$) is described by the factor K. This latter is obtained through the solution of a *boundary-value elasticity problem*. Examples of analytical computations are given in Appendix 4.

Given the behaviours (7.20) and (7.22), we might consider the introduction of a simple crack-propagation *criterion*. This is Irwin and Kries's (1951) point of view. From the qualitative point of view we must 'pull rather strongly' so as to get crack propagation. Let us introduce the *tenacity or fracture toughness* K_c characteristic of brittle materials. We can then state the following *criterion*:

$$\left.\begin{array}{ll}\text{(i)} & \text{If } K < K_c\text{, there is } no \text{ propagation: } \dot{l} = 0.\\ \text{(ii)} & \text{If } K = K_c\text{, propagation is } possible\text{: } \dot{l} \geqslant 0\text{, but we do}\\ & \text{not know whether it will actually happen or}\\ & \text{not (so that } \dot{l} = 0 \text{ cannot be excluded).}\end{array}\right\} \tag{7.23}$$

The characteristic K_c is easily determined in *mode I*. But modes II and III are difficult to realize experimentally. For materials such as plexiglass and steels we have $(K_c)_I$ of an order of 10^7 Pa $\times \sqrt{m}$.

7.3 The energy aspect of brittle fracture

The idea is to solve the problem given in a schematic form of Fig. 7.5 and to compute $\mathbf{u}$, $\boldsymbol{\sigma}$ and K. Evidently, although we can work out the computation in elasticity, the problem under study is in *irreversible evolution* and we should rather proceed with a *dissipation analysis* (in order to simplify we follow the hypothesis of an isothermal evolution, which is generally wrong – see Chapter 12 where the thermal effects are taken into account).

Total dissipation analysis The total dissipation Φ of the system is the power of the 'applied forces', minus the system's reversibly stored power (see, for example, eqn (2.54)). In the absence of volume forces, the applied-force power is written

$$\mathscr{P}_{(\mathrm{ext})} = \int_{\partial\Omega} \dot{\mathbf{u}} \cdot (\boldsymbol{\sigma} \cdot \mathbf{n})\, \mathrm{d}s, \tag{7.24}$$

whereas the stored power is

$$\mathscr{P}_r = \frac{\mathrm{d}}{\mathrm{d}t} \int_{\Omega} W(\boldsymbol{\varepsilon}(\mathbf{u}))\, \mathrm{d}\Omega; \tag{7.25}$$

therefore

$$\Phi = \int_{\partial\Omega} \dot{\mathbf{u}}(\boldsymbol{\sigma} \cdot \mathbf{n})\, \mathrm{d}s - \frac{\mathrm{d}}{\mathrm{d}t} \int_{\Omega} W(\boldsymbol{\varepsilon}(\mathbf{u}))\, \mathrm{d}\Omega. \tag{7.26}$$

We see that, when the crack is propagating (Ω evolves), the problem lies in the computation of the second contribution. Actually, *if the crack is not propagating*, $\dot{l} = 0$, and we obviously have

$$\frac{\mathrm{d}}{\mathrm{d}t}\int_\Omega W(\boldsymbol{\varepsilon}(\mathbf{u}))\,\mathrm{d}\Omega = \int_\Omega \frac{\partial W}{\partial \boldsymbol{\varepsilon}} : \dot{\boldsymbol{\varepsilon}}(\mathbf{u})\,\mathrm{d}\Omega = \int_\Omega \boldsymbol{\sigma} : \dot{\boldsymbol{\varepsilon}}\,\mathrm{d}\Omega. \tag{7.27}$$

This is *wrong* if $\dot{l} \neq 0$, because in this case the last integral is diverging. According to Section 7.2, $\varepsilon \sim K\sqrt{(1/r)}\cdot\varepsilon(\theta) = \varepsilon(X, Y)$ and we get then $\sigma \sim K\sqrt{(1/r)}\cdot\sigma(\theta)$, where (X, Y) is a frame moving with the crack (Fig. 7.8). In order to compute $\dot{\boldsymbol{\varepsilon}}$ we must keep in mind that

$$\dot{\boldsymbol{\varepsilon}} = \dot{\boldsymbol{\varepsilon}}_t(X, Y) - \boldsymbol{\varepsilon}_{t,1}(X, Y)\dot{l}, \tag{7.28}$$

because $X = x - l(t)$ and consequently $\boldsymbol{\varepsilon}_{t,1} \sim r^{-3/2}$, whence the divergence announced in the last integral of (7.27). So we must do *once more* the computation of $\mathscr{P}_r$, much more carefully as far as $\dot{l} \neq 0$ is concerned. In order to do this, we *isolate* the singularity by following a contour Γ delimiting the volume V_Γ (Fig. 7.8) which contains the *moving* tip of the crack and we write in an obvious manner that

$$\frac{\mathrm{d}}{\mathrm{d}t}\int_\Omega W(\boldsymbol{\varepsilon}(\mathbf{u}))\,\mathrm{d}\Omega = \frac{\mathrm{d}}{\mathrm{d}t}\int_{V_\Gamma} W(\boldsymbol{\varepsilon}(\mathbf{u}))\,\mathrm{d}\Omega + \frac{\mathrm{d}}{\mathrm{d}t}\int_{\Omega - V_\Gamma} W(\boldsymbol{\varepsilon}(\mathbf{u}))\,\mathrm{d}\Omega. \tag{7.29}$$

We notice that Γ is fixed within the moving frame. Therefore,

$$\frac{\mathrm{d}}{\mathrm{d}t}\int_{V_\Gamma} W(\boldsymbol{\varepsilon}(\mathbf{u}))\,\mathrm{d}\Omega = \int_{V_\Gamma} \boldsymbol{\sigma} : \dot{\boldsymbol{\varepsilon}}(X, Y)\,\mathrm{d}\Omega, \qquad (X, Y \text{ fixed}) \tag{7.30}$$

Fig. 7.8. Frame moving with the crack

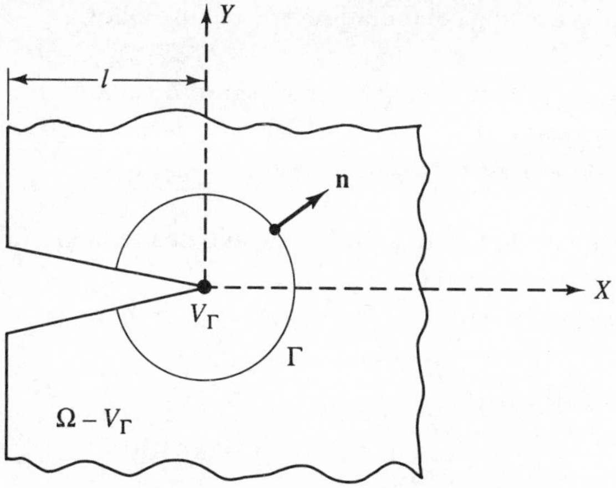

where $\dot{\boldsymbol{\varepsilon}}_t$ is in r^{-1} hence the integrability in (7.30). Besides, by using the formula for the material derivative of a nonmaterial volume integral (for example Germain, 1973, p. 37), we have

$$\frac{\mathrm{d}}{\mathrm{d}t}\int_{\Omega - V_\Gamma} W(\boldsymbol{\varepsilon}(\mathbf{u}))\,\mathrm{d}\Omega = \int_{\Omega - V_\Gamma} \boldsymbol{\sigma} : \dot{\boldsymbol{\varepsilon}}\,\mathrm{d}\Omega - \int_\Gamma W(\boldsymbol{\varepsilon}(\mathbf{u}))\dot{l}(\mathbf{e}_1 \cdot \mathbf{n})\mathrm{d}\Gamma, \tag{7.31}$$

where the second term is a flux term and $\mathbf{e}_1$ is a unit vector in the crack's propagation direction (here X). We note $n_1 = \mathbf{e}_1 \cdot \mathbf{n}$, hence

$$\frac{\mathrm{d}}{\mathrm{d}t}\int_{\Omega - V_\Gamma} W(\boldsymbol{\varepsilon}(\mathbf{u}))\,\partial\Omega = \int_{\Omega - V_\Gamma} \boldsymbol{\sigma} : \dot{\boldsymbol{\varepsilon}}\,\mathrm{d}\Omega - \int_\Gamma W\dot{l}\mathbf{n}_1\,\mathrm{d}\Gamma. \tag{7.32}$$

Let us now write the energy balance for the region $\Omega - V_\Gamma(t)$, keeping in mind that $\mathbf{n}(\Gamma) = -\mathbf{n}(\partial\bar{\Omega})$ on Γ (Fig. 7.8). We have

$$\int_{\bar{\Omega} = \Omega - V_\Gamma} \boldsymbol{\sigma} : \dot{\boldsymbol{\varepsilon}}\,\mathrm{d}\Omega = \int_{\partial\bar{\Omega}} \dot{\mathbf{u}}(\boldsymbol{\sigma}\cdot\mathbf{n})\,\mathrm{d}s = \int_{\partial\Omega} \dot{\mathbf{u}}(\boldsymbol{\sigma}\cdot\mathbf{n})\,\mathrm{d}s - \int_\Gamma \dot{\mathbf{u}}(\boldsymbol{\sigma}\cdot\mathbf{n})\,\mathrm{d}\Gamma \tag{7.33}$$

at time t. This given, as well as (7.32), (7.26) may be expressed as

$$\Phi = \int_\Gamma (W\dot{l}n_1 + \dot{\mathbf{u}}\cdot\boldsymbol{\sigma}\cdot\mathbf{n})\,\mathrm{d}\Gamma + \int_{V_\Gamma} \boldsymbol{\sigma} : \dot{\boldsymbol{\varepsilon}}_t(X, Y)\,\mathrm{d}\Omega, \tag{7.34}$$

where the integrand of the last contribution is integrable. If we take now the limit where we make the contour Γ tend towards the crack tip ($\Gamma \to 0$) and if we take into account this integrability, there remains

$$\Phi = \lim_{\Gamma\to 0}\int_\Gamma (W\dot{l}n_1 + \dot{\mathbf{u}}\cdot\boldsymbol{\sigma}\cdot\mathbf{n})\,\mathrm{d}\Gamma, \tag{7.35}$$

since the second contribution tends towards zero.

Given the preceding computation we note the following

Lemma *For any physical entity f attached to a material point, $\dot{f}$ has the same singularity as $(-\dot{l}f_{,1})$ at point O where '$_{,1}$' indicates the gradient in the direction of the crack's propagation.*

So, $\dot{\boldsymbol{\sigma}}$ is the sum of a regular contribution and of a term in $-\dot{l}\partial\boldsymbol{\sigma}/\partial X$; $\dot{\mathbf{u}}$ has a singular contribution in $(-\dot{l}\mathbf{u}_{,1})$.

Given this last remark we can rewrite (7.34) in the form

$$\left.\begin{aligned} &\Phi = G\dot{l}, \\ &G = \lim_{\Gamma\to 0}\int_\Gamma (Wn_1 - \mathbf{n}\cdot\boldsymbol{\sigma}\cdot\mathbf{u}_{,1})\,\mathrm{d}\Gamma. \end{aligned}\right\} \tag{7.36}$$

We see that G is the *thermodynamic dual* of l: it is the *force due to the singularity at the crack's tip. Thus, we are able to present a crack-propagation criterion* with a threshold, on a thermodynamic basis, in the form of Griffith's criterion (in terms of surface energy).

Criterion (Griffith) *There exists a threshold* G_c *such that*

$$\left.\begin{array}{ll} G \in C = \{G^* \leqslant G_c\}, & \\ \text{(i)}\ \dot{l} = 0, \textit{no progress if} & G < G_c, \\ \text{(ii)}\ \dot{l} \geqslant 0, \text{possible } \textit{progress if} & G = G_c, \end{array}\right\} \tag{7.37}$$

and this can be rewritten in the form of a standard law. *Actually, the dissipation potential method leads us to postulate, the same as in elastoplasticity, the existence of a potential* $\varphi(G)$ *such that*

$$\dot{l} \in \partial\varphi(G) \tag{7.38}$$

with – for Griffith's criterion in brittle fracture –

$$\varphi(G) = \left\{\begin{array}{ll} 0 & \textit{if } G \leqslant G_c, \\ +\infty & \textit{if } G > G_c; \end{array}\right\} \tag{7.39}$$

in other words, $\varphi(G)$ *is the* indicator *of the convex* $\{G^* \leqslant G_c\}$ *– see Appendix 2.*

Let us point out that G is a *surface energy* since Φ is a power and $\dim[\dot{l}] = L/T$. Therefore G_c is a *critical dissipated energy.*

Note For materials such as polymers or rubber, the *fracture is of the viscous type.* We must then make the same hypotheses as in elastoplasticity: we must regularize the above law by taking, for example

$$\varphi(G) = \frac{1}{2\eta} G^2, \tag{7.40}$$

hence a law for $\dot{l}$, $\dot{l} = G/\eta$, which is of the type of a viscous evolution law.

Relation between *K* and *G* As an exercise, if we consider homogeneous and isotropic linear elasticity, for which $W = \frac{1}{2}\lambda(\varepsilon_{kk})^2 + \mu\varepsilon_{ij}\varepsilon_{ij}$, and if we take for Γ a circle of radius R tending towards zero, we can show that

$$G = \lim_{R \to 0} \int_\Gamma \ldots = \frac{1 - \nu^2}{2E}(K_I^2 + K_{II}^2) + \frac{1}{2\mu} K_{\text{III}}^2. \tag{7.41}$$

In pure *mode I*, in particular,

$$G = \frac{1-\nu^2}{2E} K_I^2. \tag{7.42}$$

By measuring $(K_c)_I$ we obtain then

$$G_c = \frac{1-\nu^2}{2E} (K_c)_I^2, \tag{7.43}$$

which shows that the Irwin and Griffith criteria are identical. Some typical values of G are as follows: for glass $G_c \approx 3\ \mathrm{J/m^2}$; for heat-resistant ceramics $G_c \approx 60\ \mathrm{J/m^2}$ and for steels $G_c \approx 10^4 - 10^5\ \mathrm{J/m^2}$. Like electrical resistivity, the material property represented by G_c, the *critical* energy-release rate, is a property that may vary by several orders of magnitude depending on the material: the *more brittle* is a material, the smaller is G_c. The same holds true of the fracture toughness K_c.

7.4 The Rice–Eshelby–Cherepanov integral

We call the *Rice–Eshelby–Cherepanov integral* (this is not very fair as Eshelby (1951), Cherepanov (1967, 1968), and Rice (1968a,b) came in that order) the *invariant*

$$J_\Gamma = \int_\Gamma (W n_1 - \mathbf{n} \cdot \boldsymbol{\sigma} \cdot \mathbf{u}_{,1})\, \mathrm{d}\Gamma \tag{7.44}$$

such that, according to $(7.36)_2$,

$$G = \lim_{\Gamma \to 0} J_\Gamma. \tag{7.45}$$

We have thus the following.

Proposition *If a material is elastic (linear or not) and homogeneous, then* J_Γ *does not depend upon* Γ.

As a consequence of this proposition we can estimate G rather accurately, provided that Γ is rather large. This invariance property is lost if there is not perfect contact between the lips of Σ (friction in particular is excluded; reactions do not work).

Principle of proof. We consider two contours Γ_1 and Γ_2 and, with the help of the Gauss formula and the equilibrium equation $\operatorname{div} \boldsymbol{\sigma} = \mathbf{0}$, we transform the surface integrals to volume integrals. Since there are no *forces* acting upon Σ we obtain

$$J_\Gamma = \int_{\Gamma_\alpha} (W\mathbf{n}\cdot\mathbf{e}_1 - \mathbf{n}\cdot\boldsymbol{\sigma}\cdot\mathbf{u}_{,1})\,\mathrm{d}\Gamma \tag{7.46}$$

with

$$J_{\Gamma_1} - J_{\Gamma_2} = 0. \tag{7.47}$$

This result is only valid if there are neither volume forces (obvious) nor forces applied upon Σ (which is *not* the case in hydraulic cracking problems).

7.5 Global potential, generalized Rice integral, energy-release rate

Let us now consider a different way, *global* this time, permitting us to arrive at the notion represented by G. In order to do that, we must consider the total *potential energy* of the cracked elastic solid, subject to surface forces $\mathbf{T}^{\mathrm{d}}$ over $\partial\Omega_T$ (with $\mathbf{T}^{\mathrm{d}} = \lambda\mathbf{T}_0$ over $\partial\Omega_T$). We have (see eqn (4.12) with $\mathbf{g} = \mathbf{0}$)

$$I(\mathbf{u}) = \int_{\Omega_l} W(\boldsymbol{\varepsilon}(\mathbf{u}))\partial\Omega - \int_{\partial\Omega} \lambda\mathbf{T}_0\cdot\mathbf{u}\,\mathrm{d}s, \tag{7.48}$$

where $\mathbf{u}$ and λ are the parameters and $\Omega = \Omega_l$ is Ω for a length l of the crack. We define the set of admissible kinematic fields by

$$\mathcal{U}_{\mathrm{ad}} = \{\mathbf{u}\,|\,\mathbf{u} = \lambda\mathbf{u}_0 \text{ on } \partial\Omega_u\} = \mathcal{U}_{\mathrm{ad}}(\lambda). \tag{7.49}$$

We write

$$P = I(\mathbf{u}, l, \lambda), \tag{7.50}$$

the total potential energy at equilibrium (therefore $\mathbf{u}$ is *the* solution of the problem). If W is convex, which is usually the case, then we have the minimum property

$$P = \min_{\mathbf{u}^*\in\mathcal{U}_{\mathrm{ad}}} I(\mathbf{u}^*, l, \lambda) = P(l, \lambda), \tag{7.51}$$

and we have the following proposition, concerning the global potential:

$$G = -\left.\frac{\partial P(l,\lambda)}{\partial l}\right|_{\lambda\text{ fixed}}. \tag{7.52}$$

Proof We consider a fictitious problem at moment t (Fig. 7.9). In this problem we can write

$$\lim_{\Delta l\to 0}\frac{\Delta P}{\Delta l} = \cdots, \tag{7.53}$$

but this fictitious process corresponds to a particular action (belonging to reality) or to the real evolution studied above. We calculate

$$+G\Delta l^* = +\Delta(\text{external energy}) - \Delta(\text{stored energy}), \tag{7.54}$$

i.e.

$$G = \frac{\Delta(\text{external energy}) - \Delta(\text{stored energy})}{\Delta l^*}, \tag{7.55}$$

hence

$$G = -\frac{\partial P(l, \lambda)}{\partial l},$$

that is, G is evaluated as a generalized force by making it work in a *virtual* transformation following the usual method in continuum mechanics. Owing to this result, G is called the *energy-release rate.*

The result (7.52) is called the *generalized Rice theorem* and may be extended to the case of *several cracks* (without interaction among them). For example, for two cracks of lengths l_1 and l_2,

$$\left.\begin{aligned} &P = P(l_1, l_2, \lambda), \\ &G_1 = -\frac{\partial P}{\partial l_1}, \qquad G_2 = -\frac{\partial P}{\partial l_2}. \end{aligned}\right\} \tag{7.56}$$

7.6 Quasi-static evolution of a crack system in an elastic solid in brittle fracture

Let $\lambda(t)$ be the loading and $\{l_1(t), l_2(t)\}$ the evolution for the crack system. We are led to a formulation of the *generalized standard material* type (see

Fig. 7.9. Fictitious processes

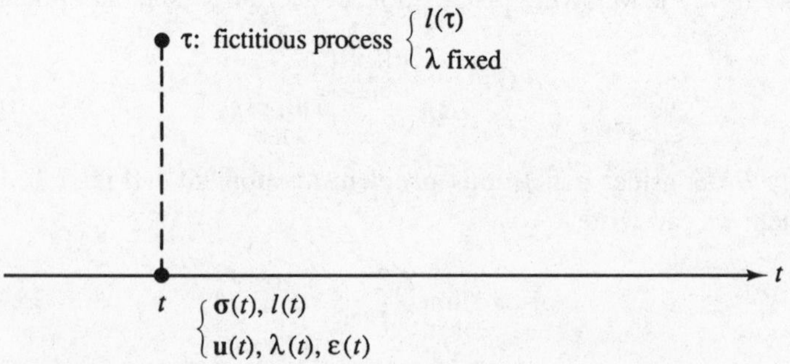

Section 5.1):

at $t = 0$ we have $\vec{l}_0 = \vec{l}\{l_1(0), l_2(0)\}$;
we may know

$$I(\mathbf{u}, \vec{l}, \lambda) = \int_{\Omega_l} W \, d\Omega - \lambda \int_{\partial\Omega} \tilde{\mathbf{T}} \cdot \mathbf{u} \, ds; \tag{5.57}$$

let us assume that we know or that we are able to compute

$$\left.\begin{aligned} P(\vec{l}, \lambda) &= \min_{u^* \in \mathscr{U}_{ad}} I(\mathbf{u}^*, \vec{l}, \lambda)_1, \\ &\text{i.e., } (\vec{l}, \lambda) \to P(\vec{l}, \lambda). \end{aligned}\right\} \tag{7.58}$$

Then we have the following.

Evolution equation *With the Griffith criterion we have*

$$\left.\begin{aligned} G_i &= -\frac{\partial P(\vec{l}, \lambda)}{\partial l_i}, \\ \dot{l}_i &= 0 \quad \text{if } G_i < G_c, \\ \dot{l}_i &\geqslant 0 \quad \text{if } G_i = G_c. \end{aligned}\right\} \tag{7.59}$$

For one crack, $G \leqslant G_c$ is a convex in $\mathbb{R}$. For two cracks we have $C = \{\vec{G} \,|\, G_i \leqslant G_i\}$ in $\mathbb{R}^2$ and we can write the normality law

$$\vec{\dot{l}} \cdot (\vec{G} - \vec{G}^*) \geqslant 0, \qquad \forall \vec{G}^* \in C, \tag{7.60}$$

that is, more explicitly,

$$\vec{\dot{l}}_1(G_1 - G_1^*) + \dot{l}_2(G_2 - G_2^*) \geqslant 0, \qquad \forall G_i \leqslant G_c.$$

The evolution problem is then of a type that we meet with in plasticity and we sum it up by writing

$$(\mathscr{P}) \left\{\begin{aligned} &\vec{G} = -\partial P/\partial \vec{l}, \\ &\vec{\dot{l}} \cdot (G - \vec{G}^*) \geqslant 0, \qquad \forall \vec{G}^* \in C, \\ &\vec{l}(0) = \vec{l}_0, \\ &\lambda(t) \to \vec{l}(t). \end{aligned}\right\} \tag{7.61}$$

Remark $P(l, \lambda)$ is not generally quadratic. If P is quadratic, then we know how to solve the problem ($\mathscr{P}$) (according to Brezis' works). Otherwise, there are technical difficulties. It can be shown that the solution $\vec{l}(t)$ is *piecewise regular*.

According to the formalism (7.60)–(7.61) we have the following.

Proposition *As in elastoplasticity, where the superimposed dot indicates the right-hand derivative, we have the orthogonality relation*

$$\dot{l}_i \dot{G}_i = 0 \tag{7.62}$$

(*of the same form as* $\dot{\boldsymbol{\sigma}} : \dot{\boldsymbol{\varepsilon}}^{\mathrm{p}} = 0$ *in perfect elastoplasticity*).

The result (7.62) is true *with* or *without* summing over i in such a way that, for a two-crack system, we also have

$$\dot{l}_1 \dot{G}_1 = 0, \qquad \dot{l}_2 \dot{G}_2 = 0. \tag{7.63}$$

So we can, as in plasticity, *pose the problem in terms of velocities*, that is: assuming that the actual state is known, and given $\dot{\lambda}$, calculate the associated $\dot{l}_i$. So, for two cracks, by $(7.63)_1$, we have

either $\dot{l}_1 = 0$, that is, we have *the* solution,
or $\dot{G}_1 = 0$, hence

$$\dot{G}_1 = -\frac{\mathrm{d}}{\mathrm{d}t}\left[\frac{\partial P}{\partial l_1}(\vec{l}, \lambda)\right] = 0, \tag{7.64}$$

which is written

$$-\frac{\partial^2 P}{\partial l_1^2}\dot{l}_1 - \frac{\partial^2 P}{\partial l_1 \partial l_2}\dot{l}_2 - \frac{\partial^2 P}{\partial l_1 \partial \lambda}\dot{\lambda} = 0. \tag{7.65}$$

If we reason likewise about $\dot{l}_2 \dot{G}_2$, we obtain

$$-\frac{\partial^2 P}{\partial l_2^2}\dot{l}_2 - \frac{\partial^2 P}{\partial l_2 \partial l_1}\dot{l}_1 - \frac{\partial^2 P}{\partial l_2 \partial \lambda}\dot{\lambda} = 0. \tag{7.66}$$

This can be summed up in a variational form (exercise; this result is due to Nguyen Quoc Son),

$$\begin{bmatrix} \dot{l}_1 - \overset{*}{l}_1 \\ \dot{l}_2 - \overset{*}{l}_2 \end{bmatrix}^{\mathrm{T}} \begin{bmatrix} \dfrac{\partial^2 P}{\partial l_1^2} & \dfrac{\partial^2 P}{\partial l_1 \partial l_2} \\ \dfrac{\partial^2 P}{\partial l_1 \partial l_2} & \dfrac{\partial^2 P}{\partial l_2^2} \end{bmatrix} \begin{bmatrix} \dot{l}_1 \\ \dot{l}_2 \end{bmatrix} + \begin{bmatrix} \dot{l}_1 - \overset{*}{l}_1 \\ \dot{l}_2 - \overset{*}{l}_2 \end{bmatrix}^{\mathrm{T}} \begin{bmatrix} \dfrac{\partial^2 P}{\partial l_1 \partial \lambda} \\ \dfrac{\partial^2 P}{\partial l_2 \partial \lambda} \end{bmatrix} \dot{\lambda} \geqslant 0, \tag{7.67}$$

with $\dot{l}_1 \geqslant 0$, $\dot{l}_2 \geqslant 0$, $\forall \overset{*}{l}_1 \geqslant 0$, $\forall \overset{*}{l}_2 \geqslant 0$,

If we set

$$N_C(\vec{G}) = \{\vec{\dot{l}} | \dot{l}_1 \geqslant 0, \dot{l}_2 \geqslant 0\}, \tag{7.68}$$

the *evolution problem* is written as follows.

Problem Find $\vec{\dot{l}} \in N_C(\vec{G})$ such that (7.67) is satisfied for all $\vec{\overset{*}{l}} \in N_C(\vec{G})$. (7.69)

We have then the following demonstrable propositions (Nguyen Quoc Son).

Proposition 1 *If the second derivative matrix $\partial^2 P/\partial l_i \partial l_j$ is strictly positive definite over N_C $(\vec{G})$, then there is* at least one solution *to the problem* (7.69).

Proposition 2 *If the matrix of second derivatives $\partial^2 P/\partial l_i \partial l_j$ is strictly positive, $\forall \vec{l}$, then there is only* one unique solution *to the problem* (7.69).

We have already noticed a strong (formal) similarity between the thermodynamic formulation of elastoplasticity and that of the brittle-fracture theory. In the first case we were able to give an abstract formulation. Let us show now that there exists a common abstract formulation, for both plasticity and fracture.

7.7 Similarity between plasticity and fracture

In fracture (see above, Sections 7.5 and 7.6), the displacement is eliminated. The similarity between fracture and elastoplasticity then is exact only if we can take $\boldsymbol{\varepsilon}^p$ as *the main variable* in plasticity. Is this possible? The answer is, yes. In fact, in perfect plasticity we have

$$\left.\begin{aligned} &\boldsymbol{\varepsilon} = \boldsymbol{\varepsilon}^e + \boldsymbol{\varepsilon}^p \\ &\boldsymbol{\sigma} = \mathbf{E} : \boldsymbol{\varepsilon}^e, \\ &\boldsymbol{\sigma} \in C = \{f(\boldsymbol{\sigma}) \leqslant 0\}. \end{aligned}\right\} \tag{7.70}$$

If $\boldsymbol{\varepsilon}^p$ is known, then we have an elasticity with known residual strains (as in thermoelasticity), so we can have a plasticity *formulation* with $\boldsymbol{\varepsilon}^p$ as the only variable. In order to do this we must proceed in two steps.

1st step: we take $\boldsymbol{\varepsilon}^p$,
we compute $\mathbf{u}(\boldsymbol{\varepsilon}^p, \lambda)$ and $\boldsymbol{\sigma}$,
we eliminate $\mathbf{u}$.
2nd step: we write the evolution equation for $\boldsymbol{\varepsilon}^p$.

First step We have the problem

$$\left.\begin{aligned} \boldsymbol{\varepsilon} &= \boldsymbol{\varepsilon}^{\mathrm{e}} + \boldsymbol{\varepsilon}^{\mathrm{p}}, \\ \boldsymbol{\sigma} &= \mathbf{E} : \boldsymbol{\varepsilon}^{\mathrm{e}}, \\ \operatorname{div} \boldsymbol{\sigma} &= \mathbf{0} \qquad \text{in } \Omega, \\ \boldsymbol{\varepsilon} &= \boldsymbol{\varepsilon}(\mathbf{u}), \\ \boldsymbol{\sigma} \cdot \mathbf{n} = \mathbf{T}^{\mathrm{d}} &= \lambda \mathbf{T}_0 \qquad \text{at } \partial\Omega_{\mathrm{T}}, \\ \mathbf{u} = \mathbf{u}^{\mathrm{d}} &= \lambda \mathbf{u}_0 \qquad \text{at } \partial\Omega_u. \end{aligned}\right\} \tag{7.71}$$

We will have

$$\boldsymbol{\sigma} = -\mathbf{Z}(\boldsymbol{\varepsilon}^{\mathrm{p}}) + \boldsymbol{\sigma}^{\mathrm{el}}(\lambda), \tag{7.72}$$

where

$$\boldsymbol{\sigma}^{\mathrm{R}} = -\mathbf{Z}(\boldsymbol{\varepsilon}^{\mathrm{p}}) \tag{7.73}$$

is a residual stress and, in the absence of $\boldsymbol{\varepsilon}^{\mathrm{p}}$, $\boldsymbol{\sigma}^{\mathrm{el}}$ is the solution. We can show that the operator $\mathbf{Z}$ such that $\boldsymbol{\varepsilon}^{\mathrm{p}} \to \mathbf{Z}(\boldsymbol{\varepsilon}^{\mathrm{p}})$ is a self-adjoint linear application, which is positive (i.e. $\mathbf{Z}(\boldsymbol{\varepsilon}^{\mathrm{p}}) : \boldsymbol{\varepsilon}^{\mathrm{p}} \geqslant 0$) but not strictly positive (since $\mathbf{Z}$ is a projection operator).

Second step We have the normality law

$$(\boldsymbol{\sigma} - \boldsymbol{\sigma}^*) : \dot{\boldsymbol{\varepsilon}}^{\mathrm{p}} \geqslant 0, \qquad \forall \boldsymbol{\sigma}^* \in C,$$

i.e, with (7.72)

$$\left.\begin{aligned} \dot{\boldsymbol{\varepsilon}}^{\mathrm{p}} : (-\mathbf{Z}(\boldsymbol{\varepsilon}^{\mathrm{p}})) + \boldsymbol{\sigma}^{\mathrm{el}}(\lambda) - \boldsymbol{\sigma}^*) \geqslant 0, \qquad \forall \boldsymbol{\sigma}^* \in C, \\ \boldsymbol{\varepsilon}^{\mathrm{p}}(0) = \boldsymbol{\varepsilon}^{\mathrm{p}}_0. \end{aligned}\right\} \tag{7.74}$$

This formulation is strictly analogous to the fracture problem, except that, instead of n-vectors (for n cracks), we have tensors. In the present case,

$$I(\boldsymbol{\varepsilon}^{\mathrm{p}}, \mathbf{u}, \lambda) = \int_\Omega \tfrac{1}{2}(\boldsymbol{\varepsilon}(\mathbf{u}) - \boldsymbol{\varepsilon}^{\mathrm{p}}) : \mathbf{E} : (\boldsymbol{\varepsilon}(\mathbf{u}) - \boldsymbol{\varepsilon}^{\mathrm{p}})\, \mathrm{d}\Omega - \int_{\partial\Omega_T} \lambda \mathbf{T}_0 \cdot \mathbf{u}\, \mathrm{d}s \tag{7.75}$$

and

$$P(\boldsymbol{\varepsilon}^{\mathrm{p}}, \lambda) = \min_{\mathbf{u}^* \in \mathscr{U}_{\mathrm{ad}}} I(\boldsymbol{\varepsilon}^{\mathrm{p}}, \mathbf{u}, \lambda). \tag{7.76}$$

If we choose the correct duality we can write

$$\langle P_{,\boldsymbol{\varepsilon}^{\mathrm{p}}}, \delta\boldsymbol{\varepsilon}^{\mathrm{p}} \rangle = \int (\ldots) : \delta\boldsymbol{\varepsilon}^{\mathrm{p}}\, \mathrm{d}\Omega, \tag{7.77}$$

where the quantity within parentheses is none other than

$$\boldsymbol{\sigma} = -\mathbf{Z}(\boldsymbol{\varepsilon}^{\mathrm{p}}) + \boldsymbol{\sigma}^{\mathrm{el}}(\lambda)$$

and then the similarity is complete. We have then the *general abstract* problem

$$(\mathscr{A}\mathscr{P})\left\{\begin{array}{l}\text{Let } P(\alpha, \lambda),\\[4pt] \mathbf{A} = -\partial P/\partial\boldsymbol{\alpha}, \quad \mathbf{A} \in C,\\[4pt] \dot{\boldsymbol{\alpha}}\cdot(\mathbf{A} - \mathbf{A}^*) \geqslant 0, \quad \forall \mathbf{A}^* \in C,\\[4pt] \boldsymbol{\alpha}(0) = \boldsymbol{\alpha}_0,\end{array}\right\} \tag{7.78}$$

with $\boldsymbol{\alpha} = (\boldsymbol{\varepsilon}^{\mathrm{p}} \text{ or } \vec{l})$ and $\mathbf{A} = (\boldsymbol{\sigma} \text{ or } \vec{G})$.

7.8 The Barenblatt theory

This is an *elastic theory of cohesive forces* that we may compare to the Irwin theory. Following Barenblatt (1972), we accept that ahead of the crack front there is a zone in which the 'atoms' can be pushed aside at a variable distance δ (see Fig. 7.10), and that this separation leads to *cohesion stresses* which are opposed to a clear separation. These cohesion stresses vary from zero where $\delta = 0$, as a function $\sigma(\delta)$ according to a law characteristic of the material. The zone of loss of cohesion is *independent* of the change and remains equal to itself, during the crack's extension. Rice has shown that

Fig. 7.10. Barenblatt's theory

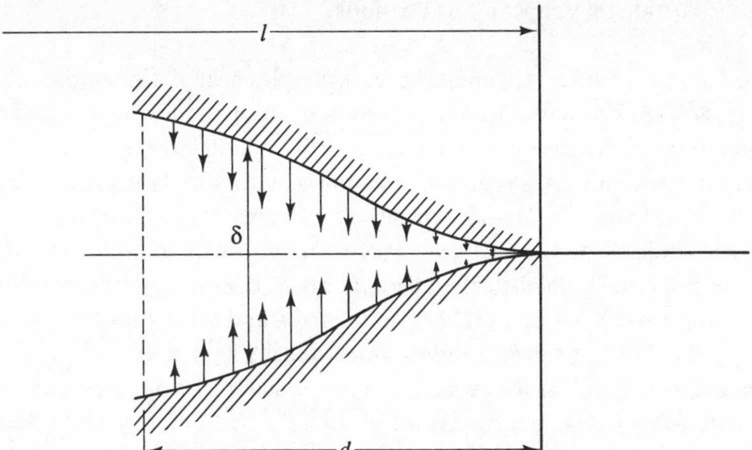

$$-\frac{\partial P}{\partial l} = \int_0^{\delta_t} \sigma(\delta)\, d\delta. \tag{7.79}$$

So we have an equivalence between Griffith's results and Barenblatt's cohesive stresses.

7.9 Introduction of a plastic zone

It is obvious that infinite stresses cannot physically exist, whereas elasticity allows for singularity in $1/\sqrt{r}$ concerning the stresses at the crack's tip. We may then think that most materials in the vicinity of the crack's tip are subject to a plastic strain that *smooths out* this point and limits the stresses by treating strains as in plasticity theory. Assuming that the plastic strain is not too far extended, we can admit, with sufficient approximation, that the existence of a plastic zone be taken into consideration, by giving to the crack an effective length l^{eff} for the calculations:

$$l^{\text{eff}} = l + r_e. \tag{7.80}$$

Obviously we must calculate r_e, where the singularity has been fictitiously displaced (see Fig. 7.11). When the external loading is augmented the plastic zone increases, the singular elastic zone disappears and a strong *non-linearity* appears at the point of the force application. What we need then is an *elastoplastic analysis*. We may consult Labbens (1980) for the computation of the correction of the plastic zone in stresses or plane strains, as well as for the models of crack opening (for example, the Dugdale model). This is outside the scope of this textbook.

The reader will find in Appendix 4 a few examples of analytic computations of stress-intensity factors, as well as some indications concerning numerical computations. As the corresponding practical implementations now use computer evaluations we can only mention handbooks listing results (Sih, 1973, Murakami, 1987) and expositions of experimental and analytical methods (Sih, 1981). General references on fracture include Atkins and May (1985), Bui (1977), Barthélémy (1980), Labbens (1980), Le Nizhery (1980), Paris and Sih (1965), Freund (1990), and Parton and Morozov (1978). Apart from some hints at further studies (three-dimensional case, kinked crack, Barenblatt crack), the following problems concern the most efficient approach, using path-independent integrals (e.g., the J-integral and others). Relating to this last type of approach we mention the general variational treatment accounting for *inhomogeneities* in a *nonlinear* deformable medium by, e.g., Casal (1978), Golebiewska-Herrmann (1981), and Epstein and Maugin (1990a, b), which allow a direct generalization of J-integrals to electromechanical fields in brittle ceramics (Maugin and Epstein, 1991).

Fig. 7.11. Plastic zone at the crack tip (after Labbens, 1980). (a) Plane stresses. $r_e = (1/2\pi)(K/R_e)^2 = \lambda$. (b) Plane strains. $r_e = (1/6\pi)(K/R_e)^2 = \lambda$.

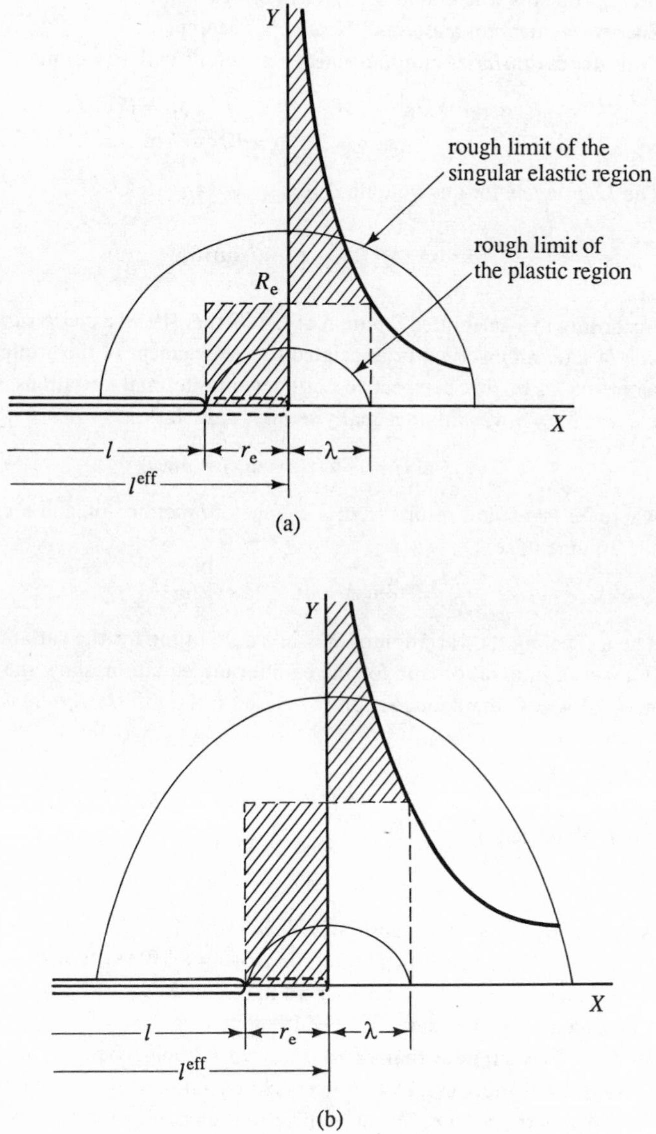

Problems for Chapter 7

7.1. Energy-release rate G
Prove the result (7.41) in homogeneous isotropic linear elasticity (Hooke's law).

7.2. Rice–Eshelby integral
Prove the path independence of the J-integral. What happens in hydraulic crack problems where fluid occupies the crack void?

7.3. Energy–momentum tensor and Noether's Theorem
Consider *elastostatics* in homogeneous materials with equations

$$\left.\begin{array}{lll} \boldsymbol{\sigma} = \partial W/\partial \boldsymbol{\varepsilon}, & W = W(\boldsymbol{\varepsilon}), & \boldsymbol{\varepsilon}(\mathbf{u}) = (\nabla \mathbf{u})_s, \\ \operatorname{div} \boldsymbol{\sigma} = \mathbf{0} \quad \text{in}\, \Omega, & \mathbf{n}\cdot\boldsymbol{\sigma} = \mathbf{T}^d \quad \text{on}\, \partial\Omega_T, & \mathbf{u} = \mathbf{u}^d \quad \text{on}\, \partial\Omega_u. \end{array}\right\} \tag{1}$$

The *Lagrangian* for any volume D reads

$$\mathscr{L}[D] = \int_D W(\boldsymbol{\varepsilon}(\mathbf{u}))\,\mathrm{d}v. \tag{2}$$

According to a celebrated theorem of E. Noether (1918) a *conservation equation or integral invariant* is associated to each element of the group of symmetry of $\mathscr{L}$. In this perspective consider infinitesimal variations $\mathbf{x} \to \mathbf{x}' = \mathbf{x} + \delta\mathbf{x}$, $\delta\mathbf{x} = \eta\mathbf{w}$, η infinitesimally small, $\forall\mathbf{w}$, such that

$$\mathbf{u}(\mathbf{x}) = \rightarrow \mathbf{u}'(\mathbf{x}') = \mathbf{u}(\mathbf{x}) + \delta\mathbf{u}(\mathbf{x}) \tag{3}$$

where $\delta\mathbf{u}$ in general results from a change of function, $\bar{\delta}\mathbf{u}$, and a change in the argument, i.e.,

$$\delta\mathbf{u}(\mathbf{x}) = \bar{\delta}\mathbf{u} + (\delta\mathbf{x}\cdot\nabla)\mathbf{u}. \tag{4}$$

Then, effecting the variation of $\mathscr{L}$ while accounting for the variation in the domain of integration and for the equilibrium equation, show that for any $\mathbf{w} \neq 0$, $\nabla\mathbf{w} = 0$, there follows for any closed surface the *integral invariant*

$$\int_{\partial D} \mathbf{n}\cdot\mathbf{P}\,\mathrm{d}s = \mathbf{0} \tag{5}$$

where $\mathbf{P}$ defined by

$$P_{ki} = W\delta_{ki} - \sigma_{kj}u_{j,i} \tag{6}$$

is the *energy–momentum tensor* of Eshelby (1951).
[*Hint*: Note that $\delta\mathscr{L} = \int_D (\partial W/\partial\boldsymbol{\varepsilon}) : \delta\boldsymbol{\varepsilon}\,\mathrm{d}v + \int_D \nabla\cdot(W\delta\mathbf{x})\,\mathrm{d}v$ and the fact that elementary energy ($W\mathrm{d}v$) is conserved in the process.]

7.4. Force on an inhomogeneity (P7.3 continued)
In D of P7.3 suppose that there exists an inhomogeneity or *inclusion* $\mathscr{F}$ (cavity, dislocation, crack). Let S be a closed surface surrounding this inclusion. Apply the balance law (5) of P7.3 to the annulus bounded by S and S'

in Fig. 7.12 (a region of homogeneous isotropic material) to show that

$$J_i = \int_S n_k P_{ki} \, \mathrm{d}s \tag{1}$$

is an integral invariant (i.e., it does not depend on S). In particular, projection of this onto the x_1-axis yields the J-integral.

7.5. Other integral invariants

Proceeding as in 7.3 and 7.4 but considering infinitesimal variations of the type (infinitesimal rotation)

$$x_i' = \eta \varepsilon_{ijk} w_j x_k, \qquad u_i' = u_i + \eta \varepsilon_{ijk} w_j x_k \quad (\forall \mathbf{w}, \nabla \mathbf{w}) \tag{1}$$

or (change of scale)

$$x_i' = (1 + \eta k) x_i, \qquad u_i' = (1 - \tfrac{1}{2}\eta k) u_i \quad (\forall k, \nabla k = 0), \tag{2}$$

show that there exist the following vector (axial) and scalar integral invariants (due to Knowles and Sternberg, 1972):

$$L_i = \int_S \varepsilon_{ijk} (W x_j n_k + n_p \sigma_{pj} u_k - n_p \sigma_{pq} u_{q,j} x_k) \, \mathrm{d}s \tag{3}$$

and

$$M = \int_S (W \mathbf{x} \cdot \mathbf{n} - n_k \sigma_{kj} u_{j,p} x_p - \tfrac{1}{2} \mathbf{n} \cdot \boldsymbol{\sigma} \cdot \mathbf{u}) \, \mathrm{d}s \tag{4}$$

which also prove to be useful in studying cracks.

7.6. Dual path-independent integral

H.D. Bui (1977) has introduced the dual of the Rice–Eshelby integral, I, by

$$I_\Gamma = \int_\Gamma \left[-W^*(\boldsymbol{\sigma}) n_1 + \mathbf{u} \cdot \frac{\partial \boldsymbol{\sigma}}{\partial x_1} \cdot \mathbf{n} \right] \mathrm{d}\Gamma. \tag{1}$$

Fig. 7.12. Figure for Problem 7.4

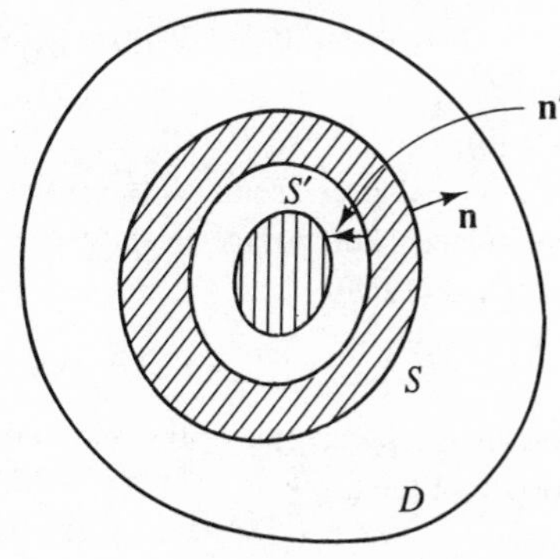

(a) Prove directly that this is also a path-independent integral.
(b) Prove that $I = J$ in modes I and II of crack opening.
(c) Show that

$$I_\Gamma = -\frac{\partial P^*}{\partial l}, \qquad P^* = -\int_\Omega W^*(\boldsymbol{\sigma})\,\mathrm{d}v + \int_{\partial\Omega_u} (\mathbf{n}\cdot\boldsymbol{\sigma}\cdot\mathbf{u}^{\mathrm{d}})\,\mathrm{d}s, \tag{2}$$

where P^* is complementary energy.

7.7. Special case of *J*- and *I*-integrals

A test specimen is cut leaving only a segment b of matter. Suppose that this specimen is subjected to a flexure moment M and its rotation θ is a function $\theta = f(M/b^2)$ only. This function is supposed to be invertible i.e., there exists g such that $M = b^2 g(\theta)$. Show that per unit thickness one has

$$J = I = (2/b)\int M\,\mathrm{d}\theta, \tag{1}$$

where J and I are the Rice–Eshelby integral and its dual integral (as in 7.6). [*Hint*: The crack has length $l = w - b$ if w is the width of the specimen; with imposed torque M and at imposed rotation the potential and complementary energies read

$$P = -\int_0^M \theta\,\mathrm{d}M \qquad \text{and} \qquad P^* = \int M\,\mathrm{d}\theta; \tag{2}$$

one obtains J and I by using the global formulas

$$J = -\partial P/\partial l, \qquad I = -\partial P^*/\partial l.]$$

7.8. *J*-integral in antiplane mode

Write down the expressions, in terms of real variables, of the J- and I-integrals for an antiplane mode.
[*Answer*:

$$J_\Gamma = \int_\Gamma \left[\frac{1}{2\mu}(\sigma_{31}^2 + \sigma_{32}^2)\,\mathrm{d}x_2 - u_{3,1}(\sigma_{31}n_1 + \sigma_{32}n_2)\,\mathrm{d}s\right]$$

$$I_\Gamma = \int_\Gamma \left[-\frac{1}{2\mu}(\sigma_{31}^2 + \sigma_{32}^2)\,\mathrm{d}x_2 + u_3(n_1\sigma_{13,1} + n_2\sigma_{23,1})\,\mathrm{d}s\right].]$$

7.9. Energy-release rate (b)

Applying the first law of thermodynamics to a solid region Ω with a crack growing at a linear velocity $\dot{l}$ or surface velocity $\dot{A} = e\dot{l}$, i.e. (see Appendix 1; this follows Cherepanov (1968))

$$\dot{K} + \dot{E} = \mathscr{P}_{(\mathrm{ext})} + \dot{Q} - 2\gamma\dot{A}, \tag{1}$$

where γ is a characteristic parameter of the material. Show that the energy available for a growth of the crack is

$$\int_{\partial\Omega} \mathbf{T} \cdot \frac{\partial \mathbf{u}}{\partial A} \mathrm{d}s - \frac{\partial W_e}{\partial A} \leqslant 2\gamma \tag{2}$$

and prove the inequality. Here $\partial\Omega$ is the boundary of Ω, and $\mathbf{T}$ is the density of surface forces on $\partial\Omega$. W_e is the elastic energy. Deduce that the energy-release rate G is none other than the left-hand side of eqn. (2). Then $G = 2\gamma$ provides the threshold of the Griffith fracture criterion.

7.10. Two-crack system

Show that for a system of *two noninteracting* cracks the evolution problem (7.61) can be given the variational formulation (7.67)

$$\forall \overset{*}{l}_1, \overset{*}{l}_2 \geqslant 0.$$

7.11. Crack in three-dimensional medium

Make a literature search and write a report on the problem of the definition of stress-intensity factors in a three-dimensional elastic medium. Study more particularly the case of a circular crack under constant pressure in a three-dimensional medium.

7.12. Kinked crack (term paper)

Make a literature search and write a short report on the problem (definition of energy-release rate in Griffith's theory, time evolution) of kinked cracks (see Fig. 7.13).

7.13. Barenblatt crack (term paper)

Search the technical literature and write a short report on Barenblatt's theory of cohesive forces at a crack. In particular, establish the relationship with stress intensity factors by introducing the notion of weight functions of Bueckner and, following J.R. Willis (*Int. J. Fracture*, **II**, p. 151, 1975), relate the work performed by cohesive forces to the stress-intensity factor.

Fig. 7.13. Figure for Problem 7.11

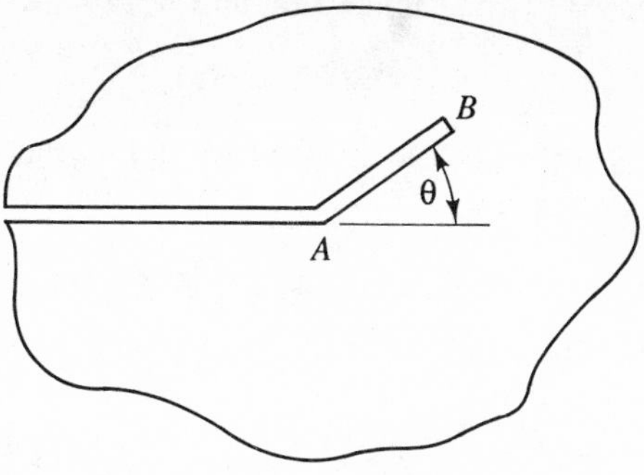

8

Elastoplasticity with finite strains

The object of the chapter When the hypothesis of small perturbations is no longer valid one has to consider the possibility of an elastoplasticity with *finite* strains. This may be important at the *monocrystal* as well as the polycrystal scale. The notion of configuration then plays an essential part and one must introduce *finite* measures of strains. The main questions, however, are (i) in what way do we distinguish between elastic and plastic strains and (ii) what is the corresponding statement of plasticity evolution and normality rule? The *multiplicative* decomposition of strains seems natural at (i) while the notion of objective time derivative comes into play at (ii). This is briefly treated by way of an introduction to the subject matter.

8.1 Decomposition of elastoplastic strains

Until now only *infinitesimally small* transformations have been considered between instants of time t and $t + \mathrm{d}t$, while assuming that the position of the matter element, the elasticity tensor and the plastic potential were all known at time t. One can say that we were only interested in the *partial* problem represented by the small-strain tensor

$$\boldsymbol{\varepsilon} = \boldsymbol{\varepsilon}^{\mathrm{e}} + \boldsymbol{\varepsilon}^{\mathrm{p}} = (\nabla \mathbf{u})_{\mathrm{S}}, \tag{8.1}$$

where neither $\boldsymbol{\varepsilon}^{\mathrm{e}}$ nor $\boldsymbol{\varepsilon}^{\mathrm{p}}$ is a true gradient. With internal variables present, the *intrinsic dissipation* read

$$\phi_{\mathrm{intr}} = \boldsymbol{\sigma} : \dot{\boldsymbol{\varepsilon}}^{\mathrm{p}} + A_i \dot{\alpha}_i \tag{8.2}$$

along with the laws of state

$$\boldsymbol{\sigma} = \frac{\partial W}{\partial \boldsymbol{\varepsilon}^{\mathrm{e}}}, \qquad A_i = -\frac{\partial W}{\partial \alpha_i}, \qquad W = \rho_{\mathrm{R}} \Psi, \tag{8.3}$$

where W was the free energy per unit *volume*.

Now, if we want to solve the full problem of evolution of an elastoplastic body in the course of time, we must study the *finite* transformation that the element of matter experiences between the initial time t_0 and the present time t. Two problems then arise. The first concerns the eventual generalization of the decomposition (8.1) to finite strains. The other is what is the most convenient form of the intrinsic dissipation in order to write down the required complementary laws, i.e. the normality rules. The latter point crucially depends on the solution of the former, which, in a sense, provides a hint to the expression of the plastic strain rate which generalizes $\dot{\boldsymbol{\varepsilon}}^{\mathrm{p}}$ to the case of finite transformations.

The general deformation of a material body between a reference configuration K_{R}, free of loads and strains (in a so-called *natural* state) and a current configuration K_t at time t is described by the *diffeomorphism* (where t is a parameter)

$$\mathbf{x} = \mathscr{X}(\mathbf{X}, t), \quad \text{i.e.,} \quad x_i = \mathscr{X}_i(X_K, t), \tag{8.4}$$

where, using Cartesian tensor notation, we denote by x_i and X_K respectively the coordinates of the *same* material point at t and t_0 (see Fig. 8.1)

The motion gradient, its inverse and its Jacobian determinant are defined by

$$\left.\begin{aligned} &\mathbf{F} = \{\partial \mathscr{X}_i / \partial X_K\}, \\ &\mathbf{G} = \mathbf{F}^{-1} = (\mathbf{F})^{-1} = \{\partial X_K / \partial \mathscr{X}_i\}, \\ &J = \det \mathbf{F}, \end{aligned}\right\} \tag{8.5}$$

The *Lagrange tensor* of finite strains is defined by (T = transpose)

$$\mathbf{E} = \tfrac{1}{2}(\mathbf{F}^{\mathrm{T}}\mathbf{F} - \mathbf{1}_{\mathrm{R}}), \quad \text{i.e.,} \quad E_{KL} = \tfrac{1}{2}(F_{iK}F_{iL} - \delta_{KL}), \tag{8.6}$$

while J is simply the ratio of densities (so-called *conservation* of mass)

$$J = \rho_{\mathrm{R}}/\rho, \tag{8.7}$$

Fig. 8.1. Finite transformation

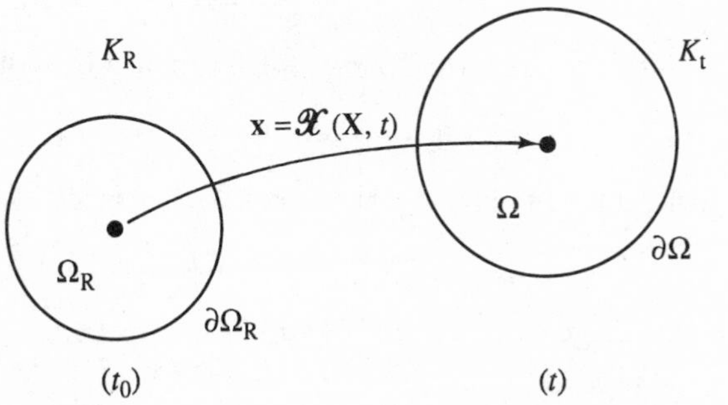

where ρ_R and ρ are the mass densities at times t_0 and t respectively, at the points $\mathbf{X}$ and $\mathscr{X}$ related by the motion (8.4) The *velocity-gradient tensor* is given by

$$\mathbf{L} = \dot{\mathbf{F}}\mathbf{F}^{-1} = \nabla\mathbf{v}, \qquad \mathbf{v} = \partial\mathscr{X}/\partial t|_{\mathbf{X}\,\text{fixed}}, \tag{8.8}$$

while the *rate-of-strain* (or strain-rate) *tensor* is the symmetric part of the latter:

$$\mathbf{D} = (\mathbf{L})_s = \tfrac{1}{2}[(\nabla\mathbf{v} + \nabla\mathbf{v})^{\mathrm{T}}] = \mathbf{D}^{\mathrm{T}}. \tag{8.9}$$

The following formulas expressing *convection* are useful:

$$\mathbf{D} = (\mathbf{F}^{-1})^{\mathrm{T}}\dot{\mathbf{E}}\mathbf{F}^{-1}, \qquad \dot{\mathbf{E}} = \mathbf{F}^{\mathrm{T}}\mathbf{D}\mathbf{F}, \tag{8.10}$$

where the superimposed dot indicates the *material time* derivative (i.e., the partial time derivative at X_K fixed).

The *second Piola–Kirchhoff tensor* $\mathbb{S}$ referred to the configuration K_R is the *symmetric* tensor defined from the symmetric Cauchy stress tensor $\boldsymbol{\sigma}$ (this one is referred to K_t) by the convection operation back to K_R:

$$\mathbb{S} = J\mathbf{F}^{-1}\boldsymbol{\sigma}(\mathbf{F}^{-1})^{\mathrm{T}}, \tag{8.11}$$

The *first Piola–Kirchhoff tensor* $\boldsymbol{\tau}$ is *not* properly speaking a tensor. It is rather a two-point $(\mathbf{x}, \mathbf{X})$ dependent object defined by partial convection (compared to (8.11)) such that

$$\boldsymbol{\tau} = J\mathbf{F}^{-1}\boldsymbol{\sigma}, \quad \text{i.e.,} \quad \tau_{iK} = JF_{Kj}^{-1}\sigma_{ij}. \tag{8.12}$$

The following identities are demonstrable:

$$(J^{-1}F_{iK})_{,i} = 0, \qquad (JF_{Ki}^{-1})_{,K} = 0 \tag{8.13}$$

It follows that

$$\begin{aligned}(\operatorname{div}\boldsymbol{\sigma})_i = \sigma_{ij,j} &= (J^{-1}F_{jk}\tau_{iK})_{,j} \\ &= J^{-1}F_{jk}\tau_{iK,j} = J^{-1}\tau_{iK,K},\end{aligned} \tag{8.14}$$

hence the equilibrium equation $\operatorname{div}\boldsymbol{\sigma} + \mathbf{g} = \mathbf{0}$ can also be written as

$$\tau_{iK,K} + Jg_i = 0 \quad \text{in} \quad \Omega_R. \tag{8.15}$$

Simultaneously, the boundary condition $\boldsymbol{\sigma}\cdot\mathbf{n} = \mathbf{T}^{\mathrm{d}}$ translates to

$$\tau_{iK}N_K = \mathscr{J}T_i^{\mathrm{d}} \quad \text{on} \quad \partial\Omega_R, \tag{8.16}$$

where N_K is the image of n_i back in K_R and

$$\begin{aligned}\mathscr{J} &= J(N_P C_{PQ}^{-1} N_Q)^{1/2},\\ N_K &= (N_P C_{PQ}^{-1} N_Q)^{1/2} n_i F_{iK}, \qquad (8.17)\\ C_{PQ}^{-1} &= \frac{\partial X_P}{\partial x_i}\frac{\partial X_Q}{\partial x_i},\end{aligned}$$

so that both n_i and N_K are the components of *unit* vectors. These simple reminders on the general, nonlinear theory of continua are sufficient for our purpose. We should note that the operations in eqns (8.10) and (8.11) are *convections* such that

$$\begin{aligned}\mathbb{S} : \dot{\mathbf{E}} &= J\mathbf{F}^{-1}\boldsymbol{\sigma}(\mathbf{F}^{-1})^{\mathrm{T}} : \mathbf{F}^{\mathrm{T}}\mathbf{D}\mathbf{F}\\ &= J\mathbf{F}^{-1}\boldsymbol{\sigma} : \mathbf{D}\mathbf{F} = J\boldsymbol{\sigma} : \mathbf{D}.\end{aligned} \qquad (8.18)$$

If we note that the power developed *per unit mass* in K_t by the internal forces (stresses) is usually given by

$$\bar{p}_{(\mathrm{i})} = \frac{1}{\rho} p_{(\mathrm{i})} = -\frac{1}{\rho}\boldsymbol{\sigma} : \mathbf{D} \qquad (8.19)$$

then with (8.7) and (8.18) we have thus

$$\bar{p}_{(\mathrm{i})} = -\frac{1}{\rho}\boldsymbol{\sigma} : \mathbf{D} = -\frac{1}{\rho_{\mathrm{R}}}\mathbb{S} : \dot{\mathbf{E}}. \qquad (8.20)$$

The quantity $\bar{p}_{(\mathrm{i})}$ therefore is *form-invariant* under convection. (As a matter of fact, it is the invariance of $\bar{p}_{(\mathrm{i})}$ that defines the *dual* convections of stress and strain rate). However, *per unit volume* in K_{R} we have

$$P_{(\mathrm{i})} = -\mathbb{S} : \dot{\mathbf{E}}; \qquad (8.21)$$

obviously

$$P_{(\mathrm{i})}\,\mathrm{d}V = p_{(\mathrm{i})}\,\mathrm{d}v = p_{(\mathrm{i})} J\,\mathrm{d}V = \rho\bar{p}_{(\mathrm{i})} J\,\mathrm{d}V = \rho_{\mathrm{R}}\bar{p}_{(\mathrm{i})}\,\mathrm{d}V$$

since

$$\mathrm{d}v = J^{-1}\,\mathrm{d}V.$$

Equations (8.20) and (8.21) tell us that $\mathbf{E}$ and $\mathbb{S}$ play the same part as $\boldsymbol{\varepsilon}$ and $\boldsymbol{\sigma}$ for finite strains, in so far as energy aspects and thermodynamics are concerned. As a matter of fact, $\mathbb{S}$ is sometimes called the *thermodynamic stress*, while $\boldsymbol{\tau}$ is called the *nominal* (or engineering) *stress*, since it is referred to the area element in the reference configuration K_{R} – see eqn (8.16).

8.2 Green–Naghdi decomposition

In 1965, A.E. Green and P.M. Naghdi proposed a *formal additive* decomposition of the finite Lagrangian strain tensor (8.6) in total analogy with the decomposition (8.1) of small-strain theory:

$$\mathbf{E} = \mathbf{E}^{e} + \mathbf{E}^{p}. \tag{8.22}$$

Then (8.21) yields

$$-P_{(i)} = \mathbb{S} : \dot{\mathbf{E}}^{e} + \mathbb{S} : \dot{\mathbf{E}}^{p}. \tag{8.23}$$

Taking now $\mathbf{E}^e$, θ and a certain set of internal variables α_i $(i = 1, .., n)$ for state variables, we can write the intrinsic *dissipation* in $K_{\mathbf{R}}$ per unit volume as

$$\Phi_{\text{intr}} = J^{-1}\phi_{\text{intr}} = \mathbb{S} : \dot{\mathbf{E}}^{p} + \mathscr{A} \cdot \dot{\boldsymbol{\alpha}}, \tag{8.24}$$

while the laws of state read

$$\mathbb{S} = \frac{\partial W}{\partial \mathbf{E}^{e}}, \qquad \mathscr{A} = -\frac{\partial W}{\partial \boldsymbol{\alpha}}, \tag{8.25}$$

where $W = \rho_{\mathbf{R}}\Psi$ is the free energy per unit volume in $K_{\mathbf{R}}$.

There is thus a complete correspondence between the expressions (8.2)–(8.3) and (8.24)–(8.25). The expression (8.24) which represents the *anelastic power* can also be re-written on account of the additive decomposition of the strain-rate tensor:

$$\mathbf{D} = \mathbf{D}^{e} + \mathbf{D}^{p}, \tag{8.26}$$

where

$$\mathbf{D}^{e} = (\mathbf{F}^{-1})^{\mathrm{T}}\dot{\mathbf{E}}^{e}\mathbf{F}^{-1}, \qquad \mathbf{D}^{p} = (\mathbf{F}^{-1})^{\mathrm{T}}\dot{\mathbf{E}}^{p}\mathbf{F}^{-1}, \tag{8.27}$$

The *intrinsic dissipation* per unit volume in K_t thus reads

$$\phi_{\text{intr}} = \boldsymbol{\sigma} : \dot{\mathbf{D}}^{p} - \rho\frac{\partial \Psi}{\partial \alpha_i}\dot{\alpha}_i \geqslant 0, \tag{8.28}$$

It is clear that, without any additional information, the *formal* nature of (8.22) and, a *fortiori*, of (8.26), hinders any measurement of the plastic strains and, thus, of any computation of the anelastic power $\boldsymbol{\sigma} : \dot{\mathbf{D}}^{p}$. It is only by constructing a nonlinear kinematics based on a stronger hypothesis than (8.22) and relying directly from the start on (8.5) that we can make some progress. The *multiplicative* decomposition of finite strains just fulfils this need.

8.3 Lee decomposition

Lee (1969) has introduced the notion of *released intermediate configuration* or state. The 'released' refers to a relaxation of elastic strains. The gradient $\mathbf{F}$ is thus decomposed in a *multiplicative* way (see Fig. 8.2)

$$\mathbf{F} = \mathbf{F}^{e}\mathbf{F}^{p}, \quad \text{i.e.,} \quad F_{iK} = F^{e}_{i\alpha}F^{p}_{\alpha K} \tag{8.29}$$

which, in fact, is the natural *composition* law for successive finite motions.

However, here, neither $\mathbf{F}^{e}$ nor $\mathbf{F}^{p}$ is a true gradient (they are Pfaff forms in differential geometry – see Choquet-Bruhat, 1968). Nonetheless, as in (8.8), velocity-'gradient' tensors can be defined

$$\mathbf{L}^{e} = \dot{\mathbf{F}}^{e}(\mathbf{F}^{e})^{-1}, \qquad \mathbf{L}^{p} = \dot{\mathbf{F}}^{p}(\mathbf{F}^{p})^{-1} \tag{8.30}$$

so that on using (8.29) and (8.8) we can write

$$\mathbf{L} = \mathbf{L}^{e} + \mathbf{F}^{e}\mathbf{L}^{p}(\mathbf{F}^{e})^{-1}. \tag{8.31}$$

Therefore, $\mathbf{L}$ is the sum of $\mathbf{L}^{e}$ and of the 'convect' of $\mathbf{L}^{p}$ from $K_{\mathscr{R}}$ into K_t, where $K_{\mathscr{R}}$ denotes the released intermediate configuration. The latter can only be defined *up to a rotation.* Actually we obviously have

$$\mathbf{F} = \mathbf{F}^{e}\mathbf{F}^{p} = \hat{\mathbf{F}}^{e}\hat{\mathbf{F}}^{p} \tag{8.32}$$

Fig. 8.2. Total deformation with intermediate configuration

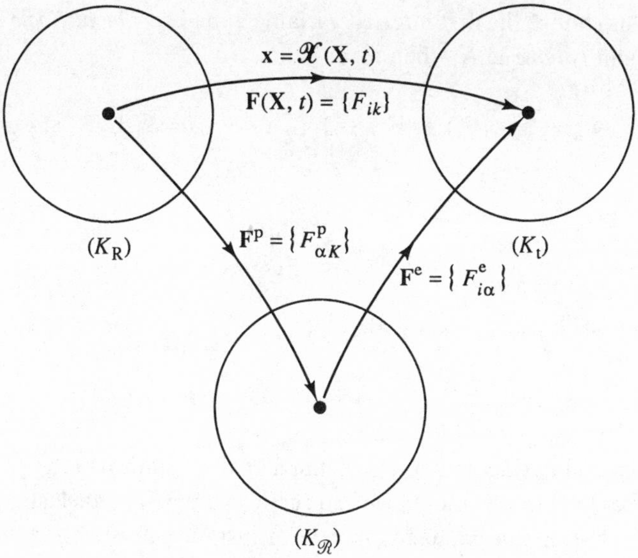

with

$$\left.\begin{aligned}\hat{\mathbf{F}}^{\mathrm{e}} &= \mathbf{F}^{\mathrm{e}}\mathbf{Q}, \quad \text{i.e.,} \quad \hat{F}^{\mathrm{e}}_{i\hat{\beta}} = F^{\mathrm{e}}_{i\alpha}Q_{\alpha\hat{\beta}},\\ \hat{\mathbf{F}}^{\mathrm{p}} &= \mathbf{Q}^{\mathrm{T}}\mathbf{F}^{\mathrm{p}}, \quad \text{i.e.,} \quad \hat{F}^{\mathrm{p}}_{\hat{\beta}K} = Q_{\hat{\beta}\alpha}F^{\mathrm{p}}_{\alpha K},\end{aligned}\right\} \tag{8.33}$$

where $\mathbf{Q}$ is an *orthogonal transformation* (representing a finite rotation in $\mathbb{E}^3$) such that

$$\mathbf{Q}\mathbf{Q}^{\mathrm{T}} = \mathbf{Q}^{\mathrm{T}}\mathbf{Q} = \mathbf{1}, \qquad \mathbf{Q}^{\mathrm{T}} = \mathbf{Q}^{-1}, \qquad \det \mathbf{Q} = \pm 1. \tag{8.34}$$

Let us now introduce the *symmetric* Piola–Kirchhoff stress tensor relative to the configuration $K_{\mathscr{R}}$ (compare with eqn (8.11)) by

$$\mathbb{S}_{\mathscr{R}} = J^{\mathrm{e}}(\mathbf{F}^{\mathrm{e}})^{-1}\boldsymbol{\sigma}[(\mathbf{F}^{\mathrm{e}})^{-1}]^{\mathrm{T}}, \tag{8.35}$$

where

$$J^{\mathrm{e}} = \rho_{\mathscr{R}}/\rho, \tag{8.36}$$

$$\frac{\rho_{\mathrm{R}}}{\rho} = J = \left(\frac{\rho_{\mathrm{R}}}{\rho_{\mathscr{R}}}\right)\left(\frac{\rho_{\mathscr{R}}}{\rho}\right) = J^{\mathrm{p}}J^{\mathrm{e}}. \tag{8.37}$$

It is immediately verified that

$$P_{(\mathrm{i})} = -J^{-1}(\mathbb{S}_{\mathscr{R}} : (\mathbf{F}^{\mathrm{e}})^{\mathrm{T}}\mathbf{F}^{\mathrm{e}}\mathbf{L}^{\mathrm{p}} + \mathbb{S}_{\mathscr{R}} : \dot{\mathbf{E}}^{\mathrm{e}}_{\mathscr{R}}), \tag{8.38}$$

where we have set

$$\mathbf{E}^{\mathrm{e}}_{\mathscr{R}} = \tfrac{1}{2}((\mathbf{F}^{\mathrm{e}})^{\mathrm{T}}\mathbf{F}^{\mathrm{e}} - \mathbf{1}_{\mathscr{R}}). \tag{8.39}$$

Now let us choose the first internal variable α_1 as $\mathbf{E}^{\mathrm{e}}_{\mathscr{R}}$. *The intrinsic dissipation per unit volume* in $K_{\mathscr{R}}$ then reads

$$\Phi_{\mathscr{R}} = \mathbb{S}_{\mathscr{R}} : (\mathbf{F}^{\mathrm{e}})^{\mathrm{T}}\mathbf{L}^{\mathrm{p}}\mathbf{F}^{\mathrm{e}})_{\mathrm{s}} - \rho\frac{\partial\Psi}{\partial\alpha_j}\dot{\alpha}_j, \qquad j = 2, 3, \ldots, n, \tag{8.40}$$

with

$$\mathbb{S}_{\mathscr{R}} = \rho_{\mathscr{R}}\frac{\partial\Psi}{\partial\mathbf{E}^{\mathrm{e}}_{\mathscr{R}}}. \tag{8.41}$$

Using then (8.31) we can show that

$$\phi_{\mathrm{intr}} = \boldsymbol{\sigma} : \mathbf{D}^{\mathrm{p}} - \rho\frac{\partial\Psi}{\partial\alpha_j}\dot{\alpha}_j, \qquad \mathbf{D}^{\mathrm{p}} = (\mathbf{F}^{\mathrm{e}}\mathbf{L}^{\mathrm{p}}(\mathbf{F}^{\mathrm{e}})^{\mathrm{T}})_{\mathrm{s}}. \tag{8.42}$$

If the released intermediate state is defined up to a rotation that is *physically observable*, then in this model we can really evaluate the anelastic power.

Other choices can be made for α_1. For instance, if we choose α_1 as $\mathbf{E}^{\mathrm{e}}$, defined by (8.22), then

$$\mathbf{E}^{\mathrm{e}} = (\mathbf{F}^{\mathrm{p}})^{\mathrm{T}}\mathbf{E}^{\mathrm{e}}_{\mathscr{R}}\mathbf{F}^{\mathrm{p}},$$
$$\mathbf{E}^{\mathrm{p}} = \tfrac{1}{2}[(\mathbf{F}^{\mathrm{p}})^{\mathrm{T}}\mathbf{F}^{\mathrm{p}} - \mathbf{1}_{\mathrm{R}}], \tag{8.43}$$

and eqn (8.42) is replaced by

$$\phi_{\mathrm{intr}} = \boldsymbol{\sigma} : \hat{\mathbf{D}}^{\mathrm{p}}, \qquad \hat{\mathbf{D}}^{\mathrm{p}} = \{[(\mathbf{F}^{\mathrm{e}})^{-1}]^{\mathrm{T}}\mathbf{L}^{\mathrm{p}}(\mathbf{F}^{\mathrm{e}})^{-1}\}_{\mathrm{S}}. \tag{8.44}$$

There still exist other possible decompositions, one example being the *additive* one of the gradient $\mathbf{H} = \mathbf{F} - \mathbf{1}$ by (see Nemat-Nasser, 1979)

$$\mathbf{H} = \mathbf{E} + \mathbf{P}. \tag{8.45}$$

Or else one can start from an additive decomposition of $\mathbf{D}$ in the form

$$\mathbf{D} = \tilde{\mathbf{D}}^{\mathrm{e}} + \tilde{\mathbf{D}}^{\mathrm{p}}, \tag{8.46}$$

where $\tilde{\mathbf{D}}^{\mathrm{e}}$ is the Jaumann time derivative of an internal variable $\tilde{\mathbf{E}}^{\mathrm{e}}$ (which is a second-order symmetric tensor) in such a way that

$$\tilde{\mathbf{D}}^{\mathrm{e}} = D_J\tilde{\mathbf{E}}^{\mathrm{e}}, \qquad D_J\mathbf{A} \equiv \dot{\mathbf{A}} - (\boldsymbol{\Omega}\mathbf{A} - \mathbf{A}\boldsymbol{\Omega}), \tag{8.47}$$

where $\boldsymbol{\Omega} = (\nabla\mathbf{v})_A$ is the skew-symmetric part of $\mathbf{L}$, whence (show this by way of exercise)

$$\phi_{\mathrm{intr}} = \boldsymbol{\sigma} : \tilde{\mathbf{D}}^{\mathrm{p}} - \rho\frac{\partial\Psi}{\partial\alpha_j}\dot{\alpha}_j, \qquad \tilde{\mathbf{D}}^{\mathrm{p}} = \mathbf{D} - \tilde{\mathbf{D}}^{\mathrm{e}}. \tag{8.48}$$

In *certain simple cases* (simple traction, small elastic strains) it can be shown that the *same* plastic strain-rate tensor obtains whatever the model (i.e., the decomposition) considered. In these cases the interpretation of energy balance does not, fortunately, depend on the model selected.

8.4 Evolution equation (normality rule)

After Mandel, the hypothesis concerning hardening variables *implies* the existence of a privileged sequence of released configurations, or at least of one privileged frame for these, called the *director frame* because it specifies the orientation of the matter element. In a *crystal*, an obvious *director frame* is the frame attached to the crystalline lattice in its released state. The elastic deformation then is nothing but the lattice deformation. In a *polycrystal* one would take as director frame any one of the frames attached to the constituent crystals (Mandel, 1971).

Typically, a flow-surface equation relative to the configuration $K_{\mathscr{R}}$ will be written down in the form

$$f_{\mathscr{R}}(\mathbb{S}_{\mathscr{R}}, \theta, \boldsymbol{\alpha}) = 0, \tag{8.49}$$

where $\boldsymbol{\alpha}$ represents the hardening variables. The evolution equation must provide $(\mathbf{L}^{\mathrm{p}})_{\mathrm{s}} = \cdots$, for instance,

$$[(\mathscr{D}_{\omega}\mathbf{F}^{\mathrm{p}})(\mathbf{F}^{\mathrm{p}})^{-1}]_{\mathrm{s}} = \dot{\lambda}\mathbb{B}(\mathbb{S}_{\mathscr{R}}, \theta, \mathbf{A}\cdot\boldsymbol{\beta}), \tag{8.50}$$

where

$$\left.\begin{aligned} &\mathscr{D}_{\omega}\mathbf{F}^{\mathrm{p}} \equiv \dot{\mathbf{F}}^{\mathrm{p}} - \boldsymbol{\omega}_{\mathrm{D}}\cdot\mathbf{F}^{\mathrm{p}}, \\ &\boldsymbol{\omega}_{\mathrm{D}} \equiv \dot{\boldsymbol{\beta}}\boldsymbol{\beta}^{\mathrm{T}}, \qquad \mathbb{B} = \frac{\partial f}{\partial \mathbb{S}_{\mathscr{R}}}, \end{aligned}\right\} \tag{8.51}$$

where $\boldsymbol{\omega}_{\mathrm{D}}$ is the angular velocity of the director frame and $\mathscr{D}_{\omega}$ is the time derivative following that frame in rotation. Simultaneously one will have the following normality law for the internal variables:

$$\dot{\boldsymbol{\alpha}} = \dot{\lambda}\frac{\partial f}{\partial \mathbf{A}}, \quad \begin{Bmatrix} \dot{\lambda} \geqslant 0 & \text{if } f = 0 \text{ and } \dot{f} = 0, \\ \dot{\lambda} = 0 & \text{if } f < 0 \text{ or } \quad \dot{f} < 0. \end{Bmatrix} \tag{8.52}$$

We shall not dwell any longer on this finite-strain generalization. The elastoplasticity theory in finite transformations (or strains) must be considered in the case of the propagation of *shock waves* throughout the material (for instance, in a crystal). The interested reader will find highly technical, but very fruitful, developments in Lee (1969), Rice (1970, 1971), Clifton (1972), Naghdi (1990), Mandel (1971, 1973a, b, 1978, 1980), Loret (1983) and Stolz (1982, 1987). For the *monocrystal*, relevant references are Mandel (1978, 1980), Zarka (1972), Havner (1987) and Asaro (1983), while for the *polycrystal* the crucial work is Zaoui (1987). Mandel (1971) must be recommended as a general reference full of excellent remarks. Stolz (1987) deals extensively with the various models and problems of stability. Van der Giessen (1989a, b) presents an overview of great interest. He emphasizes the concept of plastically induced orientational structures in finite-strain elastoplasticity. Lee, who pioneered the subject matter by proposing the decomposition (8.29), stresses the importance of strain-induced plastic anisotropy (Lee and Agah-Tehrani, 1987), which has a major influence on computed stress distributions. This anisotropy question becomes more and more relevant with computer programs being applied in connection with important technological projects using finite-strain schemes. Finally, Naghdi (1990), in a brilliant but very critical review, emphasizes the role of a yield function in *strain space* (space of E_{KL}'s).

To conclude, we should emphasize that the notion of *intermediate configurations* is a fruitful one in the description of many rheological behaviours, including *viscoelasticity* and viscoplasticity (for these, see Sidoroff, 1976; Teodosiu and Sidoroff, 1976; Maugin and Drouot, 1988).

The following problems are supposed to lead the reader through the formulation of frame-independent flow rules in finite-strain elastoplasticity.

Problems for Chapter 8

8.1. Kinematic identities

Prove the two identities (8.13) [*Hint*: account for the differentiation of a determinant and introduce the adjoint of an element).] As a direct consequence, prove eqn (8.15).

8.2. Objectivity

A tensorial object of order n is said to be *objective* if and only if it transforms tensorially under *time-dependent* rotation of frames, i.e., $\mathbf{x}^* = \mathbf{Q}(t)\cdot\mathbf{x}$. These are space transformations in the current configuration K. They are *not* Galilean transformations. By direct computation show that the strain-rate tensor $\mathbf{D}$ is objective while the rotation-rate tensor $\mathbf{\Omega}$ and the acceleration $\dot{\mathbf{v}}$ are *not* objective.

8.3. Jaumann time derivative

In continuum mechanics one usually requires the frame independence or *objectivity* of constitutive equations while balance laws are only Galilean-invariant (see Maugin, 1988, Chapter 2). If constitutive equations include time rates, then the question of building objective time derivatives is raised. The *Jaumann* co-rotational derivative, denoted by $\mathrm{D_J}$, is such a derivative. For a second-order symmetric tensor $\boldsymbol{\sigma}$, it is defined by

$$\mathrm{D_J}\boldsymbol{\sigma} = \dot{\boldsymbol{\sigma}} - \mathbf{\Omega}\boldsymbol{\sigma} + \boldsymbol{\sigma}\mathbf{\Omega}, \qquad \Omega_{ij} = \tfrac{1}{2}(v_{i,j} - v_{j,i}). \tag{1}$$

(a) By direct calculation show that $\mathrm{D_J}\boldsymbol{\sigma}$ is an objective tensor if $\boldsymbol{\sigma}$ is an objective tensor.

(b) Can you give an abstract definition of $\mathrm{D_J}$ in terms of transport of tensorial objects by *proper* rotation? [See Maugin (1988), Chapter 2.]

8.4. Convected time derivative

(a) Consider a second-order tensor σ_{ij} at $(\mathbf{x}, t)$. Convect it back to the reference configuration K_R by performing the operation

$$T_{KL} = \sigma_{ij} X_{K,i} X_{L,j}, \tag{1}$$

Evaluate $\mathrm{d}(X_{K,i})/\mathrm{d}t$ in terms of the velocity gradient and prove that

$$(\mathrm{D_C}\boldsymbol{\sigma})_{ij} = \dot{\sigma}_{ij} - v_{i,k}\sigma_{kj} - v_{j,k}\sigma_{ik} \tag{2}$$

is such that

$$\dot{T}_{KL} = (\mathrm{D_C}\boldsymbol{\sigma})_{ij} X_{K,i} X_{L,j}. \tag{3}$$

Deduce that $(\mathrm{D_C}\boldsymbol{\sigma})$ is an objective time derivative (the so-called convected time derivative) of $\boldsymbol{\sigma}$ if $\boldsymbol{\sigma}$ itself is objective.

(b) Establish the relation between $D_C\boldsymbol{\sigma}$ and $D_J\boldsymbol{\sigma}$.

(c) Is the new time derivative

$$\overset{*}{\boldsymbol{\sigma}} \equiv D_C\boldsymbol{\sigma} + \boldsymbol{\sigma}(\nabla \cdot \mathbf{v}) \tag{4}$$

also objective? How can you introduce this through a convection operation?

8.5. Angular derivative (e.g., in a director frame)

Let $\mathbf{e}^{(\alpha)}$, $\alpha = 1, 2, 3$, define at $(\mathbf{x}, t)$ a triad of unit orthogonal vectors or so-called reference frame. What are the two conditions satisfied by the $\mathbf{e}^{(\alpha)}$'s? Let

$$\sigma^{\alpha\beta} = e_i^{(\alpha)} \sigma_{ij} e_j^{(\beta)} \tag{1}$$

be the components of a second-order tensor $\boldsymbol{\sigma}$ on the frame $\{\mathbf{e}^{(\alpha)}\}$ at time t, and $\boldsymbol{\omega}$ the absolute angular velocity of the frame, i.e., $\dot{\mathbf{e}}^{(\alpha)} = \boldsymbol{\omega} \cdot \mathbf{e}^{(\alpha)}$. If $\boldsymbol{\sigma}$ is objective, then show that the derivative

$$(\mathscr{D}_\omega \boldsymbol{\sigma})_{ij} = \dot{\sigma}_{ij} - \omega_{ik}\sigma_{kj} + \sigma_{ik}\omega_{kj} \tag{2}$$

is also objective and

$$\dot{\sigma}^{\alpha\beta} = (\mathscr{D}_\omega \boldsymbol{\sigma})_{ij} e_i^{(\alpha)} e_{ij}^{(\beta)}, \tag{3}$$

The definition (2) reduces to that of the Jaumann derivative (i.e., $\mathscr{D}_\omega = D_J$) when $\omega = \Omega$. For a vector field (compare eqn (8.51)) we have

$$\mathscr{D}_\omega \mathbf{v} = \dot{\mathbf{v}} - \boldsymbol{\omega} \cdot \mathbf{v}. \tag{4}$$

8.6. Nonlinear tensor form of the Prandtl–Reuss equations

Start with

$$\mathbf{D} = \mathbf{D}(\boldsymbol{\sigma}, D_J\boldsymbol{\sigma}) \tag{1}$$

where $\mathbf{D}$ is the strain rate and D_J denotes the Jaumann derivative (see 8.3). Then show that

$$D_J\boldsymbol{\sigma} = 2\mu\left[\hat{\mathbf{D}} - \left(\frac{\mathrm{tr}(\mathbf{s}\hat{\mathbf{D}})}{2k^2}\right)\mathbf{s}\right] \tag{2}$$

is a possible *objective* generalization of the Prandtl–Reuss equations.

8.7. Finite-strain elastoplasticity (Green–Naghdi formulation)

Starting from the multiplicative decomposition (8.29) prove the results (8.31) and show that the intrinsic dissipation per unit volume in the *intermediate* configuration takes on the form (8.40).

8.8. Another multiplicative decomposition (Clifton)

R.J. Clifton (1972) has proposed a multiplicative decomposition of the total motion gradient as $\mathbf{F} = \mathbf{F}^p\mathbf{F}^e$. Study the implications of this decomposition as regards the writing of the intrinsic dissipation for an elastic – plastic body in finite strains.

8.9. Finite-strain generalization of elastic–plastic constitutive equation

Let T_{ij} denote the *Kirchhoff stress tensor* in a description where the current configuration K_t is taken as the reference configuration $K_{\mathbf{R}}$. Then a possible elastic–plastic constitutive equation is

$$\mathrm{D_J} T_{ij} = E^{\mathrm{ep}}_{ijkl} D_{kl}, \qquad T_{ij} = J\sigma_{pq} X_{K,p} X_{L,q} \delta_{iK} \delta_{jL}, \tag{1}$$

where $\mathrm{D_J}$ denotes the Jaumann derivative (see 8.3) and D_{kl} is the strain-rate tensor. Consider the convected time derivative $\mathrm{D_C} T_{ij}$ (see 8.4).

(a) Show that the elastic–plastic constitutive equation reads

$$(\mathrm{D_C} T)_{ij} = \tilde{E}^{\mathrm{ep}}_{ijkl} D_{kl}. \tag{2}$$

What is the expression for $\tilde{E}^{\mathrm{ep}}_{ijkl}$? Is this an objective relation?

(b) Eqn (2) can also be re-written as

$$\overset{*}{\sigma}_{ij} = \tilde{E}^{\mathrm{ep}}_{ijkl} D_{kl}, \tag{3}$$

where σ is the *Cauchy stress* and the superimposed asterisk denotes the convected derivative introduced in 8.4 to account for volume changes [*Hint*: make the computation in $K_{\mathbf{R}} \neq K_t$ and then make $K_{\mathbf{R}}$ coincide with K_t.]

(c) Show that in terms of *finite increments* the above finite-strain generalization can be written as

$$\Delta\sigma_{ij} = \frac{\partial W}{\partial(\Delta\varepsilon_{ij})}, \qquad W(\Delta\boldsymbol{\varepsilon}) = \frac{1}{2}\tilde{E}^{\mathrm{ep}}_{ijkl} \Delta\varepsilon_{ij} \Delta\varepsilon_{kl} \tag{4}$$

Remark. In the above-presented generalization the choice of a good stress-rate measure leads to introducing a strain-rate potential with a pleasant symmetry (of the 6×6 operator E^{ep}_{ijkl}).

8.10. Flow rule in finite-strain elastoplasticity (term paper)

Make a literature search and write a short *critical* report on the various proposals of objective flow rules in finite-strain elastoplasticity with special attention to the case of crystals.

9

Homogenization of elastoplastic composites

The object of the chapter This chapter provides a short introduction to the notion of *homogenization* (i.e., determining the parameters of a *unique* fictitious material that 'best' represents the real heterogeneous material or *composite*) and then, at some length, its application to the case where all or some of the constitutive components have an elastoplastic behaviour. The essential notions are those of *representative* volume element, procedure of *localization*, and the representation of some microscopic effects by means of internal variables. Composites with unidirectional fibres, polycrystals and cracked media provide examples of application.

9.1 Notion of homogenization

Homogenization is the modelling of a heterogeneous medium by means of a *unique* continuous medium. A heterogeneous medium is a medium of which material properties (e.g., elasticity coefficients) vary pointwise in a continuous or discontinuous manner, in a periodic or nonperiodic way, deterministically or randomly. While, obviously, homogenization is a modelling technique that applies to all fields of macroscopic physics governed by nice partial differential equations, we focus more particularly on the mechanics of deformable bodies with a special emphasis on *composite materials* (as used in aeronautics) and *polycrystals* (representing many alloys.) Most of the composite materials developed during the past three decades present a *brittle*, rather than *ductile* behaviour. As emphasized in Chapter 7, the elastic behaviour then prevails and there is no need to consider the homogenization of an elasto*plastic* behaviour. However, such an approach is needed for composites having a metallic matrix and for metallic polycrystals where the plastic behaviour necessarily comes into play. This leads to some difficulty since homogenization assumes that we are able to solve a problem at the macroscopic scale while we know by experience that the solution of a problem of elastoplasticity at this scale in exact form is most often a formidable task owing to nonlinearities. Only very simple cases are amenable to a closed-form solution.

9.2 Notion of representative volume element and localization

9.2.1 Representative volume element (RVE)

Two different scales are used in the description of heterogeneous media. One of these is a *macroscopic* (x) scale at which homogeneities are weak. The other one is the scale of inhomogeneities and is referred to as the *microscopic* (y) scale. The latter defines the size of the representative volume element (for short, RVE; see Fig. 9.1). The basic cell of a periodic composite is an example of RVE.

From the experimental point of view, we can say that there exists a kind of *statistical homogeneity* in the sense that any RVE at a specific point looks very much like any other RVE taken at random at another point.

The *mechanical* problem presents itself in the following manner. Let $\boldsymbol{\sigma}(y)$ and $\boldsymbol{\varepsilon}(y)$ be the stress and strain at the *micro* scale in the framework of SPH. We denote by $\boldsymbol{\Sigma}$ and $\mathscr{E}$ the same notions at the *macro* scale. Let $\langle \cdots \rangle$ indicate the *averaging operator*. For a volume averaging we have

$$\left.\begin{aligned} \boldsymbol{\Sigma}(\mathbf{x}) &= \langle \boldsymbol{\sigma} \rangle \\ &\equiv \frac{1}{|V|}\int_V \boldsymbol{\sigma}(x, y)\,\mathrm{d}y, \\ \mathscr{E}(\mathbf{x}) &= \langle \boldsymbol{\varepsilon} \rangle \\ &\equiv \frac{1}{|V|}\int_V \boldsymbol{\varepsilon}(x, y)\,\mathrm{d}y, \end{aligned}\right\} \tag{9.1}$$

where V is the volume of the RVE.

It is important to notice that any quantity that is an *additive* (i.e.,

Fig. 9.1. Representative volume element RVE

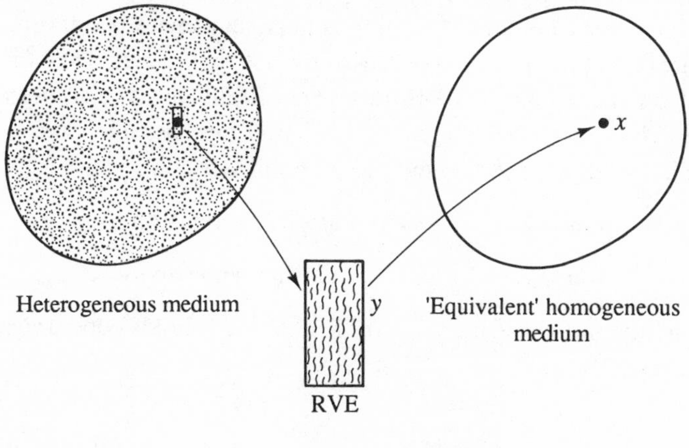

extensive, in the thermodynamical language of Appendix 1) function is *averaged* in the micro–macro transition. Thus, if $\bar{\rho} = \langle \rho \rangle$ denotes the averaged density, then we have

$$\left.\begin{aligned} \bar{\rho}E &= \langle \rho e \rangle, && \text{internal energy,} \\ \bar{\rho}S &= \langle \rho \eta \rangle, && \text{entropy,} \\ \Phi &= \langle \phi \rangle, && \text{dissipation,} \end{aligned}\right\} \tag{9.2}$$

9.2.2 Localization process

We can state the following

(a) the process that relates $(\boldsymbol{\Sigma}, \boldsymbol{\mathscr{E}})$ by means of eqns (9.1)–(9.2) and the microscopic constitutive equations is called *homogenization*;

(b) the 'inverse' process that consists in determining $\boldsymbol{\sigma}(y)$ and $\boldsymbol{\varepsilon}(y)$ from $\boldsymbol{\Sigma}$ and $\boldsymbol{\mathscr{E}}$ is called *localization.*

Therefore, the *data* are $\boldsymbol{\Sigma}$ and $\boldsymbol{\mathscr{E}}$ in the localization process which corresponds to the following problem:

$$(\mathscr{P}\mathscr{L}) \left\{\begin{array}{ll} \langle \boldsymbol{\sigma} \rangle = \boldsymbol{\Sigma}, & \langle \boldsymbol{\varepsilon} \rangle = \boldsymbol{\mathscr{E}}, \\ \operatorname{div} \boldsymbol{\sigma} = \mathbf{0}, & \text{'micro' equilibrium,} \\ \text{the 'micro' behaviour is known.} & \end{array}\right\} \tag{9.3}$$

This problem is original, because of the following two reasons:

(i) the load is the *averaged value* of a field and not a prescription at points in the bulk or at a limiting surface;

(ii) *there are no boundary conditions!*

It follows from (ii) that the problem (9.3) is *ill-posed.* How can we remedy this? The missing boundary conditions must, in some way, reproduce the internal state of the RVE in the most satisfactory manner. They therefore depend on the choice of RVE, more specifically on its size. As a rule, different choices of RVE will provide different macroscopic laws.

The following give some examples of boundary conditions:

$$\boldsymbol{\sigma} \cdot \mathbf{n} = \boldsymbol{\Sigma} \cdot \mathbf{n} \quad \text{on } \partial V \text{ – uniform tractions on } \partial V; \tag{9.4}$$

$$\mathbf{u} = \boldsymbol{\mathscr{E}} \cdot \mathbf{y} \quad \text{on } \partial V \text{ – uniform strains on } \partial V. \tag{9.5}$$

With this and $\operatorname{div} \boldsymbol{\sigma} = \mathbf{0}$, in V, it is verified that (9.1) holds good. Indeed, for (9.5) we have

$$\frac{1}{2}\int_V \left(\frac{\partial u_i}{\partial y_j} + \frac{\partial u_j}{\partial y_i}\right) \mathrm{d}v = \frac{1}{2}\int_{\partial V} (u_i n_j + u_j n_i)\,\mathrm{d}s$$
$$= \frac{1}{2}\int_{\partial V} (\mathscr{E}_{ik} y_k n_j + \mathscr{E}_{jk} y_k n_i)\,\mathrm{d}s \tag{9.6}$$

or

$$\langle \boldsymbol{\varepsilon}(\mathbf{u}) \rangle = \mathscr{E}. \tag{9.7}$$

The proof for (9.4) is self-evident.

The above reasoning does *not* apply to the case of a *periodic structure.* In that case, $\boldsymbol{\sigma}$ and $\boldsymbol{\varepsilon}$ are *locally periodic* (they are only quasi-periodic for a large sample) and the *periodicity conditions* read as follows:

the *tractions* $\boldsymbol{\sigma}\cdot\mathbf{n}$ are opposite on opposite faces of ∂V (where $\mathbf{n}$ corresponds to $-\mathbf{n}$);

the *local* strain $\boldsymbol{\varepsilon}(\mathbf{u})$ is made of two parts, the *mean* $\mathscr{E}$ and a *fluctucation* part $\boldsymbol{\varepsilon}(\mathbf{u}^*)$ such that

$$\boldsymbol{\varepsilon}(\mathbf{u}) = \mathscr{E} + \boldsymbol{\varepsilon}(\mathbf{u}^*), \qquad \langle \boldsymbol{\varepsilon}(\mathbf{u}^*) \rangle = 0, \tag{9.8}$$

where $\mathbf{u}^*$ can be shown to be periodic. Therefore, the conditions are

$$\left.\begin{array}{ll} \boldsymbol{\sigma}\cdot\mathbf{n} & \text{is antiperiodic,} \\ \mathbf{u} = \mathscr{E}\cdot\mathbf{y} + \mathbf{u}^*, \mathbf{u}^* & \text{periodic.} \end{array}\right\} \tag{9.9}$$

On account of (9.4), (9.5) *or* (9.9), the problem (9.3) now is theoretically *well-posed*, but this must be verified for *each* constitutive behaviour.

9.2.3 The Hill–Mandel principle of macrohomogeneity

Let $\bar{\boldsymbol{\sigma}}$ and $\bar{\mathbf{u}}$ be, respectively, a *statically admissible* (SA) stress field and a *kinematically admissible* (KA) displacement field. Then we have the remarkable equality

$$\langle \bar{\boldsymbol{\sigma}} : \boldsymbol{\varepsilon}(\bar{\mathbf{u}}) \rangle = \bar{\boldsymbol{\Sigma}} : \bar{\mathscr{E}}. \tag{9.10}$$

Proof. This must be proved for the three types of boundary conditions, (9.4), (9.5) and (9.9). To that purpose we introduce $\bar{\boldsymbol{\sigma}}^*$ and $\bar{\mathbf{u}}^*$ such that we can write

$$\left.\begin{array}{c} \bar{\boldsymbol{\sigma}} = \bar{\boldsymbol{\Sigma}} + \bar{\boldsymbol{\sigma}}^* \quad \text{s.t.} \quad \langle \bar{\boldsymbol{\sigma}}^* \rangle = \mathbf{0}, \quad \operatorname{div} \bar{\boldsymbol{\sigma}}^* = \mathbf{0}, \\ \boldsymbol{\varepsilon}(\bar{\mathbf{u}}) = \bar{\mathscr{E}} + \boldsymbol{\varepsilon}(\bar{\mathbf{u}}^*) \quad \text{s.t.} \quad \langle \boldsymbol{\varepsilon}(\bar{\mathbf{u}}^*) \rangle = 0. \end{array}\right\} \tag{9.11}$$

By virtue of eqns (9.11) we have then by direct computation

$$\begin{aligned}\langle\bar{\sigma}:\varepsilon(\bar{\mathbf{u}})\rangle &= \langle(\bar{\Sigma}+\bar{\sigma}^*):\varepsilon(\bar{\mathbf{u}})\rangle \\ &= \bar{\Sigma}:\bar{\mathscr{E}} + \langle\bar{\sigma}^*\rangle:\bar{\mathscr{E}} + \bar{\Sigma}:\langle\varepsilon(\bar{\mathbf{u}}^*)\rangle + \langle\bar{\sigma}:\varepsilon(\bar{\mathbf{u}}^*)\rangle \\ &= \bar{\Sigma}:\bar{\mathscr{E}} + \langle\bar{\sigma}^*:\varepsilon(\bar{\mathbf{u}}^*)\rangle.\end{aligned}$$

This also reads

$$\begin{aligned}\langle\bar{\sigma}:\varepsilon(\bar{\mathbf{u}})\rangle &= \bar{\Sigma}:\bar{\mathscr{E}} + \langle\bar{\sigma}^*:\varepsilon(\bar{\mathbf{u}})\rangle \\ &= \bar{\Sigma}:\bar{\mathscr{E}} + \langle\bar{\sigma}:\varepsilon(\bar{\mathbf{u}}^*)\rangle.\end{aligned}$$

But from eqns (9.4) and (9.5)

$$\langle\bar{\sigma}^*:\varepsilon(\bar{\mathbf{u}})\rangle = \frac{1}{|V|}\int_{\partial V}(\bar{\sigma}^*\cdot\mathbf{n})\cdot\bar{\mathbf{u}}\,\mathrm{d}s = 0,$$

while from (9.9)

$$\langle\bar{\sigma}:\varepsilon(\bar{\mathbf{u}}^*)\rangle = \frac{1}{|V|}\int_{\partial V}(\bar{\sigma}\cdot\mathbf{n})\cdot\bar{\mathbf{u}}^*\,\mathrm{d}s = 0,$$

so that (9.10) follows in all three cases of boundary conditions. The remarkable expression (9.10) is called the *principle of macrohomogeneity* of *Hill and Mandel* (after R. Hill, 1965a, and J. Mandel, 1971) or the *Hill–Mandel* relation between micro and macro scales. In statistical theories (e.g., E. Kröner, 1972) this condition is viewed as an *ergodic hypothesis*. This condition, in fact, plays in the end a much more important role than the boundary conditions applied at the RVE.

9.2.4 Functional notation

This notation is useful in variational formulations. We rewrite the conditions (9.4)–(9.5) in the compact form

$$\left.\begin{aligned}\mathbf{u} = \mathscr{E}\cdot\mathbf{y} + \mathbf{u}^*, \qquad \mathbf{u}^* \in \mathscr{U}_0,\\ \sigma \in S_0 = \mathscr{E}(\mathscr{U}_0)^{\perp},\end{aligned}\right\} \tag{9.12}$$

where the set $\mathscr{U}_0$ must be specified. For instance, for the boundary condition (9.4) we will take

$$\mathscr{U}_0 = \{\mathbf{u}^* \in H^1(V)^3 | \langle\varepsilon(\mathbf{u}^*)\rangle = \mathbf{0}\}, \tag{9.13}$$

while for the boundary conditions (9.9) we will choose

$$\mathscr{U}_0 = \mathscr{U}_{\text{per}} = \{\mathbf{u}^* \in H^1(V)^3 | \mathbf{u}^* \text{ periodic on } \partial V\}. \tag{9.14}$$

Of some use also is the set of *self-equilibrated* (SE) stress fields defined by

$$\mathrm{SE} = \{\boldsymbol{\sigma}^* \in \mathscr{E}(\mathscr{U}_0)^{\perp} | \langle \boldsymbol{\sigma} \rangle = 0\}, \tag{9.15}$$

As a consequence of (9.10) we note that

$$\left.\begin{aligned} &\boldsymbol{\sigma}\ \text{admissible} \Leftrightarrow \langle \boldsymbol{\sigma} : \boldsymbol{\varepsilon}(\mathbf{u}^*) \rangle = \mathbf{0}, \qquad \forall \mathbf{u}^* \in \mathscr{U}_0, \\ &\text{thus}\ \boldsymbol{\sigma} \in \mathscr{E}(\mathscr{U}_0)^{\perp}. \end{aligned}\right\} \tag{9.16}$$

9.3 The example of pure elasticity

Although this is of great interest in itself, the case of purely elastic components here is examined as a prerequisite for the case of elastoplastic components treated in Section 9.4. We look at *anisotropic linear-elastic* components.

9.3.1 The localization problem

This problem is written in the following form (here $\mathbf{E}(y)$ is the tensor of elasticity coefficients at the 'micro' scale):

$$\left.\begin{aligned} &\boldsymbol{\sigma}(y) = \mathbf{E}(y) : \boldsymbol{\varepsilon}(y) = \mathbf{E}(y) : [\mathscr{E} + \boldsymbol{\varepsilon}(\mathbf{u}^*(y))], \\ &\operatorname{div} \boldsymbol{\sigma} = \mathbf{0} \\ &+ \text{boundary conditions}, \end{aligned}\right\} \tag{9.17}$$

where $\mathscr{E}$ or $\boldsymbol{\Sigma}$ is prescribed. Accordingly, the *fluctuation* displacement $\mathbf{u}^*$ is the solution of the following problem:

$$\left.\begin{aligned} &\operatorname{div}(\mathbf{E} : \boldsymbol{\varepsilon}(\mathbf{u}^*)) = -\operatorname{div}(\mathbf{E} : \mathscr{E}) \\ &+ \text{boundary conditions}. \end{aligned}\right\} \tag{9.18}$$

Whenever $\mathbf{E}$ is constant for each constituent component, it can be shown that

$$\operatorname{div}(\mathbf{E} : \mathscr{E}) = ([\![\mathbf{E}]\!] : \mathscr{E})\mathbf{n}\delta(S), \tag{9.19}$$

where $[\![\mathbf{E}]\!] = \mathbf{E}^+ - \mathbf{E}^-$, $\delta(S)$ is Dirac's distribution, and $\mathbf{n}$ is the unit normal oriented from the '$-$' to the '$+$' side of the surface S separating components. Then we can state the following.

Proposition *Under classical working hypotheses applying to* $\mathbf{E}$ (*see eqns (3.30)–(3.31), the problem* (9.18) *admits a* unique *solution for all three types of boundary condition.*

To prove this we must distinguish whether it is $\mathscr{E}$ or $\boldsymbol{\Sigma}$ which is prescribed.

9.3.2 *Case where $\mathscr{E}$ is prescribed*

For the existence and uniqueness proofs one can see Suquet (1981b). We shall only give the *representation of the solution.* As the problem is linear, the solution $\boldsymbol{\varepsilon}(\mathbf{u}^*)$ depends linearly on the prescribed field $\mathscr{E}$. The latter can be decomposed into *six* elementary states of macroscopic strains (stretch in three directions and three shears). Let $\boldsymbol{\varepsilon}(\chi_{kl})$ be the fluctuation strain field induced by these six elementary states at the microscopic level. The solution $\boldsymbol{\varepsilon}(\mathbf{u}^*)$ for a general macrostrain $\mathscr{E}$ is the superposition of the six elementary solutions, so that we can write (summation over k and l)

$$\boldsymbol{\varepsilon}(\mathbf{u}^*) = \mathscr{E}_{kl}\boldsymbol{\varepsilon}(\chi_{kl}). \tag{9.20}$$

In all we have

$$\boldsymbol{\varepsilon}(\mathbf{u}) = \mathscr{E} + \boldsymbol{\varepsilon}(\mathbf{u}^*) = \mathscr{E}(\mathbf{I} + \boldsymbol{\varepsilon}(\chi))$$

or, in components,

$$\varepsilon_{ij}(\mathbf{u}) = \mathscr{D}_{ijkl}\mathscr{E}_{kl} = (\mathscr{D} : \mathscr{E})_{ij}, \tag{9.21}$$

where

$$\mathscr{D}_{ijkl} = I_{ijkl} + \varepsilon_{ij}(\chi_{kl}). \tag{9.22}$$

Here $I_{klij} = \frac{1}{2}(\delta_{ik}\delta_{jl} + \delta_{il}\delta_{jk})$ is the tensorial representation in $\mathbb{R}^3$ of the unity of $\mathbb{R}^6$ – see eqn (3.33) – and $\mathscr{D}_{ijkl}$ is called, depending on the author, the *tensor of strain localization*, or the *tensor of concentrations* (Mandel, 1971) or the *tensor of influence* (Hill, 1967).

Homogenization We can write in an obvious manner

$$\begin{aligned}\boldsymbol{\Sigma} = \langle \boldsymbol{\sigma} \rangle &= \langle \mathbf{E} : \boldsymbol{\varepsilon}(\mathbf{u}) \rangle \\ &= \langle \mathbf{E} : \mathscr{D} : \mathscr{E} \rangle = \langle \mathbf{E} : \mathscr{D} \rangle : \mathscr{E},\end{aligned}$$

so that

$$\boldsymbol{\Sigma} = \mathbf{E}^{\text{hom}} : \mathscr{E}, \qquad \mathbf{E}^{\text{hom}} = \langle \mathbf{E} : \mathscr{D} \rangle. \tag{9.23}$$

We note that (T = transpose)

$$\langle \mathscr{D} \rangle = \mathbf{I}, \qquad \langle \mathscr{D}^{\mathrm{T}} \rangle = \mathbf{I}$$

Equation $(9.23)_2$ shows that the tensor of 'macro' elasticity coefficients is obtained by taking the average of 'micro' elasticity coefficients, the latter being *weighted* by the tensor of strain localization. The tensor $\mathbf{E}^{\text{hom}}$ is symmetric. This can be proved in two ways. For a *direct* proof we compute

$\langle \mathscr{D}^{\mathrm{T}} : \bar{\boldsymbol{\sigma}} \rangle$ for an *admissible* field $\bar{\boldsymbol{\sigma}}$, obtaining thus

$$\langle \mathscr{D}^{\mathrm{T}} : \bar{\boldsymbol{\sigma}} \rangle_{ij} = \langle \mathscr{D}^{\mathrm{T}}_{ijkl} \sigma_{kl} \rangle$$
$$= \langle [I_{ijkl} + \varepsilon_{kl}(\chi_{ij})] \bar{\sigma}_{kl} \rangle = \bar{\Sigma}_{ij},$$

i.e.,

$$\boldsymbol{\Sigma} = \langle \mathscr{D}^{\mathrm{T}} : \boldsymbol{\sigma} \rangle = \langle \mathscr{D}^{\mathrm{T}} : \mathbf{E} : \boldsymbol{\varepsilon}(\mathbf{u}) \rangle = \langle \mathscr{D}^{\mathrm{T}} : \mathbf{E} : \mathscr{D} \rangle : \mathscr{E},$$

so that

$$\mathbf{E}^{\mathrm{hom}} = \langle \mathscr{D}^{\mathrm{T}} : \mathbf{E} : \mathscr{D} \rangle, \tag{9.24}$$

which is symmetric, *QED*.

For a proof based on energy arguments we use the fact that $\bar{\rho} E = \langle \rho e \rangle$, so that

$$\begin{aligned} \bar{\rho} E &= \tfrac{1}{2} \mathscr{E} : \mathbf{E}^{\mathrm{hom}} : \mathscr{E} = \langle \rho e \rangle \\ &= \tfrac{1}{2} \langle \boldsymbol{\varepsilon}(\mathbf{u}) : \mathbf{E} : \boldsymbol{\varepsilon}(\mathbf{u}) \rangle \\ &= \tfrac{1}{2} \mathscr{E} : \langle \mathscr{D}^{\mathrm{T}} : \mathbf{E} : \mathscr{D} \rangle : \mathscr{E}, \qquad QED. \end{aligned} \tag{9.25}$$

9.3.3 Case where $\boldsymbol{\Sigma}$ *is prescribed*

The localization problem then reads

$$\left. \begin{aligned} &\boldsymbol{\varepsilon}(\mathbf{u}) = \boldsymbol{\varepsilon}(\mathbf{u}^*) + \mathscr{E} = \mathbf{S} : \boldsymbol{\sigma}, \\ &\operatorname{div} \boldsymbol{\sigma} = \mathbf{0}, \\ &\langle \boldsymbol{\sigma} \rangle = \boldsymbol{\Sigma}, \\ &+ \text{boundary conditions}, \end{aligned} \right\} \tag{9.26}$$

where $\mathbf{S}$ is the tensor of the 'micro' elastic compliances and $\mathscr{E}$ is an unknown. Here again, the proof of existence and uniqueness can be found in Suquet (1981b) and we content ourselves with a *representation* of the solution. Thus we assume that a unique solution $\boldsymbol{\sigma}$ exists. This solution depends linearly on data by virtue of the linearity of the problem. Let us call C_{kl} the solution of the problem (9.26) for the datum $\boldsymbol{\Sigma} = \mathbf{I}_{kl}$ – note that $I_{ijkl} = (\mathbf{I}_{kl})_{ij}$.

Then the general solution, obtained by superposition, is written

$$\left. \begin{aligned} &\boldsymbol{\sigma} = \mathscr{C} : \boldsymbol{\Sigma}, \quad \text{i.e.,} \quad \boldsymbol{\sigma}(y) = \Sigma_{kl} \mathscr{C}_{kl}(y), \\ &\text{or} \quad \sigma_{ij} = \mathscr{C}_{ijkl} \Sigma_{kl}, \quad \mathscr{C}_{ijkl} = (\mathscr{C}_{kl})_{ij}, \end{aligned} \right\} \tag{9.27}$$

where $\mathscr{C}$ is the *tensor of stress localization.*

The homogenized compliance tensor $\mathbf{S}^{\mathrm{hom}}$ is evaluated thus. We have directly

$$\mathscr{E} = \langle \boldsymbol{\varepsilon}(\mathbf{u}) \rangle = \langle \mathbf{S} : \boldsymbol{\sigma} \rangle = \langle \mathbf{S} : \mathscr{C} \rangle : \boldsymbol{\Sigma} = \mathbf{S}^{\text{hom}} : \boldsymbol{\Sigma}, \tag{9.28}$$

whence

$$\mathbf{S}^{\text{hom}} = \langle \mathbf{S} : \mathscr{C} \rangle. \tag{9.29}$$

We note that

$$\langle \mathscr{C}^{\mathrm{T}} \rangle = \mathbf{I}, \tag{9.30}$$

and for any admissible field $\boldsymbol{\varepsilon}(\overline{\mathbf{u}})$ we can write

$$\begin{aligned} \langle \mathscr{C}^{\mathrm{T}} : \boldsymbol{\varepsilon}(\overline{\mathbf{u}}) \rangle_{ij} &= \langle \mathscr{C}^{\mathrm{T}}_{ijkl} \varepsilon_{kl}(\overline{\mathbf{u}}) \rangle = \langle (\mathscr{C}_{ij})_{kl} \varepsilon_{kl}(\overline{\mathbf{u}}) \rangle \\ &= \langle (\mathscr{C}_{ij})_{kl} \rangle \langle \varepsilon_{kl}(\overline{\mathbf{u}}) \rangle = \overline{\mathscr{E}}_{ij}, \end{aligned}$$

so that

$$\mathscr{E} = \langle \mathscr{C}^{\mathrm{T}} : \boldsymbol{\varepsilon}(\mathbf{u}) \rangle = \langle \mathscr{C}^{\mathrm{T}} : \mathbf{S} : \boldsymbol{\sigma} \rangle = \langle \mathscr{C}^{\mathrm{T}} : \mathbf{S} : \mathscr{C} \rangle : \boldsymbol{\Sigma},$$

whence

$$\mathbf{S}^{\text{hom}} = \langle \mathscr{C}^{\mathrm{T}} : \mathbf{S} : \mathscr{C} \rangle \tag{9.31}$$

and thus $\mathbf{S}^{\text{hom}}$ is *symmetric.*

9.3.4 Equivalence between 'prescribed stress' and 'prescribed strain'

First we note that $\mathbf{E}^{\text{hom}}$ and $\mathbf{S}^{\text{hom}}$ are *inverse* tensors (in $\mathbb{R}^6$) of one another if they correspond to the *same* choice of boundary conditions in the localization problem. Indeed, using the symmetry of $\mathbf{E}^{\text{hom}}$ we can write

$$\mathbf{E}^{\text{hom}} : \mathbf{S}^{\text{hom}} = (\mathbf{E}^{\text{hom}})^{\mathrm{T}} : \mathbf{S}^{\text{hom}} = \langle \mathscr{D}^{\mathrm{T}} : \mathbf{E} \rangle : \langle \mathbf{S} : \mathscr{C} \rangle, \tag{9.32}$$

in which the first factor is an *admissible stress* field (from the definition of $\mathscr{D}$ and $\mathscr{C}$) and the second factor is an *admissible strain* field. The principle (9.10) therefore applies and we can write ($\mathbf{E} : \mathbf{S} = \mathbf{I}$)

$$\begin{aligned} \mathbf{E}^{\text{hom}} : \mathbf{S}^{\text{hom}} &= \langle \mathscr{D}^{\mathrm{T}} : \mathbf{E} : \mathbf{S} : \mathscr{C} \rangle = \langle \mathscr{D}^{\mathrm{T}} : \mathscr{C} \rangle \\ &= \langle \mathscr{D}^{\mathrm{T}} \rangle : \langle \mathscr{C} \rangle = \mathbf{I}, \quad \textit{QED.} \end{aligned} \tag{9.33}$$

However, if *different* boundary conditions are used, one then has the estimate of Hill (1967) and Mandel (1971),

$$\mathbf{E}^{\text{hom}} : \mathbf{S}^{\text{hom}} = \mathbf{I} + \mathrm{O}((d/l)^3), \tag{9.34}$$

where $\mathbf{E}^{\text{hom}}$ is evaluated by using the condition (9.5), while $\mathbf{S}^{\text{hom}}$ is computed through use of the condition (9.4), d is a characteristic size of an inhomogeneity and l is the typical size of the RVE. If $l \gg d$, then the choice of boundary conditions is hardly important. For *periodic* media where $d/l = O(1)$, *this choice is most important.*

9.4 Elastoplastic constituents[†]

We now consider the case of *elastoplastic* (perfectly plastic, i.e., without hardening) constituents. It is expected that the macroscopic constitutive behaviour will exhibit some kind of hardening since some energy can be stored at the microscopic level. We assume that the constituents have a mechanical behaviour described by the basic equations of Chapter 4:

$$\left.\begin{aligned} &\boldsymbol{\varepsilon}(\mathbf{u}) = (\nabla \mathbf{u})_S = \boldsymbol{\varepsilon}^e + \boldsymbol{\varepsilon}^p && \text{(decomposition of strain)},\\ &\boldsymbol{\varepsilon}^e = \mathbf{S} : \boldsymbol{\sigma} && \text{(law of state)},\\ &(\boldsymbol{\sigma}(y) - \boldsymbol{\sigma}^*) : \dot{\boldsymbol{\varepsilon}}^p(y) \geqslant 0 \quad \forall \boldsymbol{\sigma}^* \in C(y) && \text{(normality rule)}, \end{aligned}\right\} \tag{9.35}$$

where the normality rule is expressed in variational form and $C(y)$ is the convex of plasticity at the 'micro' level.

9.4.1 Macroscopic potentials

Starting from the second of (9.35) we write

$$\begin{aligned} \langle \mathscr{C}^T : \boldsymbol{\varepsilon} \rangle &= \langle \mathscr{C}^T : \mathbf{S} : \boldsymbol{\sigma} \rangle + \langle \mathscr{C}^T : \boldsymbol{\varepsilon}^p \rangle \\ &= \langle \boldsymbol{\sigma} : \mathbf{S} : \mathscr{C} \rangle + \langle \mathscr{C}^T : \boldsymbol{\varepsilon}^p \rangle, \end{aligned} \tag{9.36}$$

where $\boldsymbol{\sigma}$ and $\mathbf{S} : \mathscr{C}$ are *admissible* stress and strain fields, respectively.

Thus applying eqn (9.10) we can transform eqn (9.36) to

$$\begin{aligned} \mathscr{E} &= \langle \boldsymbol{\sigma} \rangle : \langle \mathbf{S} : \mathscr{C} \rangle + \langle \mathscr{C}^T : \boldsymbol{\varepsilon}^p \rangle \\ &= \mathbf{S}^{\text{hom}} : \boldsymbol{\Sigma} + \langle \mathscr{C}^T : \boldsymbol{\varepsilon}^p \rangle. \end{aligned} \tag{9.37}$$

We can define the elastic part $\mathscr{E}^e$ or $\mathscr{E}$ by

$$\mathscr{E}^e = \mathbf{S}^{\text{hom}} : \boldsymbol{\Sigma}, \tag{9.38}$$

so that eqn (9.37) implies that

$$\mathscr{E}^p = \mathscr{E} - \mathscr{E}^e = \langle \mathscr{C}^T : \boldsymbol{\varepsilon}^p \rangle, \tag{9.39}$$

Remarkably enough, neither $\mathscr{E}^e$ nor $\mathscr{E}^p$ is a direct average of the corresponding microscopic quantity.

† We follow Suquet (1987)

Macroscopic internal energy We compute

$$\bar{\rho}E = \langle \rho e \rangle = \tfrac{1}{2}\langle (\boldsymbol{\varepsilon}(\mathbf{u}) - \boldsymbol{\varepsilon}^{\mathrm{p}}) : \mathbf{E} : (\boldsymbol{\varepsilon}(\mathbf{u}) - \boldsymbol{\varepsilon}^{\mathrm{p}}) \rangle$$
$$= \tfrac{1}{2}\langle \boldsymbol{\sigma} : \mathbf{S} : \boldsymbol{\sigma} \rangle. \tag{9.40}$$

As in classical elastoplasticity (see eqn (4.26)), at the microscopic level we can introduce a field of elastic stresses $\boldsymbol{\sigma}^{\mathrm{E}}$ and a field of self-equilibrated residual stresses $\boldsymbol{\sigma}^{\mathrm{r}}$ in such a way that

$$\boldsymbol{\sigma}(y) = \boldsymbol{\sigma}^{\mathrm{E}}(y) + \boldsymbol{\sigma}^{\mathrm{r}}(y), \tag{9.41}$$

where $\boldsymbol{\sigma}^{\mathrm{E}}(y) = \mathscr{C} : \boldsymbol{\Sigma}$ according to (9.27). On account of this eqn (9.40) is rewritten as

$$\bar{\rho}E = \tfrac{1}{2}\langle \boldsymbol{\sigma} : \mathbf{S} : \boldsymbol{\sigma} \rangle$$
$$= \tfrac{1}{2}\boldsymbol{\Sigma} : \langle \mathscr{C}^{\mathrm{T}} : \mathbf{S} : \mathscr{C} \rangle : \boldsymbol{\Sigma} + \tfrac{1}{2}\langle \boldsymbol{\sigma}^{\mathrm{r}} : \mathbf{S} : \mathscr{C} \rangle : \boldsymbol{\Sigma} + \tfrac{1}{2}\boldsymbol{\Sigma} : \langle \mathscr{C}^{\mathrm{T}} : \mathbf{S} : \boldsymbol{\sigma}^{\mathrm{r}} \rangle$$
$$+ \tfrac{1}{2}\langle \boldsymbol{\sigma}^{\mathrm{r}} : \mathbf{S} : \boldsymbol{\sigma}^{\mathrm{r}} \rangle,$$

or, with the Hill–Mandel condition and the fact that $\langle \boldsymbol{\sigma}^{\mathrm{r}} \rangle = \mathbf{0}$,

$$\bar{\rho}E = \tfrac{1}{2}\boldsymbol{\Sigma} : \mathbf{S}^{\mathrm{hom}} : \boldsymbol{\Sigma} + \tfrac{1}{2}\langle \boldsymbol{\sigma}^{\mathrm{r}} : \mathbf{S} : \boldsymbol{\sigma}^{\mathrm{r}} \rangle, \tag{9.42}$$

in which the second contribution may be considered as an (always positive) *stored energy*.

Plastic dissipation As it is an additive quantity, accounting for the decomposition (9.41), we can write for the *dissipation*

$$\Phi = \langle \phi \rangle = \langle \boldsymbol{\sigma} : \dot{\boldsymbol{\varepsilon}}^{\mathrm{p}} \rangle = \langle \boldsymbol{\sigma}^{\mathrm{E}} : \dot{\boldsymbol{\varepsilon}}^{\mathrm{p}} \rangle + \langle \boldsymbol{\sigma}^{\mathrm{r}} : \dot{\boldsymbol{\varepsilon}}^{\mathrm{p}} \rangle$$
$$= \langle (\mathscr{C} : \Sigma) : \dot{\boldsymbol{\varepsilon}}^{\mathrm{p}} \rangle + \langle \boldsymbol{\sigma}^{\mathrm{r}} : \dot{\boldsymbol{\varepsilon}}^{\mathrm{p}} \rangle$$
$$= \boldsymbol{\Sigma} : \langle \mathscr{C}^{\mathrm{T}} : \dot{\boldsymbol{\varepsilon}}^{\mathrm{p}} \rangle + \langle \boldsymbol{\sigma}^{\mathrm{r}} : \dot{\boldsymbol{\varepsilon}}^{\mathrm{p}} \rangle,$$

so that

$$\Phi = \boldsymbol{\Sigma} : \dot{\mathscr{E}}^{\mathrm{p}} + \langle \boldsymbol{\sigma}^{\mathrm{r}} : \dot{\boldsymbol{\varepsilon}}^{\mathrm{p}} \rangle, \tag{9.43}$$

where

$$\left.\begin{aligned} \boldsymbol{\sigma}^{\mathrm{r}} \in \mathrm{SE}, \quad \mathbf{S} : \boldsymbol{\sigma}^{\mathrm{r}} + \boldsymbol{\varepsilon}^{\mathrm{p}} = \boldsymbol{\varepsilon}(\mathbf{u}^{\mathrm{r}}), \\ \boldsymbol{\varepsilon}(\mathbf{u}) = \boldsymbol{\varepsilon}(\mathbf{u}) - \mathbf{S} : \mathscr{C} : \boldsymbol{\Sigma} \quad \text{is admissible.} \end{aligned}\right\} \tag{9.44}$$

As $\boldsymbol{\sigma}^{\mathrm{r}}$ is self-equilibrated we can write

$$\Phi = \boldsymbol{\Sigma} : \dot{\mathscr{E}}^{\mathrm{p}} + \underbrace{\langle \boldsymbol{\sigma}^{\mathrm{r}} : \boldsymbol{\varepsilon}(\dot{\mathbf{u}}^{\mathrm{r}}) \rangle}_{=0} - \langle \boldsymbol{\sigma}^{\mathrm{r}} : \mathbf{S} : \dot{\boldsymbol{\sigma}}^{\mathrm{r}} \rangle$$

i.e.,

$$\mathbf{\Phi} = \mathbf{\Sigma} : \dot{\mathscr{E}}^{\mathrm{p}} - \langle \boldsymbol{\sigma}^{\mathrm{r}} : \mathbf{S} : \dot{\boldsymbol{\sigma}}^{\mathrm{r}} \rangle = \langle \boldsymbol{\sigma} : \dot{\boldsymbol{\varepsilon}}^{\mathrm{p}} \rangle. \tag{9.45}$$

Obviously

$$\mathbf{\Sigma} : \dot{\mathscr{E}}^{\mathrm{p}} \neq \langle \boldsymbol{\sigma} : \dot{\boldsymbol{\varepsilon}}^{\mathrm{p}} \rangle$$

and the difference is none other than the rate of elastic energy caused by the developed residual stresses. While the plastic power is *entirely dissipated* at the microscopic level, it appears at the macroscopic level that one part is dissipated and another part is stored (see Fig. 9.2).

9.4.2 Stability in the sense of Drucker

As there is no hardening at the 'micro' scale, the following orthogonality relation holds true (compare to eqn (3.45)):

$$\dot{\boldsymbol{\sigma}} : \dot{\boldsymbol{\varepsilon}}^{\mathrm{p}} = 0. \tag{9.46}$$

By using (9.41) while taking the average we obtain

$$\langle \dot{\boldsymbol{\sigma}} : \dot{\boldsymbol{\varepsilon}}^{\mathrm{p}} \rangle = \langle (\mathscr{C} : \dot{\mathbf{\Sigma}}) : \dot{\boldsymbol{\varepsilon}}^{\mathrm{p}} \rangle + \langle \dot{\boldsymbol{\sigma}}^{\mathrm{r}} : \dot{\boldsymbol{\varepsilon}}^{\mathrm{p}} \rangle.$$

But

$$\langle (\mathscr{C} : \dot{\mathbf{\Sigma}}) : \dot{\boldsymbol{\varepsilon}}^{\mathrm{p}} \rangle = \dot{\mathbf{\Sigma}} : \langle \mathscr{C}^{\mathrm{T}} : \dot{\boldsymbol{\varepsilon}}^{\mathrm{T}} \rangle = \dot{\mathbf{\Sigma}} : \dot{\mathscr{E}}^{\mathrm{p}},$$

and

$$\langle \dot{\boldsymbol{\sigma}}^{\mathrm{r}} : \dot{\boldsymbol{\varepsilon}}^{\mathrm{p}} \rangle = -\langle \dot{\boldsymbol{\sigma}}^{\mathrm{r}} : \mathbf{S} : \dot{\boldsymbol{\sigma}}^{\mathrm{r}} \rangle,$$

Fig. 9.2. Plastic dissipation of the homogenized medium

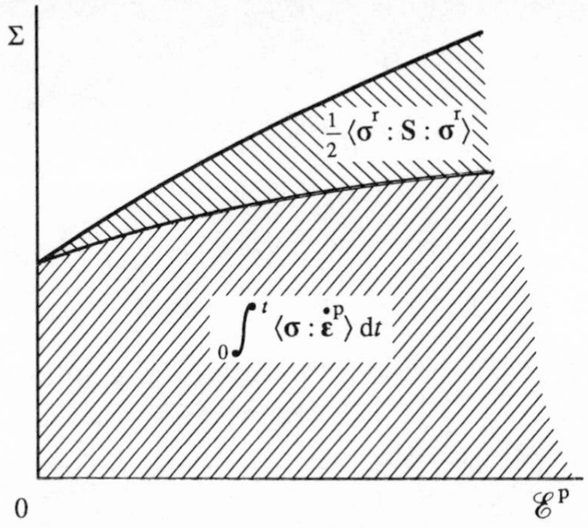

so that

$$\dot{\boldsymbol{\Sigma}} : \dot{\mathscr{E}}^{\mathrm{p}} = \underbrace{\langle \dot{\boldsymbol{\sigma}} : \dot{\boldsymbol{\varepsilon}}^{\mathrm{p}} \rangle}_{=0} + \underbrace{\langle \dot{\boldsymbol{\sigma}}^{\mathrm{r}} : \mathbf{S} : \dot{\boldsymbol{\sigma}}^{\mathrm{r}} \rangle}_{\geqslant 0},$$

and thus

$$\dot{\boldsymbol{\Sigma}} : \dot{\mathscr{E}}^{\mathrm{p}} \geqslant 0. \tag{9.47}$$

This is *Drucker's inequality* for the macroscopic fields (compare eqn (5.75)). We can say that the 'micro – macro' transition 'stabilizes the elastoplastic material'.

9.4.3 Macroscopic loading surface, macroscopic 'convex'

After (9.41) we have

$$\boldsymbol{\sigma}(y) = \mathscr{C}(y) : \boldsymbol{\Sigma} + \boldsymbol{\sigma}^{\mathrm{r}}(y), \qquad y \in V. \tag{9.48}$$

Let us introduce the following definition.

Definition The macroscopic loading surface is the set of all macroscopic stresses $\boldsymbol{\Sigma}^*$ that can be reached from the actual stress state $\boldsymbol{\Sigma}$ by following along an elastic path for which residual stresses remain unchanged.

The microscopic stress state $\boldsymbol{\sigma}^*$ corresponding to $\boldsymbol{\Sigma}^*$ on account of (9.48) then satisfies the equation

$$\boldsymbol{\sigma}^*(y) - \boldsymbol{\sigma}(y) = \mathscr{C}(y) : (\boldsymbol{\Sigma}^* - \boldsymbol{\Sigma}) \tag{9.49}$$

with

$$\boldsymbol{\sigma}^*(y) = \mathscr{C}(y) : \boldsymbol{\Sigma}^* + \boldsymbol{\sigma}^{\mathrm{r}}(y). \tag{9.50}$$

The condition

$$\boldsymbol{\sigma}^*(y) \in C(y), \qquad y \in V, \tag{9.51}$$

is then equivalent to the macroscopic condition

$$\boldsymbol{\Sigma}^* \in \mathscr{C}(y)^{-1} : (C(y) - \{\boldsymbol{\sigma}(y)\}), \qquad y \in V, \tag{9.52}$$

so that we can write

$$\boldsymbol{\Sigma}^* \in C^{\mathrm{hom}}(\{\boldsymbol{\sigma}^{\mathrm{r}}\}) = \underbrace{\underbrace{\bigcap_{y \in V} \mathscr{C}^{-1}(y) : \underbrace{(C(y) - \{\boldsymbol{\sigma}^{\mathrm{r}}(y)\})}_{1}}_{2}}_{3} \tag{9.53}$$

where the successive operations, denoted by 1, 2, 3, are needed to define $\mathbf{C}^{\text{hom}}$ which is a convex set, being the intersection of convex sets. However, this definition involves an essential difficulty in that the construction of $\mathbf{C}^{\text{hom}}$ at time t requires a knowledge of the entire set of residual stresses. In this sense, we can say that the 'micro' description is not entirely removed from the 'macro' description.

Simple properties of $\mathbf{C}^{\text{hom}}$

(i) The set $C(y) - \{\boldsymbol{\sigma}^{\text{r}}(y)\}$ is the result of a *translation* of $C(y)$ without alteration in its shape or size. This is akin to *kinematical hardening*.

(ii) The operation $\mathscr{C}^{-1}(y)$ produces a *rotation* and *anisotropic expansion* and thus is akin to *isotropic hardening* although not exactly.

(iii) The third operation in (9.53), the intersection, is a rather complex operation that mixes all types of possible alterations to a convex set. In particular, $\mathbf{C}^{\text{hom}}$ may present angular points (e.g., in a monocrystal where the $C(y)$'s may be piecewise constant, plastic flow may take place along a finite set of planes) which may be rounded off (e.g., when the RVE itself is a polycrystal containing numerous grains.

9.5 Structure of macroscopic constitutive equations

We give only a qualitative analysis of this structure which requires paying some attention to the state variables, the expression of the internal energy, the laws of state, and the loading surface.

9.5.1 State variables

We could consider as state variables (i) the macroscopic strain $\mathscr{E}$ and (ii) the *whole* set of *microscopic* plastic strains, i.e., $\{\boldsymbol{\varepsilon}^{\text{p}}(y),\ y \in V\}$, *so that, in fact, we would have an infinity* of internal variables. Let us check the validity of this assumption *a posteriori*. On using (9.37) we deduce $\boldsymbol{\Sigma}$ from $\mathscr{E}$ and $\boldsymbol{\varepsilon}^{\text{p}}$ by

$$\begin{aligned}\boldsymbol{\Sigma} &= (\mathbf{S}^{\text{hom}})^{-1} : \mathscr{E} + \mathbf{E}^{\text{hom}} : \langle \mathscr{C}^{\text{T}} : \boldsymbol{\varepsilon}^{\text{p}} \rangle \\ &= \mathbf{E}^{\text{hom}} : \mathscr{E} + \mathbf{E}^{\text{hom}} : \langle \mathscr{C}^{\text{T}} : \boldsymbol{\varepsilon}^{\text{p}} \rangle.\end{aligned} \tag{9.54}$$

Then $\boldsymbol{\sigma}^{\text{r}}$ is determined as the solution of an *elastoplastic problem* such as

$$\boldsymbol{\sigma}^{\text{r}} \in \text{SE}, \qquad \mathbf{S} : \boldsymbol{\sigma}^{\text{r}} + \boldsymbol{\varepsilon}^{\text{p}} = \boldsymbol{\varepsilon}(\mathbf{u}^{\text{r}}), \tag{9.55}$$

where $\boldsymbol{\varepsilon}^{\text{p}}$ is supposed to be known (this is quite similar to what we do with thermal stresses in thermoelasticity). Consequently, using the *Green's function technique*, from (9.55) we obtain

$$\boldsymbol{\sigma}^{\mathrm{r}}(y) = -\mathcal{R} : \boldsymbol{\varepsilon}^{\mathrm{p}} = -\int \mathbf{R}(y, y') : \boldsymbol{\varepsilon}^{\mathrm{p}}(y')\,\mathrm{d}y, \tag{9.56}$$

Here $\mathcal{R}$ is an integrodifferential operator.

We conclude that $\mathscr{E}$ and $\{\boldsymbol{\varepsilon}^{\mathrm{p}}(y)\}$ allow one to evaluate both $\boldsymbol{\Sigma}$ and the set $\{\boldsymbol{\sigma}^{\mathrm{r}}(y)\}$, whence the validity of the original assumption.

9.5.2 *Internal energy of the macroscopic material*

Eqn (9.42) reads

$$\bar{\rho}E = \tfrac{1}{2}\boldsymbol{\Sigma} : \mathbf{S}^{\mathrm{hom}} : \boldsymbol{\Sigma} + \tfrac{1}{2}\langle \boldsymbol{\sigma}^{\mathrm{r}} : \mathbf{S} : \boldsymbol{\sigma}^{\mathrm{r}} \rangle. \tag{9.57}$$

This must be expressed in terms of the state variables $(\mathscr{E}, \{\boldsymbol{\varepsilon}^{\mathrm{p}}\})$.

Taking into account eqn (9.53) we write

$$\bar{\rho}E = \tfrac{1}{2}(\mathscr{E} - \mathscr{E}^{\mathrm{p}}) : \mathbf{E}^{\mathrm{hom}} : (\mathscr{E} - \mathscr{E}^{\mathrm{p}}) + \tfrac{1}{2}\langle \boldsymbol{\sigma}^{\mathrm{r}} : (\boldsymbol{\varepsilon}(\mathbf{u}^{\mathrm{r}}) - \boldsymbol{\varepsilon}^{\mathrm{p}}) \rangle,$$

and, using both (9.56) and the fact that $\boldsymbol{\sigma}^{\mathrm{r}} \in \mathrm{SE}$, we obtain

$$\bar{\rho}E(\mathscr{E}, \{\boldsymbol{\varepsilon}^{\mathrm{p}}\}) = \tfrac{1}{2}(\mathscr{E} - \mathscr{E}^{\mathrm{p}}) : \mathbf{E}^{\mathrm{hom}} : (\mathscr{E} - \mathscr{E}^{\mathrm{p}}) + \tfrac{1}{2}\langle (\mathcal{R} : \boldsymbol{\varepsilon}^{\mathrm{p}}) : \boldsymbol{\varepsilon}^{\mathrm{p}} \rangle, \tag{9.58}$$

which is the expression sought.

9.5.3 *Equations of state*

According to the general formalism of Appendix 1, the equations of state of the homogenized material are given by the partial derivatives

$$\bar{\rho}\frac{\partial E}{\partial \mathscr{E}}, \quad -\bar{\rho}\frac{\partial E}{\partial \{\boldsymbol{\varepsilon}^{\mathrm{p}}\}}, \tag{9.59}$$

where some meaning must be granted to the second expression. As a matter of fact, the derivatives (9.59) can be evaluated from (9.58). For the first of (9.59) we have

$$\bar{\rho}\frac{\partial E}{\partial \mathscr{E}} = \mathbf{E}^{\mathrm{hom}} : (\mathscr{E} - \mathscr{E}^{\mathrm{p}}) = \boldsymbol{\Sigma}. \tag{9.60}$$

For the second of (9.59), introducing a virtual variation $\delta\boldsymbol{\varepsilon}^{\mathrm{p}}$, we can write

$$\mathscr{A} \equiv \left\langle -\bar{\rho}\frac{\partial E}{\partial \{\boldsymbol{\varepsilon}^{\mathrm{p}}\}} : \delta\boldsymbol{\varepsilon}^{\mathrm{p}} \right\rangle = \left\langle \mathbf{E}^{\mathrm{hom}} : (\mathscr{E} - \mathscr{E}^{\mathrm{p}})\frac{\partial \mathscr{E}^{\mathrm{p}}}{\partial \{\boldsymbol{\varepsilon}^{\mathrm{p}}\}} : \delta\boldsymbol{\varepsilon}^{\mathrm{p}} \right\rangle - \langle (\mathcal{R} : \boldsymbol{\varepsilon}^{\mathrm{p}}) : \delta\boldsymbol{\varepsilon}^{\mathrm{p}} \rangle \tag{9.61}$$

since $\mathscr{E}^{\mathrm{p}}$ depends on the set $\{\boldsymbol{\varepsilon}^{\mathrm{p}}\}$. We have thus after some manipulations

$$\begin{aligned} \mathscr{A} &= \langle (\boldsymbol{\Sigma} : \mathscr{C}^{\mathrm{T}}) : \delta\boldsymbol{\varepsilon}^{\mathrm{p}} \rangle - \langle (\mathcal{R} : \boldsymbol{\varepsilon}^{\mathrm{p}}) : \delta\boldsymbol{\varepsilon}^{\mathrm{p}} \rangle \\ &= \langle \delta\boldsymbol{\varepsilon}^{\mathrm{p}} : (\mathscr{C} : \boldsymbol{\Sigma} + \boldsymbol{\sigma}^{\mathrm{r}}) \rangle = \langle \delta\boldsymbol{\varepsilon}^{\mathrm{p}} : \boldsymbol{\sigma} \rangle, \end{aligned} \tag{9.62}$$

so that it is the set of *microstress* fields $\{\boldsymbol{\sigma}\}$ that is the thermodynamical force associated with $\{\boldsymbol{\varepsilon}^{\mathrm{p}}\}$.

Gathering the results enunciated in Subsections 9.5.1–9.5.3, we can formally introduce the following model of *generalized standard medium* (compare Chapter 5)

$$\left.\begin{array}{ll}\boldsymbol{\alpha} = \{\boldsymbol{\varepsilon}^{\mathrm{p}}\}, & \mathbf{A} = \{\boldsymbol{\sigma}\}, \\ C = \{\boldsymbol{\tau} \in \mathscr{E}(\mathscr{U}_0)^{\perp} \mid \boldsymbol{\tau}(y) \in C(y), y \in V\}, & \\ \boldsymbol{\Sigma} = \bar{\rho}\dfrac{\partial E}{\partial \mathscr{E}}, & \mathbf{A} = -\bar{\rho}\dfrac{\partial E}{\partial \boldsymbol{\alpha}}, \\ (\mathbf{A} - \mathbf{A}^*)\dot{\boldsymbol{\alpha}} \geqslant 0, & \mathbf{A} \in \mathbf{C}, \forall \mathbf{A}^* \in C,\end{array}\right\} \tag{9.63}$$

where the subtlety consists in the fact that $\boldsymbol{\alpha}$ represents an *infinity* of internal variables, so that the practical importance of the formulation (9.63) is purely formal. There is need for *approximate* models.

9.5.4 Example of an approximate model

This model is borrowed from Michel (1984). Here it is assumed that the condition

$$\left.\begin{array}{ll}\boldsymbol{\sigma}(y) \in C(y), & y \in V, \\ \text{or} \quad f(y, \boldsymbol{\sigma}(y)) \leqslant 0, & y \in V,\end{array}\right\} \tag{9.64}$$

is satisfied in the mean, i.e.,

$$\langle f(y, \boldsymbol{\sigma}(y)) \rangle \leqslant 0. \tag{9.65}$$

Therefore, using the notation of Chapter 4, we can write

$$\left.\begin{array}{l}C^{\mathrm{macro}} = \{\boldsymbol{\sigma} \in \mathscr{E}(\mathscr{U}_0)^{\perp} \mid \langle f(y, \boldsymbol{\sigma}(y)) \rangle \leqslant 0\}, \\ \dot{\boldsymbol{\varepsilon}}^{\mathrm{p}} = \dot{\lambda}\dfrac{\partial f}{\partial \boldsymbol{\sigma}}(y, \boldsymbol{\sigma}(y)), \qquad \boldsymbol{\sigma} \in C\end{array}\right\} \tag{9.66}$$

where there exists only *one* plastic multiplier $\dot{\lambda}$ for the RVE since we have to consider only *one* inequality, eqn (9.65).

Let us further assume, with Michel (1984), that $f(y, \boldsymbol{\sigma}(y))$ is none other than the *microscopic elastic energy*! That is,

$$f(y, \boldsymbol{\sigma}) = \tfrac{1}{2}\boldsymbol{\sigma} : \mathbf{S} : \boldsymbol{\sigma} - k, \qquad k = \text{const.} \tag{9.67}$$

(This is akin to a Mises criterion (distortion-energy theory of yielding) except that the norm of $\boldsymbol{\sigma}$ is the *energy norm* in local form.)

Eqn (9.66) yields

$$\dot{\boldsymbol{\varepsilon}}^{\mathrm{p}} = \dot{\lambda}\mathbf{S} : \boldsymbol{\sigma}. \tag{9.68}$$

The problem (9.55) concerning the evaluation of $\boldsymbol{\sigma}^{\mathrm{r}}$ can be stated in terms of velocities as

$$\boldsymbol{\sigma}^{\mathrm{r}} \in \mathrm{SE}, \mathbf{S} : \dot{\boldsymbol{\sigma}}^{\mathrm{r}} + \dot{\lambda}\mathbf{S} : \boldsymbol{\sigma}^{\mathrm{r}} = \boldsymbol{\varepsilon}(\dot{\mathbf{u}}^{\mathrm{r}}) - \dot{\lambda}(\mathbf{S} : \mathscr{C}) : \boldsymbol{\Sigma}, \tag{9.69}$$

where both $\mathbf{S} : \mathscr{C}$ and $\boldsymbol{\Sigma}$ are admissible fields. Multiplying tensorially by $\boldsymbol{\tau} \in \mathrm{SE}$, we obtain from (9.69)

$$\left.\begin{array}{l} \langle \boldsymbol{\tau} : \mathbf{S} : \dot{\boldsymbol{\sigma}}^{\mathrm{r}} \rangle + \dot{\lambda}\langle \boldsymbol{\tau} : \mathbf{S} : \boldsymbol{\sigma}^{\mathrm{r}} \rangle = 0, \\ \forall \boldsymbol{\tau} \in \mathrm{SE}, \qquad \boldsymbol{\sigma}^{\mathrm{r}} \in \mathrm{SE}, \end{array}\right\} \tag{9.70}$$

of which the $\boldsymbol{\sigma}^{\mathrm{r}}$ solution is given by

$$\boldsymbol{\sigma}^{\mathrm{r}}(t) = \boldsymbol{\sigma}^{\mathrm{r}}(0) \exp\{-[\lambda(t) - \lambda_0]\} = \boldsymbol{\sigma}^{\mathrm{r}}(0)\xi(t), \tag{9.71}$$

where *only one* parameter, $\xi(t)$, allows one to determine the whole field of residual stresses. The flow condition (9.65) reads

$$\begin{aligned} \mathscr{F} &:= \langle f \rangle = \langle \tfrac{1}{2}\boldsymbol{\sigma} : \mathbf{S} : \boldsymbol{\sigma} - k \rangle \\ &= \tfrac{1}{2}\langle (\mathscr{C} : \boldsymbol{\Sigma} + \boldsymbol{\sigma}^{\mathrm{r}}) : \mathbf{S} : (\mathscr{C} : \boldsymbol{\Sigma} + \boldsymbol{\sigma}^{\mathrm{r}}) \rangle - \langle k \rangle, \end{aligned}$$

i.e.,

$$\mathscr{F} = \tfrac{1}{2}\boldsymbol{\Sigma} : \mathbf{S}^{\mathrm{hom}} : \boldsymbol{\Sigma} + \tfrac{1}{2}h\xi^2 - \langle k \rangle = \mathscr{F}(\boldsymbol{\Sigma}, \xi), \tag{9.72}$$

where we have set

$$h = \langle \boldsymbol{\sigma}^{\mathrm{r}}(0) : \mathbf{S} : \boldsymbol{\sigma}^{\mathrm{r}}(0) \rangle. \tag{9.73}$$

Eqn (9.72) shows that the macroscopic equivalent material is subjected to an isotropic hardening with parameter ξ. Let us verify that the model obtained corresponds to a generalized standard material with variables $\mathscr{E}$, $\mathscr{E}^{\mathrm{p}}$ and ξ. Indeed, in the present case the internal energy (9.58) takes on the form

$$\bar{\rho}E(\mathscr{E}, \mathscr{E}^{\mathrm{p}}, \xi) = \tfrac{1}{2}(\mathscr{E} - \mathscr{E}^{\mathrm{p}}) : \mathbf{E}^{\mathrm{hom}} : (\mathscr{E} - \mathscr{E}^{\mathrm{p}}) + \tfrac{1}{2}h\xi^2, \tag{9.74}$$

because (this defines the hardening modulus h)

$$\langle \boldsymbol{\sigma}^{\mathrm{r}} : \mathbf{S} : \boldsymbol{\sigma}^{\mathrm{r}} \rangle = h\xi^2.$$

By direct computation we have

$$\bar{\rho}\frac{\partial E}{\partial \mathscr{E}} = \boldsymbol{\Sigma}, \qquad -\bar{\rho}\frac{\partial E}{\partial \mathscr{E}^{\mathrm{p}}} = \boldsymbol{\Sigma}, \qquad -\bar{\rho}\frac{\partial E}{\partial \xi} = -h\xi \equiv A_{\xi}. \tag{9.75}$$

We set

$$\mathscr{F}(\boldsymbol{\Sigma}, A_\xi) = \frac{1}{2}\boldsymbol{\Sigma} : \mathbf{S}^{\text{hom}} : \boldsymbol{\Sigma} + \frac{1}{2h}(A_\xi)^2 - \langle k \rangle. \tag{9.76}$$

It remains to formulate the evolution equations. First we have

$$\begin{aligned} \dot{\mathscr{E}}^{\text{p}} &= \langle \mathscr{C}^{\text{T}} : \dot{\boldsymbol{\varepsilon}}^{\text{p}} \rangle = \langle \mathscr{C}^{\text{T}} : \dot{\lambda}\mathbf{S} : \boldsymbol{\sigma} \rangle \\ &= \dot{\lambda}\langle \mathscr{C}^{\text{T}} : \mathbf{S} : \mathscr{C} \rangle : \boldsymbol{\Sigma} + \dot{\lambda}\langle \mathscr{C}^{\text{T}} : \mathbf{S} : \boldsymbol{\sigma}^{\text{r}} \rangle \end{aligned} \tag{9.77}$$

since $\boldsymbol{\sigma} = \mathscr{C} : \boldsymbol{\Sigma} + \boldsymbol{\sigma}^{\text{r}}$. But $\langle \boldsymbol{\sigma}^{\text{r}} : \mathbf{S} : \mathscr{C} \rangle = 0$ from the Hill–Mandel condition, so that (9.77) is reduced to

$$\dot{\mathscr{E}}^{\text{p}} = \dot{\lambda}\mathbf{S}^{\text{hom}} : \boldsymbol{\Sigma} = \dot{\lambda}\frac{\partial \mathscr{F}}{\partial \boldsymbol{\Sigma}}(\boldsymbol{\Sigma}, A_\xi), \tag{9.78}$$

On the other hand,

$$\dot{\xi} = \dot{\lambda}\xi = \dot{\lambda}\frac{\partial \mathscr{F}}{\partial A_\xi}(\boldsymbol{\Sigma}, A_\xi). \tag{9.79}$$

In summary, this peculiar model is a *generalized standard medium* for which the following hold.

(i) The hardening modulus is nothing but the elastic energy of initial residual stresses (see equation following (9.74)).

(ii) The 'size' of the flow surface $\mathscr{F}$ is given by $\langle k \rangle - \frac{1}{2}h\xi^2$. As $\dot{\lambda} > 0$, λ increases and ξ decreases on account of eqn (9.71) so that the flow surface expands, tending asymptotically to $\langle k \rangle$.

(iii) The stored energy decreases for *any* loading since ξ decreases *always.*

(iv) The multiplier $\dot{\lambda}$ is proportional to the 'macro' dissipation. This is directly checked by computing

$$\Phi = \langle \phi \rangle = \langle \boldsymbol{\sigma} : \dot{\boldsymbol{\varepsilon}}^{\text{p}} \rangle = \langle \boldsymbol{\sigma} : \dot{\lambda}\mathbf{S} : \boldsymbol{\sigma} \rangle = \dot{\lambda}\langle \boldsymbol{\sigma} : \mathbf{S} : \boldsymbol{\sigma} \rangle.$$

But

$$\dot{\lambda}\langle \boldsymbol{\sigma} : \mathbf{S} : \boldsymbol{\sigma} \rangle = \begin{cases} 0 & \text{if} \langle \boldsymbol{\sigma} : \mathbf{S} : \boldsymbol{\sigma} \rangle - \langle k \rangle < 0, \\ \dot{\lambda}\langle k \rangle & \text{if} \langle \boldsymbol{\sigma} : \mathbf{S} : \boldsymbol{\sigma} \rangle - \langle k \rangle = 0, \end{cases}$$

and thus

$$\Phi = \dot{\lambda}\langle k \rangle. \tag{9.80}$$

9.6 First example: composite with unidirectional fibres

This example is due to Dvorak and Rao (1976) and is older than the present formal developments (Suquet, 1982, 1987) – see Fig. 9.3.

The matrix is made of an elastic–perfectly-plastic material with plasticity criterion $f \leqslant 0$. The fibres are purely elastic. The RVE is a cylindrical element such that

$$\text{RVE}: \begin{cases} \text{fibre:} \quad 0 < r < a, \quad r = \sqrt{(y_2^2 + y_3^2)}, \\ \text{matrix:} a < r < b. \end{cases}$$

The mechanical loading consists in a traction or compression along the axis direction and an equiaxial stress in planes orthogonal to the y_1-direction. Therefore, the stress field $\mathbf{\Sigma}$ can be viewed as a two-vector of components

$$\mathbf{\Sigma} = \{\mathbf{\Sigma}_{11}, \tfrac{1}{2}(\Sigma_{22} + \Sigma_{33})\}, \tag{9.81}$$

The localization process in the *elastic* regime yields

$$\boldsymbol{\sigma}(r) = \mathscr{C}(r) : \mathbf{\Sigma}. \tag{9.82}$$

Dvorak and Rao (1976) assume that, in the *elastoplastic* regime, the maximum local stress is reached in the fibre or at the interface (this is the case in the purely elastic regime). In these conditions the macroscopic plastic flow occurs as soon as the stress at the interface has reached the yield value, i.e., for

$$f(\mathscr{C}(r = a) : \mathbf{\Sigma} + \boldsymbol{\sigma}^{\mathrm{r}}(r = a)) = 0. \tag{9.83}$$

On setting

Fig. 9.3. Composite with unidirectional fibres

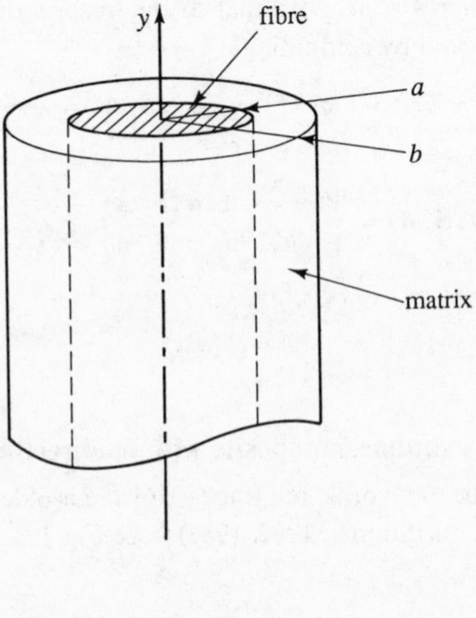

$$\mathbf{X} = -\mathscr{C}^{-1}(a) : \boldsymbol{\sigma}^{\mathrm{r}}(a), \tag{9.84}$$

we can re-write (9.83) as

$$g(\boldsymbol{\Sigma} - \mathbf{X}) = 0. \tag{9.85}$$

The very form of this expression shows that the macroscopic flow surface is subjected to *kinematic hardening* (compare eqn (1.36)) since its centre is displaced in the space of axisymmetric loads (9.81) while its shape is left unchanged.

The hardening law It is readily checked that the increment $\mathrm{d}\boldsymbol{\Sigma} - \mathrm{d}\mathbf{X}$ and the gradient of g are orthogonal to one another, i.e.,

$$0 = \mathrm{d}g = \frac{\partial g}{\partial \boldsymbol{\Sigma}_1} \cdot (\mathrm{d}\boldsymbol{\Sigma}_1 - \mathrm{d}\mathbf{X}_1) + \frac{\partial g}{\partial \boldsymbol{\Sigma}_2} \cdot (\mathrm{d}\boldsymbol{\Sigma}_2 - \mathrm{d}\mathbf{X}_2). \tag{9.86}$$

Therefore, there must exist a multiplier $\mathrm{d}\mu$ (to be determined) such that

$$\mathrm{d}\mathbf{X} = \mathrm{d}\boldsymbol{\Sigma} - \mathrm{d}\mu \left(-\frac{\partial g}{\partial \boldsymbol{\Sigma}_2}, -\frac{\partial g}{\partial \boldsymbol{\Sigma}_1} \right)^{\mathrm{T}}. \tag{9.87}$$

Determination of the multiplier $\mathrm{d}\mu$ To that purpose we suppose that the increment $\mathrm{d}\mathbf{X}$ of internal stresses is directed along the direction of the *new* stress state $\boldsymbol{\Sigma} + \mathrm{d}\boldsymbol{\Sigma}$. Hence the two-vectors $\mathrm{d}\mathbf{X}$ and $\boldsymbol{\Sigma} - \mathbf{X} + \mathrm{d}\boldsymbol{\Sigma}$ must be collinear. Neglecting contributions of the second order while using (9.87) we obtain

$$\mathrm{d}\mu = \frac{(\boldsymbol{\Sigma}_2 - \mathbf{X}_2) \cdot \mathrm{d}\boldsymbol{\Sigma}_1 - (\boldsymbol{\Sigma}_1 - \mathbf{X}_1) \cdot \mathrm{d}\boldsymbol{\Sigma}_2}{(\boldsymbol{\Sigma}_1 - \mathbf{X}_1) \cdot \dfrac{\partial g}{\partial \boldsymbol{\Sigma}_1} + (\boldsymbol{\Sigma}_2 - \mathbf{X}_2) \cdot \dfrac{\partial g}{\partial \boldsymbol{\Sigma}_2}}. \tag{9.88}$$

This, together with eqn (9.87), completely determines the law of hardening.

Flow law To obtain the full expression of this law we need to determine the macroscopic plastic multiplier $\mathrm{d}\lambda$ such that

$$(\mathrm{d}\mathscr{E}_1^{\mathrm{p}}, \mathrm{d}\mathscr{E}_2^{\mathrm{p}})^{\mathrm{T}} = \mathrm{d}\lambda \left(\frac{\partial g}{\partial \boldsymbol{\Sigma}_1}, \frac{\partial g}{\partial \boldsymbol{\Sigma}_2} \right)^{\mathrm{T}}. \tag{9.89}$$

Dvorak and Rao (1976) have obtained the following expression:

$$\mathrm{d}\lambda = (\partial g / \partial \boldsymbol{\Sigma}_1)^{-1} \cdot \left\{ \left(\frac{1}{c_{\mathrm{f}} E_{\mathrm{f}}} - S_{12}^{\mathrm{hom}} \right) \mathrm{d}X_1 + \left[\frac{-2\nu_{\mathrm{f}} + (1 - c_{\mathrm{f}})/c_{\mathrm{f}}}{E_{\mathrm{f}}} S_{22}^{\mathrm{hom}} \right] \mathrm{d}X_2 \right\}, \tag{9.90}$$

where E_f and ν_f are the elastic parameters (Young's modulus and Poisson's ratio) of fibres, c_f is the volume fraction of fibres, and S_{12}^{hom} and S_{22}^{hom} are the elastic compliances that relate $\mathscr{E}$ and Σ.

9.7 Second example: polycrystals

A polycrystal, composed of identical crystals of different orientations, may be considered as a special case of composite material. The general statements in Sections 9.1–9.5 therefore apply to the case of polycrystals with a few adjustments: we need to perform averages at the scale of the monocrystal in order to construct an elastoplasticity of polycrystals. Thus a prerequisite is a thorough examination of the monocrystal.

9.7.1 The monocrystal

In agreement with the so-called Schmid (experimental) law (Schmid, 1924) which may be justified theoretically on the basis of dislocation theory, it is accepted that the plastic deformation of the monocrystal takes place by *gliding* (without change in volume) following atomic planes, along atomic directions, when the stress vector on the glide plane, projected onto the glide direction – this is called the shear stress – has reached a value τ_g which depends on hardening. If the latter is represented by an n-vector $\boldsymbol{\alpha}$ of internal variables, an *activation criterion* is written in the form

$$\tau_r^\beta - \tau_g^\beta(\boldsymbol{\alpha}) = 0, \tag{9.91}$$

where

$$\tau_r^\beta = m_i^{(\beta)} \sigma_{ij} n_j^{(\beta)} = \mathbf{m}^{(\beta)} \cdot \boldsymbol{\sigma} \cdot \mathbf{n}^{(\beta)} \tag{9.92}$$

is the *resolved shear stress.* Only the deviatoric part $\boldsymbol{\sigma}^d$ of $\boldsymbol{\sigma}$ intervenes in eqn (9.92) where $\mathbf{n}^\beta$ is the unit normal to the glide plane β, and $\mathbf{m}^\beta$ is a unit vector in the glide direction in that plane; thus $\mathbf{m}^\beta \cdot \mathbf{n}^\beta = 0$.

Eqn (9.91) represents a *hyperplane* in the space of second-order symmetric tensors (or $\mathbb{R}^6$). The elastic domain of the monocrystal, C, is the intersection of domains $C^{(\beta)}$ defined by

$$C^{(\beta)} = \{\tau_r^\beta \leqslant \tau_g^\beta(\boldsymbol{\alpha})\}, \tag{9.93}$$

i.e.,

$$C = \bigcap_\beta C^{(\beta)}. \tag{9.94}$$

This is pictured in Fig. 9.4.

In general, there are many possible glide systems. However, in the definition (9.94) need be considered only those systems for which τ_r^β is sufficiently small. In material science (Peierls–Nabarro) it is shown that this is true for a moderate hardening for atomic planes of higher atomic densities and for the densest directions in those planes. Thus, for cubic crystals with centred face (FCC) such as Al, Ag, Cu, Ni, there are twelve pairs of glide systems that correspond to octahedral planes and directions along the diagonals of the faces (Fig. 9.5)

9.7.2 The polycrystal

In this case of an aggregate of crystals, we consider eqn (9.41) or (9.47), i.e.,

$$\boldsymbol{\sigma} = \mathscr{C} : \boldsymbol{\Sigma} + \boldsymbol{\sigma}^r, \tag{9.95}$$

where $\boldsymbol{\sigma}^r$ is a residual stress and the elastic behaviour is supposed to be

Fig. 9.4. Convex of elasticity for the monocrystal

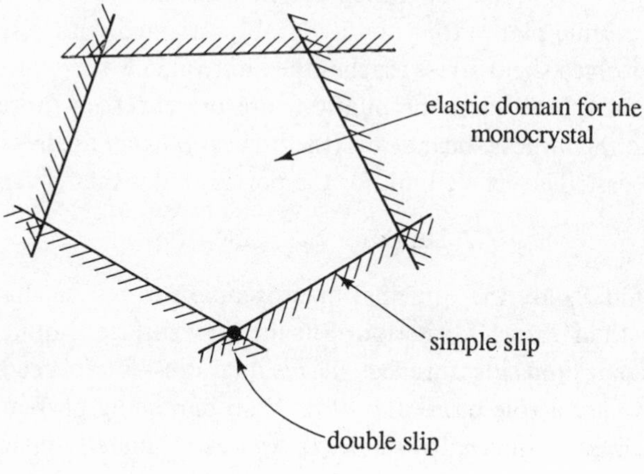

Fig. 9.5. Activation plane in an FCC crystal

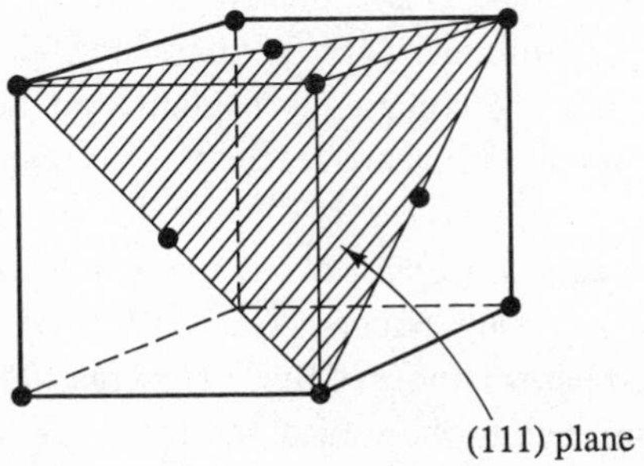

linear. In the space of macroscopic stresses $\boldsymbol{\Sigma}$, the elastic domain will be given by

$$C^{\text{hom}} = \bigcap_{\beta} \bar{C}^{(\beta)}, \tag{9.96}$$

where

$$\bar{C}^{(\beta)} = \{\boldsymbol{\Sigma} : (\mathbf{m}^{(\beta)} \otimes \mathbf{n}^{(\beta)})\mathscr{C} : \boldsymbol{\Sigma} \leqslant \tau_{\mathrm{g}}^{\beta}(\boldsymbol{\alpha}) - \tau_{\mathrm{r}}^{\beta}\}, \tag{9.97}$$

in which

$$\tau_{\mathrm{r}}^{\beta} \equiv \mathbf{m}^{\beta} \cdot (\boldsymbol{\sigma}^{\mathrm{r}} \cdot \mathbf{n}^{\beta})$$

and the intersection in eqn (9.96) is to be understood for the various possible glide systems in the different crystals.

Remarks (a) If all monocrystals are identical but for their orientation in the polycrystal, and they are also in the *same* state of hardening in a stress-free ($\boldsymbol{\sigma}^{\mathrm{r}} = \mathbf{0}$) initial state, then plastic deformation occurs in grains and along atomic planes that are favourably oriented with respect to $\boldsymbol{\Sigma}$, when the resolved shear stress reaches the limit value K (here the *same* one for all crystals). If all plane orientations are present, then those that are subjected to the largest shear stress (the highest tangential stress) must be considered first. The elastic limit for the polycrystal is then given by

$$\operatorname{Sup}(\Sigma_{\alpha} - \Sigma_{\beta}) = 2K, \tag{9.98}$$

where Σ_{α} and Σ_{β} are the principal macroscopic stresses, so that (9.98) is none other than *Tresca's* criterion! But the flow surface is quite different from this elastic limit since, in effect, the residual stresses $\boldsymbol{\sigma}^{\mathrm{r}}$ play a prevailing role, in any case a role more important than hardening proper to grains and represented by internal variables $\boldsymbol{\alpha}$. An overestimated upper bound is thus found (for example, in hexagonal crystals, the strength can reach 50 K);

(b) In practice, the volume is not rigorously constant during the plastic deformation of polycrystals. This is due to the fact that residual stresses cause elastic changes in volume. These changes are usually small (because $\langle \boldsymbol{\sigma}^{\mathrm{r}} \rangle = \mathbf{0}$ and they become completely negligible at the scale of plastic strains when the flow conditions are reached.

It is clear that *polycrystal elastoplasticity* is a very difficult field of research. We refer the reader to review papers (Asaro, 1983; Havner, 1987; Zaoui, 1987) and specialized papers (those of Bui, 1970; Hill, 1965a, b; Zaoui, 1987, *etc*) for a deeper approach to the subject matter.

9.8 Notion of limit analysis for composites*

Here we are interested in *ductile* materials, i.e., materials whose fracture and ruin occur in the *plastic* regime.

9.8.1 Extremal flow surface

Without any further information concerning the material behaviour, we assume that each constituent of the composite possesses an *extremal surface* that delineates the *convex* set $P(y)$ of statically admissible stress states (see eqns (6.3))

$$P(y) = \{\boldsymbol{\sigma}(y) | f(y, \boldsymbol{\sigma}(y)) \leqslant 0, \forall y \in V\} \tag{9.99}$$

or, $\|\ldots\|$ denoting a semi-norm,

$$P(y) = \{\boldsymbol{\sigma} | \|\boldsymbol{\sigma}\| \leqslant \sigma_0(y)\}. \tag{9.100}$$

As the microscopic stress field is 'constrained', this is also the case for its average; thus

$$\|\boldsymbol{\Sigma}\| \leqslant \langle \|\boldsymbol{\sigma}\| \rangle \leqslant \langle \sigma_0 \rangle. \tag{9.101}$$

This provides an *upper bound.*

For a macroscopic stress field $\boldsymbol{\Sigma}$ to be reached (i.e., admissible), it must be possible to determine the microscopic stress field $\boldsymbol{\sigma}$ that satisfies the following problem.

$$\left.\begin{aligned} &\langle \boldsymbol{\sigma} \rangle = \boldsymbol{\Sigma}, \\ &\operatorname{div} \boldsymbol{\sigma} = \mathbf{0}, \\ &\boldsymbol{\sigma}(y) \in P(y), \quad \forall y \in V, \\ &+\text{boundary conditions.} \end{aligned}\right\} \tag{9.102}$$

This leads us to consider a *macroscopic* stress field $\boldsymbol{\Sigma}$ such that

$$\begin{aligned} P^{\text{hom}} = \{\boldsymbol{\Sigma} | \exists \boldsymbol{\sigma} \quad \text{s.t.} \quad &\langle \boldsymbol{\sigma} \rangle = \boldsymbol{\Sigma}, \\ &\boldsymbol{\sigma} \in \mathscr{E}(\mathscr{U}_0)^{\perp}, \boldsymbol{\sigma}(y) \in P(y), \forall y \in V\}. \end{aligned} \tag{9.103}$$

It follows then (this must be proved) that, if $P(y)$ is a closed convex, then P^{hom} is also a closed convex, so that we can say that *convexity is a property that is stable under homogenization.* It is also possible to show that P^{hom} is nonempty. The following question may be raised. Are all stress states $\boldsymbol{\Sigma}$ in

* We follow Suquet (1983)

P^{hom} physically admissible? The answer is positive only if it is possible to associate a microscopic stress field $\boldsymbol{\sigma}$ satisfying the (currently nonspecified) constitutive equation, with any $\boldsymbol{\Sigma}$. Therefore, for the time being, states in P^{hom} can be referred to only as *potentially safe. However, if elastoplastic constituents obeying a normality rule are considered, then* P^{hom} is the set of *actually safe* states. It must also be noticed that numerous fibres used in composites are *brittle* so that the computation of P^{hom} will provide a serious prediction of the fracture of composites only for elastoplastic or rigid–plastic materials (otherwise a serious overestimate of the strength of composites is possible).

In what follows we assume that $P(y)$ is a closed convex and all constituents satisfy a normality rule.

Example We consider *rigid–plastic* constituents obeying a normality rule. Therefore ($\boldsymbol{\varepsilon}^e \equiv \mathbf{0}$)

$$\boldsymbol{\sigma}(y) \in P(y), \quad (\boldsymbol{\sigma}(y) - \boldsymbol{\sigma}^*) : \boldsymbol{\varepsilon}(\dot{\mathbf{u}}(y)) \geqslant 0, \quad \forall \boldsymbol{\sigma}^* \in P(y). \tag{9.104}$$

Let $\boldsymbol{\Sigma}^* \in P^{hom}$, corresponding to $\boldsymbol{\sigma}^r$ *through* (9.103). *Taking the average of* (9.104) *while taking account of the Hill–Mandel condition, we obtain*

$$\boldsymbol{\Sigma} \in P^{hom}, \quad (\boldsymbol{\Sigma} - \boldsymbol{\Sigma}^*) : \dot{\mathscr{E}} \geqslant 0, \quad \forall \boldsymbol{\Sigma}^* \in P^{hom}. \tag{9.105}$$

If $\boldsymbol{\Sigma}$ is *in* P^{hom}, we can write $\boldsymbol{\Sigma}^* = \boldsymbol{\Sigma} + \tilde{\boldsymbol{\Sigma}}$ where the norm $\|\tilde{\boldsymbol{\Sigma}}\|$ is sufficiently small. Then eqn $(9.105)_2$ yields

$$\tilde{\boldsymbol{\Sigma}} : \dot{\mathscr{E}} \leqslant 0 \tag{9.106}$$

for any $\tilde{\boldsymbol{\Sigma}}$ with a sufficiently small norm. If we apply the condition to $+\tilde{\boldsymbol{\Sigma}}$ and $-\tilde{\boldsymbol{\Sigma}}$, (9.106) provides an equality that implies that $\dot{\mathscr{E}} = \mathbf{0}$ so that the composite behaves *rigidly* if $\boldsymbol{\Sigma} \in P^{hom}$ and it can be concluded that (i) the composite behaves macroscopically like a *rigid–plastic material*, (ii) P^{hom} is exactly its domain of safe stresses, and (iii) the composite obeys a normality rule.

9.8.2 Determination of homogenized plastically admissible stresses

From the very definition (9.103), the determination of P^{hom} amounts to solving a problem of limit analysis on the RVE where loading parameters are the components of $\boldsymbol{\Sigma}$. As in Chapter 7, two methods can be used to approach this solution, i.e., either the construction of statically and plastically admissible fields (so-called *static method*), or the evaluation of the rate of dissipated plastic energy in strains yielding ruin (so-called *kinematic method*).

Static method Here one must solve the following problem of limit analysis:

$$\left.\begin{array}{l}\delta_{\mathrm{L}} = \operatorname{Sup} \delta, \quad \text{s.t.}\, \exists \boldsymbol{\sigma}, \quad \langle \boldsymbol{\sigma} \rangle = \delta \boldsymbol{\Sigma}^0, \\ \boldsymbol{\sigma} \in \mathscr{E}(\mathscr{U}_0)^{\perp}, \quad \boldsymbol{\sigma}(y) \in P(y), \quad y \in V.\end{array}\right\} \tag{9.107}$$

Then $\delta_{\mathrm{L}} \boldsymbol{\Sigma}^0$ is the boundary of P^{hom}.

Kinematic method Let (compare to eqn (6.41))

$$H(\dot{\mathscr{E}}) = \{\tilde{\boldsymbol{\Sigma}} | \tilde{\boldsymbol{\Sigma}} : \dot{\mathscr{E}} \leqslant \Phi(\dot{\mathscr{E}})\}, \tag{9.108}$$

where

$$\Phi(\dot{\mathscr{E}}) = \langle \phi(y, \boldsymbol{\varepsilon}(\dot{\mathbf{u}})) \rangle.$$

Then (Suquet, 1987)

$$\mathrm{P}^{\mathrm{hom}} = \bigcap_{\dot{\mathscr{E}}} H(\dot{\mathscr{E}}). \tag{9.109}$$

For different practical applications to reinforced materials we refer the reader to the works of Shu and Rosen (1967), McLaughlin (1970), Le Nizhery (1980) and a book by Salençon (1983). Periodic structures are examined in Marigo *et al.* (1987).

9.9 Homogenization of cracked materials

Although continuum solid matter is most often considered as purely elastic, cracked bodies obviously present a *dissipative* (irreversible) behaviour. If we want to homogenize such bodies we now know that we must adjoin to the classical global variables of strain or stress a set of variables, collectively, denoted by β, which serve to desc1ibe the irreversibilities in the structure. For example, following Suquet (1982), to obtain the free-energy density of the homogenized body corresponding to a doubly periodic structure of period Y, it is necessary to fix all the global variables. One then obtains a purely elastic problem for the basic cell Y (since all variables describing irreversibilities are kept fixed). Therefore, it follows that[†]

$$\begin{aligned} W^{\mathrm{eff}}(\boldsymbol{\varepsilon}(\mathbf{u}^0), \beta) = \operatorname*{Min}_{\mathbf{u} \in \mathrm{DP}(Y)} & \langle W(y, \boldsymbol{\varepsilon}_y(\mathbf{u}), \beta) \rangle \\ & \langle \tfrac{1}{2}(\mathbf{u} \otimes \mathbf{n} + \mathbf{n} \otimes \mathbf{u}) \rangle_Y = \boldsymbol{\varepsilon}(\mathbf{u}^0) \\ & + \text{additional conditions on } \beta \text{ (in some cases).} \end{aligned} \tag{9.110}$$

The *macroscopic dissipation potential* is given by

† DP = doubly periodic functions on the period Y

$$\Phi(\boldsymbol{\varepsilon}(\mathbf{u}^0), \beta) = \langle \phi(y, \boldsymbol{\varepsilon}_y(\mathbf{u}), \beta) \rangle. \tag{9.111}$$

Defining the generalized forces by

$$\boldsymbol{\sigma}^0 = \frac{\partial W^{\text{eff}}}{\partial \boldsymbol{\varepsilon}(\mathbf{u}^0)}, \qquad B^0 = -\frac{\partial W^{\text{eff}}}{\partial \beta}, \tag{9.112}$$

we can compute a *macroscopic force potential* by

$$\Phi^*(\boldsymbol{\sigma}^0, B^0) = \langle \varphi^*(y, \boldsymbol{\sigma}(y), B(y)) \rangle, \tag{9.113}$$

where φ^*, $\boldsymbol{\sigma}$ and B are, respectively, the force potential, the stress tensor and the thermodynamic forces, at the microscopic level.

Example of a cracked body with a dissipative contact law We call Σ the crack in the cell Y. The following *microscopic* laws are considered:

$$\left.\begin{aligned} \boldsymbol{\sigma} &= \mathbf{E} : \boldsymbol{\varepsilon}_y(\mathbf{u}) && \text{in } Y - \Sigma, \\ \operatorname{div}_y \boldsymbol{\sigma} &= \mathbf{0} && \text{in } Y - \Sigma, \\ [\![\mathbf{u}]\!] &= \mathbf{u}^+ - \mathbf{u}^- && \text{on } \Sigma, \end{aligned}\right\} \tag{9.114}$$

$$\left.\begin{aligned} [\![\mathbf{u}]\!] &= [\![\mathbf{u}]\!]^{\text{el}} + [\![\mathbf{u}]\!]^{\text{anel}} \\ [\![\mathbf{u}]\!]^{\text{el}} &= \frac{\partial W^*(\mathbf{T})}{\partial \mathbf{T}}, \quad \mathbf{T} = \boldsymbol{\sigma} \cdot \mathbf{n}, \\ [\![\dot{\mathbf{u}}]\!]^{\text{anel}} &= \frac{\partial \varphi^*(\mathbf{T})}{\partial \mathbf{T}}, \end{aligned}\right\} \tag{9.115}$$

where W^* and φ^* are convex functions of the traction vector $\mathbf{T}$, and the superscript 'el' and 'anel' refer to the 'elastic' and 'anelastic' behaviours, respectively. Let us consider a few examples of these functions,

(i) *Nondissipative crack*: here we take

$$\varphi^* = 0, \qquad W(\mathbf{v}) = \begin{Bmatrix} +\infty & \text{if } v_n = \mathbf{v} \cdot \mathbf{n} < 0, \\ 0 & \text{if } v_n \geqslant 0. \end{Bmatrix} \tag{9.116}$$

Then we have

$$\left.\begin{gathered} [\![\mathbf{u}]\!] = [\![\mathbf{u}]\!]^{\text{el}}, \\ \boldsymbol{\sigma}_t \equiv \boldsymbol{\sigma} \cdot \mathbf{n} - \boldsymbol{\sigma}_n \mathbf{n}, \quad [\![\mathbf{u}]\!]_n \geqslant 0, \quad \sigma_n \leqslant 0, \\ \sigma_n = 0 \quad \text{if } [\![\mathbf{u}]\!]_n > 0. \end{gathered}\right\} \tag{9.117}$$

This is a law of unilateral contact without friction between the lips of the crack.

(ii) *Dissipative crack*: here we take

$$W^* = 0, \qquad \varphi^*(\mathbf{T}) = \frac{1}{2\mu}\|\mathbf{T}_t\|^2. \tag{9.118}$$

It follows from (9.115) that

$$\left.\begin{array}{l}[\![\mathbf{u}]\!] = [\![\mathbf{u}]\!]^{\text{anel}}, \quad [\![\dot{\mathbf{u}}]\!]_n = 0, \quad \text{thus} \\ [\![\mathbf{u}]\!]_n = 0 \quad \text{by integration,}\end{array}\right\} \tag{9.119}$$

and

$$\boldsymbol{\sigma}_t = \mu[\![\dot{\mathbf{u}}]\!]_t. \tag{9.120}$$

We have thus a model of composite with *imperfectly bonded constituents* and *viscous joints.*

Other examples can be found in Suquet (1982)

What about the homogenization procedure? We have

$$\boldsymbol{\sigma}^0 = \frac{1}{|Y|}\int_{Y-\Sigma} \boldsymbol{\sigma}(y)\,\mathrm{d}y, \qquad \boldsymbol{\varepsilon}(\mathbf{u}^0) = \frac{1}{|Y|}\int_{\partial Y} (\mathbf{u}\otimes\mathbf{n})_s\,\mathrm{d}s. \tag{9.121}$$

Suquet (1982) has proved the following variant of the Hill–Mandel condition which accounts for the discontinuity in **u**:

$$\frac{1}{|Y|}\int_{Y-\Sigma} \boldsymbol{\tau}(y):\boldsymbol{\varepsilon}_y(\mathbf{u})\,\mathrm{d}y + \frac{1}{|Y|}\int_{\Sigma} (\boldsymbol{\tau}\cdot\mathbf{n})\cdot[\![\mathbf{u}]\!]\,\mathrm{d}s = \boldsymbol{\tau}^0:\boldsymbol{\varepsilon}(\mathbf{u}^0) \tag{9.122}$$

for $\mathbf{u} \in \mathrm{DP}(Y-\Sigma)$ and $\boldsymbol{\tau} \in S^0_{\text{per}}(Y-\Sigma)$.

Homogenized law The internal variables are none other than the set of the anelastic parts of the displacement jump at the crack, i.e., $\beta = \{[\![\mathbf{u}^{\text{anel}}]\!]\}$ applying then the principles enunciated in (9.110) through (9.113), we obtain (see Suquet, 1982)

$$\begin{aligned} W^{\text{eff}}(\boldsymbol{\varepsilon}(\mathbf{u}^0),\beta) = \underset{\substack{\mathbf{u}^*\in \mathrm{DP}(Y) \\ \langle\ldots\rangle_{\partial Y}=\boldsymbol{\varepsilon}(\mathbf{u}^0)}}{\mathrm{Min}} &\left[\frac{1}{2|Y|}\int_{Y-\Sigma} (\mathbf{E}:\boldsymbol{\varepsilon}_y(\mathbf{u}^*)):\boldsymbol{\varepsilon}_y(\mathbf{u}^*)\,\mathrm{d}v \right. \\ &\left. + \frac{1}{|Y|}\int_\Sigma W([\![\mathbf{u}^*]\!]-\beta)\,\mathrm{d}y\right], \end{aligned} \tag{9.123}$$

$$\boldsymbol{\sigma}^0 = \frac{\partial W^{\text{eff}}}{\partial\boldsymbol{\varepsilon}(\mathbf{u}^0)}, \tag{9.124}$$

$$B^0 = -\frac{\partial W^{\text{eff}}}{\partial\beta} = \frac{1}{|Y|}\boldsymbol{\sigma}\cdot\mathbf{n}, \tag{9.125}$$

which is the microscopic stress on the lips of the crack, and

$$\Phi^*(B^0) = \frac{1}{|Y|}\int_\Sigma \varphi^*(\boldsymbol{\sigma}\cdot\mathbf{n})\,\mathrm{d}y. \tag{9.126}$$

The functions W^{eff} and Φ^* define a *generalized standard medium* with an infinite set of internal variables which are the anelastic jumps in displacement on the lips of the crack. In certain cases, however, a reduction to a *finite* number of internal variables is possible. This occurs if the anelastic jumps are constant along the crack so that only *one* conjugate force is needed to describe the model and, by duality, only *one* internal variable (see an illustration in Leguillon and Sanchez-Palencia (1983) where the basic cell is made of an isotropic linear-elastic material and it contains a rectilinear crack with unilateral contact and a Coulomb type of friction law).

Problems for chapter 9

9.1. Homogenization of composites
Prove the result (9.7) for both boundary conditions (9.4) and (9.5).

9.2. Homogenization of composites – continued
Redo in detail all steps of the proof of the Hill–Mandel principle of macrohomogeneity.

9.3. Stabilization of plastic materials by homogenization
Consider a plastic composite material whose components present *no* hardening at the microscale, so that the orthogonality condition (3.45) holds good at this scale. By averaging and taking into account a decomposition of stress of the type (9.41) at the microscale, show that the 'macroscopic" material is plastic and satisfies Drucker's inequality, and is therefore a stable plastic material in the sense of Drucker.

9.4. Homogenization of a peculiar plastic material (Michel, 1984)
Consider the very special case of a yield condition such as (9.67). Then use the general formulation (9.63) to establish the *macroscopic* flow rule for the macroscopic plastic strain and the internal variable. [*Hint*: replace $\boldsymbol{\sigma}$ by its decomposition (9.41) and apply the stress localization rule to $\boldsymbol{\sigma}^{\text{E}}$.]

9.5. Self-consistent method for elastic materials
The so-called self-consistent scheme (Kröner, 1958) consists in assimilating mechanical interactions between any component of a heterogeneous aggregate and the set of all the other components to that between this component and an effective homogeneous medium of unknown behaviour with the help of which one tries to model the aggregate. Therefore, the problem is reduced to studying the interaction between an '*inclusion*' (generic name given

to *the* component in question) and an *infinite matrix* (the effective medium). It happens that the problem has an explicit solution in many cases. Here we consider the case of polycrystal elasticity.

(a) Suppose that in the inhomogeneous medium we represent the tensor of elasticity coefficients by

$$E_{ijkl}(\mathbf{x}) = E^{0}_{ijkl} + E^{*}_{ijkl}(\mathbf{x}), \tag{1}$$

where $\mathbf{E}^0$ is spatially uniform. We also have $\mathscr{E} = \langle \boldsymbol{\varepsilon} \rangle$ and (9.6) holds good. Show that local equilibrium yields

$$E^{0}_{ijkl} E^{*}_{kl,i} = -(E^{*}_{ijkl}(\mathbf{x})\varepsilon_{kl}(\mathbf{x}))_{,i}. \tag{2}$$

Introduce the Green function to write the solution ε^{*}_{kl} of the equation. [*Hint*: This is the Green function for the infinite medium of constant coefficients $\mathbf{E}^0$.] Show that there results from this an *integral equation* for $\boldsymbol{\varepsilon}$ if $E_{ijkl}(\mathbf{x})$ is given.

(b) Consider the case of elastic inclusions ω in an elastic matrix. We can write

$$\mathbf{E}(\mathbf{x}) : E_{ijkl}(\mathbf{x}) = \bar{E}_{ijkl} + \Delta E_{ijkl}\delta^{0}(\omega), \tag{3}$$

where $\delta^0(\omega) = 0$ if $\mathbf{x} \notin \omega$ and $\delta^0(\omega) = 1$ if $\mathbf{x} \in \omega$. If $\boldsymbol{\varepsilon}$ and $\boldsymbol{\sigma}$ are *uniform* in the inclusion (this occurs for ellipsoidal inclusions), show that the solution $\boldsymbol{\varepsilon}$ of (a) can be written as

$$\boldsymbol{\varepsilon} = \mathscr{E} + \mathbf{S}^{*} : (\boldsymbol{\Sigma} - \boldsymbol{\sigma}) \tag{4}$$

or

$$\boldsymbol{\sigma} = \boldsymbol{\Sigma} + \mathbf{E}^{*} : (\mathscr{E} - \boldsymbol{\varepsilon}), \tag{5}$$

where $\mathbf{E}^*$ depends on $\bar{\mathbf{E}}$ (such that $\boldsymbol{\Sigma} = \bar{\mathbf{E}} : \mathscr{E}$). Show that $\bar{\mathbf{E}}$ is given formally by

$$\bar{\mathbf{E}} = \langle \mathbf{E}_c(\mathbf{E}_c + \mathbf{E}^{*})^{-1}(\bar{\mathbf{E}} + \mathbf{E}^{*}) \rangle, \tag{6}$$

where $\mathbf{E}_c$ characterizes the local elastic behaviour ($\boldsymbol{\sigma}_c = \mathbf{E}_c : \boldsymbol{\varepsilon}_c$ for a grain) and $\langle \ldots \rangle$ denotes the average over volume elements of the polycrystal. Eqn (6) is the *self-consistent equation* (an implicit integral relation).

9.6. Self-consistent method for metallic polycrystals

Metallic polycrystals exhibit a plastic behaviour. The question therefore arises of their homogenization, e.g., through the self-consistent method. Then we will have $\boldsymbol{\varepsilon} = \boldsymbol{\varepsilon}^{e} + \boldsymbol{\varepsilon}^{p}$ locally and eqn (3) of 9.5 must by complemented by

$$\boldsymbol{\varepsilon}^{p}(\mathbf{x}) = \mathscr{E}^{p} + (\Delta\boldsymbol{\varepsilon}^{p})\delta^{0}(\omega). \tag{1}$$

Write the incremental form of eqn (5) of 9.5. Make a literature search (e.g., work by Berveiller and Zaoui, 1978) and write a short report on the application of the self-consistent method to metallic polycrystals (assume Hencky's law in loading for the plastic matrix and consider spherical inclusions).

9.7. Mathematical (variational) properties in homogenized elasticity

Show that when $\mathscr{E}$ is prescribed the solution $\mathbf{u}^*$ appearing in eqn (9.20) is such that $\boldsymbol{\varepsilon}(\mathbf{u}^*)$ minimizes the microscopic elastic energy

$$J = (\boldsymbol{\varepsilon}(\overline{\mathbf{u}}^*) + \mathscr{E}) : \mathbf{E}(\boldsymbol{\varepsilon}(\overline{\mathbf{u}}^*) + \mathscr{E}) \tag{1}$$

among all $\boldsymbol{\varepsilon}(\overline{\mathbf{u}}^*)$ in $\mathscr{E}(\mathscr{U}_0)$. This minimum is

$$\min_{\boldsymbol{\varepsilon}(\overline{\mathbf{u}}^*) \in \mathscr{E}(\mathscr{U}_0)} J = \mathscr{E} : \mathbf{E}^{\text{hom}} : \mathscr{E}. \tag{2}$$

9.8. Homogenization of materials with Maxwell viscoelastic constituents

Assume that each constituent has the constitutive equation (compare to the scalar equation (2.70))

$$\boldsymbol{\varepsilon}(\dot{\mathbf{u}}) = \mathbf{S}(y) : \dot{\boldsymbol{\sigma}}(y) + (\mathbf{ET})^{-1}(y) : \boldsymbol{\sigma}(y). \tag{1}$$

Prove that the homogenized constitutive equation reads

$$\dot{\mathscr{E}} = \mathbf{S}^{\text{hom}} : \dot{\boldsymbol{\Sigma}} + \int_0^t \mathbf{J}(t - \sigma) : \dot{\boldsymbol{\Sigma}}(s)\,\mathrm{d}s + [(\mathbf{ET})^{-1}]^{\text{hom}} : \boldsymbol{\Sigma}, \tag{2}$$

where $\mathbf{S}$ and $(\mathbf{ET})^{-1}$ are fourth-order tensors with all required nice properties (symmetry, boundedness, coercivity) and the kernel $\mathbf{J}$ has the expression

$$\mathbf{J}(\xi) = \langle \mathbf{S} : \dot{\mathscr{C}}(\xi) + (\mathbf{ET})^{-1} : \mathscr{C}(\xi) \rangle, \tag{3}$$

in which $\mathscr{C}$ is a fourth-order tensor solution of the following evolution equation:

$$\left.\begin{aligned} \langle \boldsymbol{\tau} : \mathbf{S} : \dot{\mathscr{C}} \rangle + \langle \boldsymbol{\tau} : (\mathbf{ET})^{-1} : \mathscr{C} \rangle = 0, \qquad \forall \boldsymbol{\tau} \in \mathrm{SE}, \\ \mathscr{C}(t) \in \mathrm{SE}, \qquad \mathscr{C}(0) = \mathscr{C}^{S} - \mathscr{C}^{ET}, \end{aligned}\right\} \tag{4}$$

where $\mathscr{C}^S$ and $\mathscr{C}^{ET}$ are the elastic stress localization tensors associated with $\mathbf{S}$ and $(\mathbf{ET})^{-1}$, respectively. [*Hint*: work with the Laplace transform of all equations.]

9.9. Homogenization of materials with Kelvin–Voigt viscoelastic constituents

Assume that each constituent has the constitutive equation (compare to eqn (2.65))

$$\boldsymbol{\sigma}(y) = \mathbf{E}(y) : \boldsymbol{\varepsilon}(\mathbf{u}(y)) + \mathbf{N} : \boldsymbol{\varepsilon}(\dot{\mathbf{u}}(y)). \tag{1}$$

Then prove that the homogenized constitutive equation reads

$$\boldsymbol{\Sigma}(t) = \mathbf{E}^{\text{hom}} : \mathscr{E}(t) + \int_0^t \mathbf{K}(t - s) : \dot{\mathscr{E}}(s)\,\mathrm{d}s + \mathbf{N}^{\text{hom}} : \dot{\mathscr{E}}(t), \tag{2}$$

where the kernel $\mathbf{K}$ should be specified (see Francfort *et al.* 1983).

9.10. Limit analysis of a composite material

Consider a stratified two-phase composite that is periodically inhomogeneous in the y_3-direction and infinite and homogeneous in the other two directions. Let P_1 and P_2 be the yield sets of phases 1 and 2 (notation of Section 9.8).

(i) Show that P^{hom} defined in (9.103) is such that

$$P^{\text{hom}} = \{\boldsymbol{\sigma} | \boldsymbol{\Sigma} = c_1 \boldsymbol{\Sigma}_{(1)} + c_2 \boldsymbol{\Sigma}_{(2)}, \boldsymbol{\Sigma}_{(1)} \in P_1, \boldsymbol{\Sigma}_{(2)} \in P_2, \Sigma_{(1)i3} = \Sigma_{(2)i3}\}. \quad (1)$$

(ii) Assume that

$$P_i = \left\{ \boldsymbol{\sigma} | \operatorname*{Sup}_{\alpha,\beta} |\sigma_\alpha - \sigma_\beta| \leqslant 2k_i \right\}, \quad (2)$$

i.e., both phases are Tresca materials; compare eqn (1.20). Confining to two-dimensional problems, show that P^{hom} is given by

$$P^{\text{hom}} = \left\{ \boldsymbol{\Sigma} | \operatorname*{Sup}_{\alpha,\beta} |\Sigma_\alpha - \Sigma_\beta| \leqslant 2K(\theta) \right\}, \quad (3)$$

where θ is the angle made by the y_1-axis with the direction of larger principal stress (see de Buhan and Salençon, 1983).

10

Coupling between plasticity and damage

The object of the chapter This brief chapter introduces the reader to a thermodynamic formulation of the coupling between *plasticity* (for ductile materials) and *damage* (i.e., the deterioration of the elastic behaviour as a result of microscopic effects such as the growth of microcracks and microcavities).

10.1 Notion of damage

We agree to call *damage* the decrease in elasticity property consequent on a decrease of the areas that transmit internal forces, through the appearance and subsequent growth of microcracks and microcavities: hence, in a damage process, *the elasticity modulus is decreased.* To make the notion clear, fix ideas, a simple representation of damage, viewed as some isotropic phenomenon (very often an oversimplification), is the one provided by a *scalar* quantity D that represents the surface density of intersection of microcracks and microcavities with any one plane cut in the material. This idea is due to Kachanov (1958) and Rabotnov (1963) and was further elaborated upon by Kachanov (1986) himself, Lemaître and Chaboche (1985), Chaboche (1989a, b), Krajcinovic (1989), and others. The last reference gives all the points of view on damage in a balanced manner. A totally different view is given in Grabacki (1989) and Rabier (1989). On the basis of the original concepts of Kachanov and Rabotnov, if ΔS is the area of the section by such a plane, $(1 - D)\Delta S$ is the true area that can transmit the stress. Consequently,

$$0 \leqslant D \leqslant 1, \tag{10.1}$$

where $D = 0$ corresponds to a virgin element while $D = 1$ obviously corresponds to a fully damaged element, in other words, to fracture (see Fig. 10.1).

In a real physical situation the element of material will break before D reaches the value 1 since atomic decohesion will occur. Therefore we can envisage a *criterion of fracture initiation* by

$$D = D_c \qquad \text{(say 0.2–0.8).} \tag{10.2}$$

A line segment is a convex set so that we can say that D is restrained to a *convex* $[0, D_c]$. Let Y be the thermodynamic *dual* variable to D, thus having the physical dimension of *work*. The power dissipated in a damage process, which obviously is fully irreversible, is written

$$\varphi_{\text{dam}} = Y\dot{D}. \tag{10.3}$$

It is observed that, in the course of damage, plastic hardening decreases and can even reach negative values before fracture and failure of the element occur. More complex phenomena, such as increase in plastic strain rate in so-called tertiary creep, are also related to damage. It would seem reasonable then to think of an analysis that takes simultaneously into account both *damage* and *elastoplasticity*, up to the vicinity of the fracture point.

10.2 Thermodynamic formulation in SPH

As in previous chapters, the observable variables are taken to be the total strain $\boldsymbol{\varepsilon}$ and the temperature θ. We set $\boldsymbol{\varepsilon}^p = \boldsymbol{\varepsilon} - \boldsymbol{\varepsilon}^e$, $\boldsymbol{\alpha}$ and β denote internal variables (respectively a second-order tensor and a scalar) that account for both forms of hardening: kinematic through $\boldsymbol{\alpha}$ (displacement of the loading surface), and isotropic through β (dilatation of the relevant convex). The free energy per unit volume is considered as

$$W = W(\boldsymbol{\varepsilon}^e, \theta; \beta, \boldsymbol{\alpha}, D). \tag{10.4}$$

The equations of state follow by

$$\boldsymbol{\sigma} = \frac{\partial W}{\partial \boldsymbol{\varepsilon}^e}, \quad S = -\frac{\partial W}{\partial \theta}, \quad B = -\frac{\partial W}{\partial \beta}, \quad \mathbf{A} = -\frac{\partial W}{\partial \boldsymbol{\alpha}}, \quad Y = -\frac{\partial W}{\partial D}. \tag{10.5}$$

Fig. 10.1. Notion of damage. (a) Virgin element ($D = 0$). (b) Partially damaged element. (c) Fully damaged element ($D = 1$).

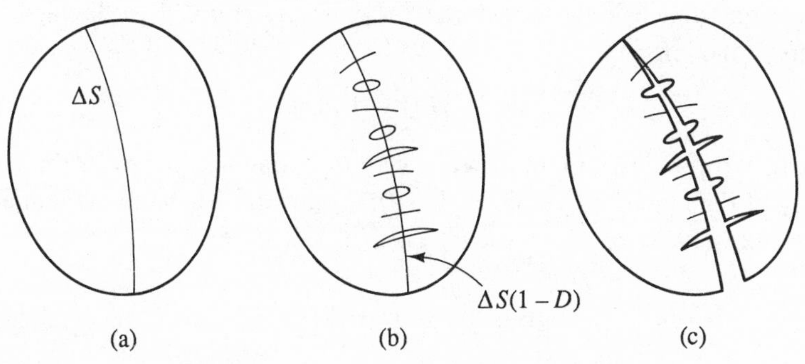

The dissipation reads

$$\phi = \phi_q + \phi_{\mathrm{intr}}, \tag{10.6}$$

in which the thermal and intrinsic dissipations are given by

$$\phi_q = \theta \mathbf{q} \cdot \nabla \left(\frac{1}{\theta} \right) \tag{10.7}$$

and

$$\phi_{\mathrm{intr}} = \boldsymbol{\sigma} : \dot{\boldsymbol{\varepsilon}}^{\mathrm{p}} + \mathbf{A} : \dot{\boldsymbol{\alpha}} + B\dot{\beta} + Y\dot{D}. \tag{10.8}$$

Taking for granted the hypothesis of generalized normality (see Chapter 3) we assume that there exists a pseudo-potential of dissipation $\mathscr{D}^*$ such that the time rates present in (10.8) derive from it according to

$$\dot{\boldsymbol{\varepsilon}}^{\mathrm{p}} = \frac{\partial \mathscr{D}^*}{\partial \boldsymbol{\sigma}}, \qquad \dot{\boldsymbol{\alpha}} = \frac{\partial \mathscr{D}^*}{\partial \mathbf{A}}, \qquad \dot{\beta} = \frac{\partial \mathscr{D}^*}{\partial B}, \qquad \dot{D} = \frac{\partial \mathscr{D}^*}{\partial Y}. \tag{10.9}$$

If the dissipation phenomena contributing to ϕ_{intr} have no time scale (thus ϕ_{intr} is positively homogeneous of degree 1 in the various time rates), $\mathscr{D}^*$ is the indicator function of a convex set C whose boundary has the equation $f = 0$. Eqns (10.9) are thus replaced by the plastic-like evolution equations

$$\dot{\boldsymbol{\varepsilon}}^{\mathrm{p}} = \dot{\lambda} \frac{\partial f}{\partial \boldsymbol{\sigma}}, \qquad \dot{\boldsymbol{\alpha}} = \dot{\lambda} \frac{\partial f}{\partial \mathbf{A}}, \qquad \dot{\beta} = \dot{\lambda} \frac{\partial f}{\partial B}, \qquad \dot{D} = \dot{\lambda} \frac{\partial f}{\partial Y}, \tag{10.10}$$

where $f(\boldsymbol{\sigma}, \mathbf{A}, B, Y) = 0$. The generality of eqns (10.10) is such as to be quite useless in practice. In order then to shed some light on the dominant phenomena we should consider some more specialized cases.

10.3 Elastoplasticity of a damaged body

10.3.1 Damage criterion

For a one-dimensional model and for *elasticity* coupled to damage the energy function

$$W^{\mathrm{e}} = \tfrac{1}{2} E(1 - D)(\varepsilon^{\mathrm{e}})^2 \tag{10.11}$$

can be written down in an intuitive way since damage *decreases* the elasticity of the material by a ratio $(1 - D)$. Following Kachanov (1958) and Rabotnov (1963), to whom the idea is due, we can write

$$\sigma = \frac{\partial W^{\mathrm{e}}}{\partial \varepsilon^{\mathrm{e}}} = (1 - D) E \varepsilon^{\mathrm{e}}, \tag{10.12}$$

hence

$$\varepsilon^{\mathrm{e}} = \frac{\sigma}{(1-D)E} = \frac{\sigma}{\overline{E}} = \frac{\overline{\sigma}}{E}, \tag{10.13}$$

where $\overline{\sigma} = \sigma/(1-D)$ is an effective stress (referred to the surface that really transmits the internal forces) and $\overline{E}$ is the effective elasticity modulus resulting from damage.

In anisotropic linear elasticity with isotropic damage (not coupled with plasticity at the free-energy level; we consider an isothermal evolution ignoring temperature effects) the following free energy provides a rather good model:

$$W = W(\boldsymbol{\varepsilon}^{\mathrm{e}}, \boldsymbol{\alpha}, \beta, D) = W^{\mathrm{e}}(\boldsymbol{\varepsilon}^{\mathrm{e}}, D) + W^{\mathrm{p}}(\boldsymbol{\alpha}, \beta) \tag{10.14}$$

with

$$W^{\mathrm{e}}(\boldsymbol{\varepsilon}^{\mathrm{e}}, D) = \tfrac{1}{2}(1-D)\boldsymbol{\varepsilon}^{\mathrm{e}} : \mathbf{E} : \boldsymbol{\varepsilon}^{\mathrm{e}}. \tag{10.15}$$

As a generalization of (10.12) we have thus

$$\left.\begin{aligned} \boldsymbol{\sigma} &= \frac{\partial W^{\mathrm{e}}}{\partial \boldsymbol{\varepsilon}^{\mathrm{e}}} = (1-D)\mathbf{E} : \boldsymbol{\varepsilon}^{\mathrm{e}}, \\ Y &= -\frac{\partial W^{\mathrm{e}}}{\partial D} = \frac{1}{2}\boldsymbol{\varepsilon}^{\mathrm{e}} : \mathbf{E} : \boldsymbol{\varepsilon}^{\mathrm{e}}, \\ B &= -\frac{\partial W^{\mathrm{p}}}{\partial \beta}, \qquad \mathbf{A} = -\frac{\partial W^{\mathrm{p}}}{\partial \boldsymbol{\alpha}}. \end{aligned}\right\} \tag{10.16}$$

The second of these equations shows that the generalized force Y, the thermodynamic dual of D, is none other than the *elastic energy* of the *non*damaged body in this simple model.

For linear isotropic elasticity we would obtain

$$Y = \frac{\sigma_Y^2}{2E(1-D)^2} R_v, \tag{10.17}$$

where

$$\sigma_Y = (\tfrac{3}{2}\sigma_{ij}^{\mathrm{d}}\sigma_{ij}^{\mathrm{d}})^{1/2},$$

$$\sigma_{ij}^{\mathrm{d}} = \sigma_{ij} - \sigma_m \delta_{ij}.$$

$$\sigma_m = \tfrac{1}{3}(\sigma_1 + \sigma_2 + \sigma_3),$$

are, respectively, the equivalent stress, the deviatoric stress, and the mean stress, and we have set

$$R_v = \frac{2}{3}(1+v) + 3(1-2v)\left(\frac{\sigma_m}{\sigma_Y}\right)^2, \tag{10.18}$$

where E and v are Young's modulus and Poisson's ratio for the non-damaged material. The dimensionless quantity R_v can be viewed as defining a *damage criterion* by (Lemaître and Chaboche, 1985)

$$\sigma_Y\sqrt{R_v} - \sigma^* = 0. \tag{10.19}$$

10.3.2 Evolution of damage parameters

For a one-dimensional model we can write (see eqn (10.13))

$$\bar{E} = (1-D)E, \qquad \text{i.e. } D = 1 - \frac{\bar{E}}{E}. \tag{10.20}$$

Both E and $\bar{E}$ can be carefully measured and experimental curves giving D as a function of the number of loading cycles in a fatigue-creep test at low frequency can be plotted (see Fig. 10.2). This exhibits a damage accumulation in fatigue process and a distinction needs to be made between low- and high-cycle fatigue (see Kachanov, 1986). Typically, we would write

$$\frac{\delta D}{\delta N} = g(\sigma_m, \sigma_M, D),$$

where N is the number of cycles, σ_m is the mean stress and σ_M is the maximum stress. In the case of low-cycle fatigue the damage increment in

Fig. 10.2. Experimental plot of D. Material: stainless steel A316. Test temperature: 550°C. N_c = 218 cycles. (Fracture at D = 1) (After Lemaître, 1985)

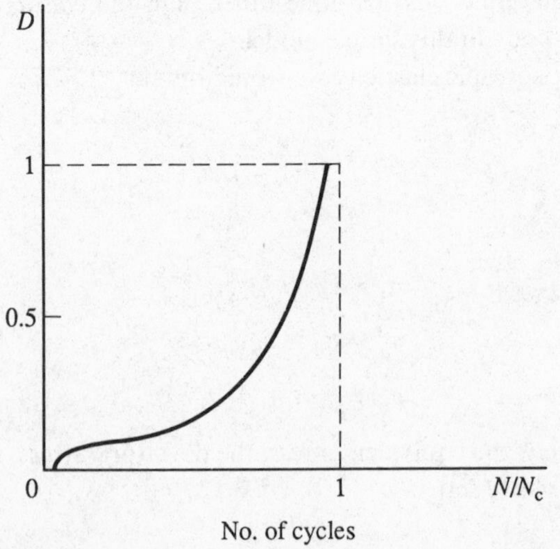

each cycle is typically related to the plastic strain accumulated during the cycle, while for high-cycle fatigue the microcrack nucleation resulting in damage is thought to be related to stress concentrations at the tips of isolated slip bands, hence to so-called *microplasticity* (see below). The choice of function g above requires some ingenuity. However, results such as those in Fig. 10.2 suggest the possibility of using a pseudo-potential $\mathscr{D}^*(Y, D, \theta)$ in the form of a *power* law in terms of Y, i.e.,

$$\mathscr{D}^*_{\text{dam}}(Y, D) = \frac{S_0}{s_0 + 1}\left(\frac{Y}{S_0}\right)^{s_0+1} \frac{1}{1 - D}, \tag{10.21}$$

in such a way that $(10.9)_4$ yields the evolution equation

$$\dot{D} = \frac{\partial \mathscr{D}^*_{\text{dam}}}{\partial Y} = \left(\frac{Y}{S_0}\right)^{s_0} \frac{1}{1 - D} \tag{10.22}$$

where only two material parameters, S_0 and s_0, are present. They possibly depend on temperature. Fracture can be defined by

$$D = D_c \leqslant 1. \tag{10.23}$$

10.3.3 Plastic microstrains

In reality, damage is always associated with an irreversible strain. This can be disregarded as compared to elastic strains in fatigue tests in the elastic range. We call $\boldsymbol{\varepsilon}^\pi$ the plastic microstrain that may become of some importance outside this range, and $\mathscr{P}$ its thermodynamic dual variable, i.e., a stress. The usual thermodynamic potential reads

$$W = W(\boldsymbol{\varepsilon}^e, D, \boldsymbol{\varepsilon}^\pi). \tag{10.24}$$

Its conjugate by partial Legendre–Fenchel transformation performed with respect to $\boldsymbol{\varepsilon}^e$ yields

$$W^* = W^*(\boldsymbol{\sigma}, D, \boldsymbol{\varepsilon}^\pi), \qquad \boldsymbol{\sigma} = \frac{\partial W}{\partial \boldsymbol{\varepsilon}^e}. \tag{10.25}$$

For instance (case with isotropic linear elasticity and $\boldsymbol{\varepsilon}^\pi$ as separate variable), we can take

$$W^* = \frac{1}{2}\left[\frac{1 + \nu}{E} \frac{\operatorname{tr} \boldsymbol{\sigma}^2}{1 - D} - \frac{\nu}{E} \frac{(\operatorname{tr} \boldsymbol{\sigma})^2}{1 - D}\right] + W^*_\pi(\boldsymbol{\varepsilon}^\pi), \tag{10.26}$$

from which follow the equations of state

$$\boldsymbol{\varepsilon}^e = \frac{\partial W^*}{\partial \boldsymbol{\sigma}}, \qquad Y = -\frac{\partial W^*}{\partial D}, \qquad \mathscr{P} = -\frac{\partial W^*_\pi}{\partial \boldsymbol{\varepsilon}^\pi}. \tag{10.27}$$

The first of these equations is coupled to (10.22) or, whenever time has no effect, to the following equation which then replaces (10.22):

$$\dot{D} = \frac{\dot{\lambda}}{(1 - D)} \left(\frac{Y}{S_0} \right)^{s_0}. \tag{10.28}$$

It remains to formulate an evolution equation for the plastic microstrains. For instance, Chaboche (1974) proposes

$$\left. \begin{aligned} \dot{\boldsymbol{\varepsilon}}^{\pi} &= \left[\frac{\Delta \sigma_Y}{k(1 - D)} \right]^{\beta - 1} \frac{|\dot{\sigma}_Y|}{1 - D}, \\ \Delta \sigma_Y &= [\tfrac{3}{2}(\sigma_{ij}^{\mathrm{d}} - \bar{\sigma}_{ij}^{\mathrm{d}})(\sigma_{ij}^{\mathrm{d}} - \bar{\sigma}_{ij}^{\mathrm{d}})]^{1/2}, \end{aligned} \right\} \tag{10.29}$$

where the stress $\bar{\sigma}^{\mathrm{d}}$ can be identified with the mean of $\boldsymbol{\sigma}^{\mathrm{d}}$ (obviously, the mean is applied to the components of $\boldsymbol{\sigma}^{\mathrm{d}}$) over a loading cycle. It is shown that

$$\dot{\lambda} = \varepsilon^{\pi}(1 - D). \tag{10.30}$$

The coefficients k and β characterize internal frictions. We have thus at our disposal a model of fatigue for a high number of cycles. It provides a nonlinear evolution of damage. The unknown coefficients are identified via the so-called *Woehler curve* which gives the number of cycles at fracture in terms of the maximum stress reached in a combined fatigue and traction–compression test. These curves go back to the nineteenth century (Woehler, 1858) and they prove to be among the oldest, extremely reliable results in experimental solid mechanics (see Bell, 1973, pp. 461–2).†

10.3.4 Coupling with plasticity

For illustrative purposes we use the working hypothesis of *generalized standard materials* (existence of the normality rule with *associated plasticity*, i.e., the plasticity criterion is used as a plasticity potential). The coupling, for instance, applies particularly well in the case of the so-called *ductile damage* during the initiation and growth of cavities from the boundary of inclusions in the presence of plastic instability, as is the case during metal forming, or in a fatigue test at low cycle number ($N < 10^4$ cycles) at relatively low temperature. The modelling is as follows:

the state variables are $\boldsymbol{\varepsilon}^{\mathrm{e}}$, $\boldsymbol{\varepsilon}^{\mathrm{p}}$, β and D where the last three dissipate;

the dissipation potential, taken as the plasticity criterion, reads

$$f(\boldsymbol{\sigma}, B, D) = 0, \tag{10.31}$$

† A Woehler (1819–1914) may be considered the Founding Father of mechanical testing laboratories (see Timoshenko, 1953, pp. 167–73).

with

$$\mathscr{D}^* = f(\boldsymbol{\sigma}, B, D) + \mathscr{D}_{\mathrm{p}}^*(Y, D) \tag{10.32}$$

from which there follow the normality rules

$$\dot{\boldsymbol{\varepsilon}}^{\mathrm{p}} = \dot{\lambda}\frac{\partial f}{\partial \boldsymbol{\sigma}}, \qquad \dot{\beta} = \dot{\lambda}\frac{\partial f}{\partial B}, \qquad \dot{D} = \dot{\lambda}\frac{\partial \mathscr{D}_{\mathrm{p}}^*}{\partial Y}. \tag{10.33}$$

For Mises plasticity criterion with hardening (compare (10.21)), we have

$$f = \overline{(\sigma_Y + B)} + K_0 \tag{10.34}$$

where $\overline{\sigma} = \sigma/(1 - D)$. Then eqns (10.33) give

$$\left.\begin{aligned} \dot{\boldsymbol{\varepsilon}}^{\mathrm{p}} &= \dot{\lambda}\frac{\partial f}{\partial \boldsymbol{\sigma}} = \frac{3}{2}\frac{\dot{\lambda}}{1-D}\frac{\boldsymbol{\sigma}^{\mathrm{d}}}{\sigma_Y}, \\ \dot{\beta} &= \dot{\lambda}\frac{\partial f}{\partial B} = \frac{\dot{\lambda}}{1-D} = \left(\frac{2}{3}\dot{\varepsilon}_{ij}^{\mathrm{p}}\dot{\varepsilon}_{ij}^{\mathrm{p}}\right)^{1/2}, \\ \dot{D} &= \dot{\lambda}\frac{\partial \mathscr{D}_{\mathrm{p}}^*}{\partial Y} = \left(\frac{Y}{S_0}\right)^{s_0}\frac{\dot{\lambda}}{1-D}, \end{aligned}\right\} \tag{10.35}$$

if the expression (10.21) is assumed for $\mathscr{D}_{\mathrm{p}}^*$.

The plastic multiplier $\dot{\lambda}$ is determined as usual by the consistency condition $\dot{f} = 0$. One obtains thus

$$\dot{\lambda} = H(f)\frac{\langle \dot{\sigma}_Y \rangle (1 - D)}{(\mathrm{d}B/\mathrm{d}\beta) - K_0(Y/S_0)^{s_0}}; \tag{10.36}$$

$\langle \cdot \rangle$ denotes the positive part and $H(f) = 0$, if $f < 0$, and $H(f) = 1$, if $f = 0$. The derivative $\mathrm{d}B/\mathrm{d}\beta$ characterizes *nonlinear hardening*. For example,

$$B = -\frac{\partial W}{\partial \beta} = K\beta^{1/M} \tag{10.37}$$

and the *hardening modulus* h is given by

$$h = \mathrm{d}B/\mathrm{d}\beta = \frac{K}{M}\beta^{(1-M)/M}; \tag{10.38}$$

the two coefficients K and M can be identified in a traction hardening test.

10.4 Example of a complete model with ductile damage

To conclude this chapter we quote in full a model proposed by Cordebois (1983). In this model devised for steel the evolution equations (10.35) take on the following precise form:

$$\left.\begin{aligned}
\dot{\boldsymbol{\varepsilon}}^{\mathrm{e}} &= \frac{1}{1-D}\left[\frac{1+\nu}{E}\dot{\boldsymbol{\sigma}} - \frac{\nu}{E}(\operatorname{tr}\dot{\boldsymbol{\sigma}})\mathbf{1}\right] \quad \text{(isotropy)},\\
\dot{\boldsymbol{\varepsilon}}^{\mathrm{p}} &= \frac{3}{2}H(f)\frac{\langle\dot{\sigma}_Y\rangle}{\dfrac{K}{M}\beta^{(1-M)/M} - K_0\left(\dfrac{D_{\mathrm{c}}R_\nu}{\varepsilon_{\mathrm{R}}-\varepsilon_D}\right)}\cdot\frac{\boldsymbol{\sigma}^{\mathrm{d}}}{\sigma_Y},\\
\dot{D} &= \frac{D_{\mathrm{c}}}{\varepsilon_{\mathrm{R}}-\varepsilon_{\mathrm{D}}}R_\nu\dot{\beta},
\end{aligned}\right\} \tag{10.39}$$

with

$$\dot{\beta} = (\tfrac{2}{3}\dot{\varepsilon}^{\mathrm{d}}_{j}\dot{\varepsilon}^{\mathrm{d}}_{ij})^{1/2} \tag{10.40}$$

and

$$\beta < \varepsilon_{\mathrm{D}} \Rightarrow D = 0, \tag{10.41}$$

where $D = D_{\mathrm{c}}$ corresponds to the initiation of *macro*cracks and $\varepsilon = \varepsilon_{\mathrm{R}}$ is the strain reached at fracture. Cordebois (1983) has identified all the parameters of this model for steel (French standard 35 NCD 16) at room temperature. These parameters are

$$E = 210\,\mathrm{MPa}, \qquad \nu = 0.3,$$

$$K_0 = 1200\,\mathrm{MPa}, \quad K = 3340\,\mathrm{MPa}, \qquad M = 3.1,$$

$$\varepsilon_{\mathrm{D}} = 2\%_0, \quad \varepsilon_{\mathrm{R}} = 0.7, \qquad D_{\mathrm{c}} = 0.34.$$

Damage can also be coupled to elasto*visco*plasticity, especially in loading tests at relatively high temperature. Like most viscous processes, this lies outside the scope of this book. Continuum damage mechanics is a blossoming field of both applied mechanics *and* applied mathematics. The reader is referred to Krajcinovic and Lemaître (Editors, 1987), Lemaître and Chaboche (1985), Krajcinovic (1989), and Kachanov (1986), for more ample developments. The last reference is of special interest as it gives many examples of evaluation of the critical time at fracture. The following problems, also, have for object to widen the horizon of readers as regards the thermodynamic formulation of damage, the choice of damage parameter and its time evolution, and the important property of *anisotropy* induced by damage.

Problems for Chapter 10

10.1. In three dimensions damage should be represented by a second-order tensor $\mathbf{D}$ (not to be mistaken for the strain-rate tensor).

(a) Show that a possible three-dimensional generalization of the notion of effective stress $\bar{\sigma} = \sigma/(1-D)$ (see eqn 10.13)) reads (Murakami and Ohno, 1981)

Ohno, 1981)

$$\bar{\boldsymbol{\sigma}} = \tfrac{1}{2}[(\mathbf{I} - \mathbf{D})^{-1}\boldsymbol{\sigma} + \boldsymbol{\sigma}(\mathbf{I} - \mathbf{D})^{-1}]. \tag{1}$$

(b) Assume that the behaviour of the damaged material is described by an elasticity tensor $\bar{\mathbf{E}}(\mathbf{D})$ such that

$$\boldsymbol{\sigma} = \bar{\mathbf{E}}(\mathbf{D}) : \boldsymbol{\varepsilon}^{\mathrm{e}}. \tag{2}$$

In this absence of damage this reads $\boldsymbol{\sigma} = \mathbf{E} : \boldsymbol{\varepsilon}^{\mathrm{e}}$. (Therefore $\mathbf{D}$ is formally an operator of order 8, as it transforms $\mathbf{E}$ into $\bar{\mathbf{E}}$.) Then the effective stress can be defined by

$$\bar{\boldsymbol{\sigma}} = \mathbf{E} : \boldsymbol{\varepsilon}^{\mathrm{e}}. \tag{3}$$

Show formally that

$$\bar{\boldsymbol{\sigma}} = \mathbf{M}(\mathbf{D}) : \boldsymbol{\sigma}, \qquad \bar{\boldsymbol{\sigma}} = \mathbf{E}\bar{\mathbf{E}}^{-1}(\mathbf{D})\boldsymbol{\sigma}. \tag{4}$$

Eqn (1) is a simplification of the last expression which does not provide an equivalence in strains as $\bar{\mathbf{E}}(\mathbf{D})$ in general is *not* a symmetric operator in $\mathbb{R}^6$, except for the case where $\mathbf{D}$ is spherical, i.e., proportional to unity.

10.2. Thermodynamical model of damage (Cordebois and Sidoroff, 1982)
For a damaged material consider the following internal energy and enthalpy (obtained by partial Legendre transformation):

$$\left.\begin{aligned} W(\boldsymbol{\varepsilon}^{\mathrm{e}}, \mathbf{D}, \boldsymbol{\alpha}) &= \tfrac{1}{2}\boldsymbol{\varepsilon}^{\mathrm{e}} : \mathbf{E}(\mathbf{D}) : \boldsymbol{\varepsilon}^{\mathrm{e}} + g(\boldsymbol{\alpha}),\\ W^*(\boldsymbol{\sigma}, \mathbf{D}, \boldsymbol{\alpha}) &= \boldsymbol{\sigma} : \boldsymbol{\varepsilon}^{\mathrm{e}} - W = \tfrac{1}{2}\boldsymbol{\sigma} : \mathbf{S}(\mathbf{D}) : \boldsymbol{\sigma} - \mathbf{g}(\boldsymbol{\alpha}),\end{aligned}\right\} \tag{1}$$

where $\mathbf{D}$ is the damage parameter and $\boldsymbol{\alpha}$ represents the plastic internal variables. For a virgin material eqns (1) reduce to

$$\begin{aligned} W(\boldsymbol{\varepsilon}^{\mathrm{e}}, \boldsymbol{\alpha}) &= \tfrac{1}{2}\boldsymbol{\varepsilon}^{\mathrm{e}} : \mathbf{E}_0 : \boldsymbol{\varepsilon}^{\mathrm{e}} + \mathbf{g}(\boldsymbol{\alpha}),\\ W^*(\boldsymbol{\sigma}, \boldsymbol{\alpha}) &= \tfrac{1}{2}\boldsymbol{\sigma} : \mathbf{S}_0 : \boldsymbol{\sigma} - \mathbf{g}(\boldsymbol{\alpha}).\end{aligned} \tag{2}$$

(a) Write the state laws for $\boldsymbol{\sigma}$ and $\boldsymbol{\varepsilon}^{\mathrm{e}}$ in the damaged regime. Show that the dissipation per unit volume takes the form

$$\phi = \boldsymbol{\sigma} : \dot{\boldsymbol{\varepsilon}}^{\mathrm{p}} + \mathbf{G}\dot{\mathbf{D}} + \mathbf{A}\dot{\boldsymbol{\alpha}} \geqslant 0 \tag{3}$$

with

$$\mathbf{G} = \tfrac{1}{2}\boldsymbol{\sigma} : \mathbf{A}'(\mathbf{D})[\boldsymbol{\sigma}]. \tag{4}$$

(b) The damaged constitutive equation (effective stress) is obtained by an energy (enthalpy) *identification*, i.e.,

$$W^*(\boldsymbol{\sigma}, \mathbf{D}, \boldsymbol{\alpha}) = W_0^*(\bar{\boldsymbol{\sigma}}, \boldsymbol{\alpha}), \tag{5}$$

so that we will obtain $\bar{\sigma} = \mathbf{M}(\mathbf{D})\boldsymbol{\sigma}$ generalizing the one-dimensional formula $\bar{\sigma} = \sigma/(1 - D)$. Then show that

$$\mathbf{S}(\mathbf{D}) = \mathbf{M}^{\mathrm{T}}(\mathbf{D})\mathbf{S}_0\mathbf{M}(\mathbf{D}), \quad \bar{\boldsymbol{\varepsilon}}^{\mathrm{e}} = (\mathbf{M}^{\mathrm{T}})^{-1}(\mathbf{D})[\boldsymbol{\varepsilon}^{\mathrm{e}}], \quad \boldsymbol{\varepsilon}^{\mathrm{e}} = \mathbf{S}_0 : \bar{\boldsymbol{\sigma}}. \tag{6}$$

10.3. Damage-induced elastic anisotropy

Assume that the virgin material of 10.2 is *isotropic* so that

$$\mathbf{S}_0 : \boldsymbol{\sigma} = \frac{1+\nu}{E}\boldsymbol{\sigma} - \frac{\nu}{E}(\operatorname{tr}\boldsymbol{\sigma})\mathbf{I}, \tag{1}$$

where ν and E are Poisson's ratio and Young's modulus. If we work in the principal axes of $\mathbf{D}$ and assume that the stress tensor and the damage tensor have the same eigendirections, eqn (1) in 10.1 may be considered as a good definition. Then show that

$$\bar{\sigma}_{11} = \frac{\sigma_{11}}{W_1}, \qquad \bar{\sigma}_{12} = \frac{\sigma_{12}}{W_{12}}, \tag{2}$$

where the Ws are *weakening coefficients* such that

$$W_1 = 1 - D_1, \qquad W_{12} = \frac{2W_1 W_2}{W_1 + W_2} = 2\left[\frac{1}{(1-D_1)} + \frac{1}{(1-D_2)}\right]^{-1}. \tag{3}$$

Show using the Voigt notation (six-dimensional vector notation) that

$$\varepsilon^e_\alpha = S_{\alpha\beta}(\mathbf{D})\sigma_\beta, \qquad \alpha, \beta = 1, 2, \ldots, 6, \tag{4}$$

with, e.g.,

$$S_{11}(\mathbf{D}) = \frac{1}{E}\frac{1}{W_1^2}, \qquad S_{12} = -\frac{\nu}{E}\frac{1}{W_1 W_2}, \tag{5}$$

so that the *damaged* material does *not* behave isotropically in its elastic regime.

10.4. Uniaxial damage (10.3 continued)

Loading in pure traction can create a uniaxial state of damage. Then $\mathbf{D} = \operatorname{diag}(D_1, D_2, D_3)$. Taking account of the results of 10.3 show that apparent Young's modulus and Poisson's coefficient are given by

$$\bar{E} = E(1 - D_1)^2, \qquad \bar{\nu} = \nu\frac{1 - D_1}{1 - D_2}.$$

The ratio D_2/D_1 will characterize the degree of induced anisotropy through damage.

10.5. Another model

Instead of eqn (1) of 10.1 consider an *effective stress* defined by

$$\bar{\boldsymbol{\sigma}} = \sqrt{\varphi(\mathbf{D})}\cdot[(\mathbf{I} - \mathbf{D})^{-1/2}\boldsymbol{\sigma}(\mathbf{I} - \mathbf{D})^{-1/2}]. \tag{1}$$

Then show that the elastic constitutive equations of the damaged material read

$$\boldsymbol{\varepsilon}^e = \varphi(\mathbf{D})\left\{\frac{1+\nu}{E}(\mathbf{I} - \mathbf{D})^{-1}\cdot\boldsymbol{\sigma}\cdot(\mathbf{I} - \mathbf{D})^{-1} - \frac{\nu}{E}\operatorname{tr}[\boldsymbol{\sigma}\cdot(\mathbf{I} - \mathbf{D})^{-1}](\mathbf{I} - \mathbf{D})^{-1}\right\}, \tag{2}$$

if ν and E are the elastic coefficients of the isotropic virgin material. For instance, one may take $\varphi(\mathbf{D}) = \varphi(D_1, D_2) = (1 - D_1)^p(1 - D_2)^q$ with $p = 7/4$ and $q = -3/4$ (D. Nouailhas, unpublished report, 1980).

10.6. Justification by means of homogenization

Make a literature search and write a report showing that if damage is due to ellipsoidal cavities or periodically spaced parallel cracks the tensor $\bar{\mathbf{E}}$ appearing in $\mathbf{M}(\mathbf{D}) = \mathbf{E}\bar{\mathbf{E}}^{-1}$ has an expression given by

$$\bar{E}_{ijkl} = \left(\frac{\bar{V}}{V}\delta_{ir}\delta_{js} - \frac{1}{V}\int_V \mathscr{C}_{irjs}\,\mathrm{d}V\right)E_{rskl},$$

where V is the apparent volume element, $\bar{V}$ is the apparent volume from which the void volume has been substracted and $\mathscr{C}_{ijkl}$ are the stress concentration coefficients introduced in Section 9.3.

10.7. Evolution of damage (simple model)

An admitted simple form is Norton's law in one dimension

$$\dot{D} = (\bar{\sigma}/\alpha)^n \tag{1}$$

(Kachanov, 1958, so-called *creep damage*), where α and n are two characteristic coefficients. Let D_c be the critical value of D (at fracture). Then show by integration in time that

$$D = 1 - \left(1 - \frac{t}{t_c}\right)^q, \qquad t_c = q(\bar{\sigma}/\alpha)^n. \tag{2}$$

What is the value of q?

10.8. Evolution of damage (thermodynamic theory)

Consider the dissipation inequality of 10.2 and a dissipation potential $\mathscr{D}^* = \mathscr{D}^*(\boldsymbol{\sigma}, \mathbf{A}, \mathbf{G}; \boldsymbol{\varepsilon}^e, \boldsymbol{\alpha}, \mathbf{D})$. We suppose that the following separation is admissible:

$$\mathscr{D}^* = \mathscr{D}_1(\boldsymbol{\sigma}, \mathbf{A}; \boldsymbol{\alpha}) + \mathscr{D}_D^*(\mathbf{G}; \boldsymbol{\varepsilon}^e, \mathbf{D}). \tag{1}$$

As a particular case we may take

$$\mathscr{D}_D^*(\mathbf{G}, \boldsymbol{\varepsilon}^e, \mathbf{D}) = -F(\boldsymbol{\varepsilon}^e, \mathbf{D})\mathbf{Q}\cdot\mathbf{G}, \tag{2}$$

where $D = c\,\mathrm{tr}\,\mathbf{D}$. Using the normality condition for generalized standard materials give the evolution equations for D and $\mathbf{D}$. In particular, show that, if $\dot{D}$ is given, then $\dot{\mathbf{D}} = \mathbf{Q}\dot{D}$, which is a convenient form in many applications.

10.9. Ductile damage

Consider

$$\dot{D} = \left(\frac{\bar{\sigma}^2}{2ES_0}\right)^n |\dot{\varepsilon}^p| \tag{1}$$

relating the damage evolution and plastic strain rate. Here S_0 and n are material parameters and $\bar{\sigma}$ is the effective stress. Considering a hardening law such as $\varepsilon^p = (\bar{\sigma}/K)^M$, where K and M are material parameters, find an expression of $\dot{D}$ in terms of the evolution in stresses. Integrate this in time to deduce $D(\sigma)$. Ductile fracture occurs for a certain value of $\sigma = \sigma_R$.

[*Answer*:
$$D = \left(1 - \left\langle \frac{\sigma - \sigma_D}{S_0} \right\rangle^n\right)^{1/n}$$

where σ_D, the threshold in stress, has been introduced, so that σ has been replaced by $\langle \sigma - \sigma_D \rangle$.]

10.10. Ductile damage (term paper)

Make a literature search and show in a short report how to relate the growth of cavities in a plastic body to the evolution of damage parameters. Compare with the modelling of brittle damage by growth of microvoids. [*Hint*: study Marigo (1985).]

11

Numerical solution of plasticity problems

The object of the chapter The technical difficulties faced in solving plasticity problems, which are free-boundary problems, are such that sooner or later one has to use a numerical implementation. While works fully devoted to numerical methods in solid mechanics give general solution techniques, here we focus on the specificity of the *incremental* or evolutionary nature of elastoplasticity problems and on Moreau's implicit scheme which is particularly well suited to this.

11.1 Introduction

Save for a few exceptions (see Appendix 3) the analytic solution of a problem of elastoplasticity is a formidable task since it involves a free *boundary* which is none other than the border between elastic and plastic domains, in general an unknown in the problem. In addition, by the very nature of elastoplasticity, the corresponding problems are *nonlinear* and the nature of certain plasticity criteria does not improve the situation. The relevant question at this point is: what is the *quasi-static evolution* of an elastoplastic structural member? The very nature of elastoplasticity and the corresponding incremental formulation are well suited to the study of general features of such a mechanical behaviour (see Chapters 4 to 6) and, indeed, via both spatial and temporal discretizations, to a numerical solution for real problems that involve complex geometries, somewhat elaborated plasticity criteria, and complex loading paths (including both loading *and* unloading). The most appropriate method for the spatial problem obviously is the one of *finite elements* (for short FEM).[†] It can be noticed that sufficiently accurate computations of the elastoplastic response of structural members are rather recent (1965, 1970s).

[†] Some textbooks are entirely devoted to elastoplastic computations although in a rather standard way: see, e.g. Owen and Hinton (1980). Here we focus on the temporal evolution which is specific to elasto*plasticity*.

11.2 Elementary notions on numerical computations

Usually we make the distinction between *explicit* and *implicit* schemes of numerical approaches. This distinction can be made clear with the following simple *evolution* problem that we state thus:

Problem Let

$$\dot{y} = A(y,t), \qquad y(t) \in V, \qquad y(0) = y_0. \tag{11.1}$$

Define a numerical approximation of $y(t)$ on the time interval $[0, T]$, knowing both y_0 and the expression for A.

We call Δt the time increment. We have $y_n \approx y(t_n)$ and $t_n = n\Delta t$.

In the *Eulerian explicit scheme* we have

$$y_{n+1} = y_n + (\Delta t)A(y_n, t_n) \Leftrightarrow \Delta y = \dot{y}_n \Delta t, \tag{11.2}$$

in the implicit scheme we have

$$y_{n+1} = y_n + (\Delta t)A(y_{n+1}, t_{n+1}) \Leftrightarrow \Delta y = \dot{y}_{n+1} \Delta t, \tag{11.3}$$

These different ways of constructing y_{n+1} lead to the following comments.

(i) With y_n known, the two schemes (11.2) and (11.3) define the same sequence y_n step by step.

(ii) Clearly, the implicit scheme (11.3) is *more involved* to solve than the explicit one, since (11.3) must be solved with respect to y_{n+1}, while (11.2) only involves a simple evaluation of its right-hand side. Furthermore, as the operator A is practically always *nonlinear*, the solution of (11.3) requires the use of *iterations*.

(iii) Both schemes (11.2) and (11.3) introduce computational errors of the type $e_n = y_n - y(t_n)$ at step n.

More sophisticated schemes than (11.2) and (11.3) can be envisaged, but this will be at the price of lengthy intermediate computations and a cost in computer memory space.

11.3 Application to elastoplasticity

11.3.1 Explicit scheme

To illustrate our concern we consider a unified formulation of the type already introduced in Section 5.1 (Example 4). Let Ω be an elastoplastic material structure in quasi-static evolution (inertia terms are disregarded). Let $\mathbf{g}^d$, $\mathbf{T}^d$ and $\mathbf{u}^d$ denote the data in Ω, on $\partial\Omega_T$ and on $\partial\Omega_u$, respectively. The temporal interval $[0, T]$ is subdivided as $0 < t_1 < t_2 < \cdots < t_n = T$ and we have to account for initial data

$$\mathbf{u}(0) = \mathbf{u}_0, \qquad \mathscr{E}(0) = \mathscr{E}_0, \qquad \mathscr{E}^p(0) = \mathscr{E}_0^p, \qquad \mathbf{\Sigma}(0) = \mathbf{\Sigma}_0. \tag{11.4}$$

We can state then the following.

Problem Knowing the increments $\Delta\mathbf{g}^d$, $\Delta\mathbf{T}^d$ and $\Delta\mathbf{u}^d$ determine the increments $\Delta\mathbf{u} = \mathbf{u}_{n+1} - \mathbf{u}_n$, $\Delta\mathscr{E} = \mathscr{E}_{n+1} - \mathscr{E}_n$ and $\Delta\mathbf{\Sigma} = \mathbf{\Sigma}_{n+1} - \mathbf{\Sigma}_n$ with the incremental laws

$$\left.\begin{aligned} \Delta\mathscr{E} &= \Delta\mathscr{E}^e + \Delta\mathscr{E}^p, \\ \Delta\mathscr{E}^p &= N_C(\mathbf{\Sigma}_n), \\ \Delta\mathbf{\Sigma} &= \mathscr{L} \cdot \Delta\mathscr{E}^e, \end{aligned}\right\} \tag{11.5}$$

and

$$\Delta\mathbf{u} = \Delta\mathbf{u}^d \quad \text{(kinematic condition at time } t = t_{n+1}),$$

$$\mathbf{\Sigma}_n + \Delta\mathbf{\Sigma} \in S_{n+1} \quad \text{(static condition at time } t = t_{n+1}).$$

This is *indeed* an explicit scheme since it is $\mathbf{\Sigma}_n$ that appears in the right-hand side of the evolution equation (compare (11.5) and (11.2)).

11.3.2 Implicit scheme

The *same* problem as the one considered above is numerically formulated by using the following incremental laws:

$$\left.\begin{aligned} \Delta\mathscr{E} &= \Delta\mathscr{E}^e + \Delta\mathscr{E}^p, \\ \Delta\mathscr{E}^p &\in N_C(\mathbf{\Sigma}_{n+1}), \\ \Delta\mathbf{\Sigma} &= \mathscr{L} \cdot \Delta\mathscr{E}^e, \end{aligned}\right\} \tag{11.6}$$

This is indeed an implicit scheme since it is $\mathbf{\Sigma}_{n+1}$ that appears in the right-hand side of the evolution equation. This is *more satisfactory* in the sense that we can then respect the normality rule at time t_{n+1} and it is $\mathbf{\Sigma}_{n+1}$ that belongs to the convex C. Several authors (among them Moreau, 1971, Johnson, 1976) have discussed the convergence of the implicit scheme. Their essential results can be stated thus

Convergence *There exists a constant γ, which is independent of h (spatial discretization) and Δt (temporal discretization) and is such that*

$$\underset{n}{\text{Max}}\ \|\Sigma_n - \Sigma_n^h\| \leqslant [\delta(h) + \sqrt{\Delta t}]\gamma \tag{11.7}$$

when $\delta(h)$ *characterizes the accuracy in the spatial discretization* $[\lim_{h\to 0}\delta(h) = 0]$.

Remarkably enough, we have not said anything about spatial discretization. A good reason for that is that spatial and temporal discretizations are *uncoupled.*

It has also been established that the implicit scheme is *unconditionally stable.* We recall that a scheme is said to be stable if its approximate solution is continuous with respect to initial data. Therefore, in spite of its at first sight complex appearance, the quasi-stable evolution of elastoplastic structural members introduces less numerical difficulties than other problems in solid mechanics (those involving high-frequency dynamics in particular).

Fig. 11.1 provides in an obvious manner an illustration of implicit and

Fig. 11.1. Illustration of explicit and implicit schemes for problem (11.8).

(a) Explicit scheme:

$$\begin{cases} y_{n+1} - y_n + (\Delta t)A(y_n) = f_{n+1} - f_n, \\ y_{n+1} = \text{Proj}|_{\text{tang}}[y_n + f_{n+1} - f_n] \end{cases} \tag{11.8$'$}$$

(after Moreau, 1971).

(b) Implicit scheme;

$$\begin{cases} y_{n+1} - y_n + (\Delta t)A(y_{n+1}) = f_{n+1} - f_n, \\ y_{n+1} = \text{Proj}|_C[y_n + f_{n+1} - f_n] \end{cases} \tag{11.8$''$}$$

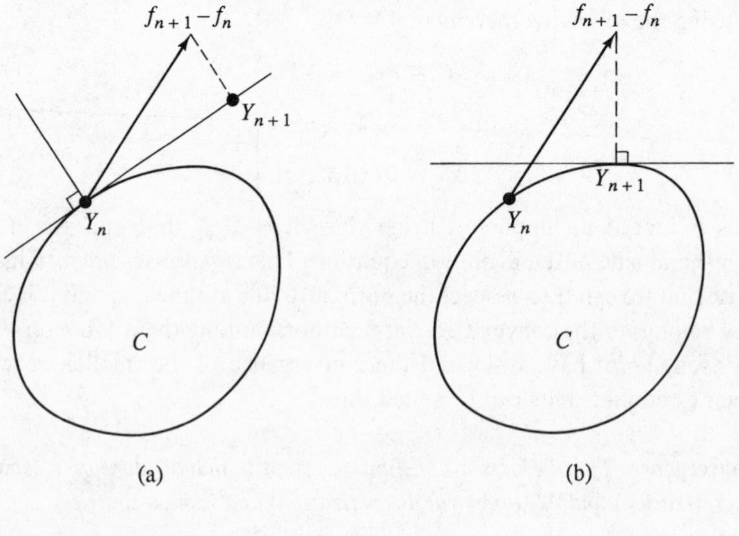

explicit schemes (after Moreau, 1971) for a problem of evolution involving a convex set C, e.g.

$$\left.\begin{aligned} \dot{Y} + A(Y) &= \dot{f}(t), \\ A &\in \partial\Psi_{\mathrm{c}}. \end{aligned}\right\} \tag{11.8}$$

11.3.3 Incremental problem for the implicit scheme

This is a nonlinear boundary-value problem which leads to principles of *minimum* and can be stated thus.

Problem Determine $\Delta\mathbf{u}$, $\Delta\mathscr{E}$ and $\Delta\boldsymbol{\Sigma}$ subject to

$$\left.\begin{aligned} \Delta\mathbf{u} = \Delta\mathbf{u}^{\mathrm{d}} \text{ on } \partial\Omega_u \quad &(\text{kinematic condition at } t = t_{n+1}), \\ \boldsymbol{\Sigma}_n + \Delta\boldsymbol{\Sigma} \in S_{n+1} \quad &(\text{static condition at } t = t_{n+1}). \end{aligned}\right\} \tag{11.9}$$

Because of nonlinearity we must have recourse to an *iterative* method, of which we give below an example.

11.3.4 Example of iterative method (Ilyushin)

The example utilizes the elasticity operator. The principle of the method is as follows. The solution $\Delta\mathbf{u}$ is approximated by a sequence of elastic solutions taking, each time, as initial condition, the result at the previous step. The basic idea goes back to Picard's method of successive approximations to nonlinear equations (see, e.g. Ince (1944)). This method can be illustrated with the help of the following simple differential equation:

$$\frac{\mathrm{d}y}{\mathrm{d}x} - y = 0, \qquad y(0) = 1. \tag{11.10}$$

Integrating (11.10) once, on account of the initial condition we have

$$y = 1 + \int_0^x y\,\mathrm{d}x. \tag{11.11}$$

As a first approximation we take $y^{(1)} = 1$ and substituting from this in (11.11) we get

$$y^{(2)} = 1 + \int_0^x \mathrm{d}x = 1 + x. \tag{11.12}$$

Substituting then from this in (11.11), and so on, we obtain the following sequence of solutions:

$$\left.\begin{aligned} y^{(3)} &= 1 + x + \frac{1}{2}x^2, \\ y^{(4)} &= 1 + x + \frac{1}{2!}x^2 + \frac{1}{3!}x^3, \\ &\vdots \\ y^{(h+1)} &= 1 + \int_0^x y^{(k)}\,\mathrm{d}x = 1 + x + \frac{1}{2!}x^2 + \cdots + \frac{1}{k!}x^k. \end{aligned}\right\} \quad (11.13)$$

As k gets larger and larger $y^{(k)}$ will approach the infinite series for $\exp(x)$, which is indeed the exact solution of (11.10).

Apparently, Ilyushin (1943) was the first to use such an iterative method in plastic-flow problems; then, the implementation of the method goes as follows. Denoting by k the iteration number at step n of the computation of Subsection 11.3.3 above, we take thus

$\Delta\mathbf{u}^0 = \Delta\mathbf{u}^{\mathrm{E}}$,

$\Delta\mathbf{u}^k$ being known, $\Delta\mathbf{u}^{k+1}$ is obtained as the displacement solution of the elasticity problem,

$\mathbf{u}_n + \Delta\mathbf{u}^{k+1} \in \mathscr{U}^{n+1}$,

$\Sigma_n + \Delta\Sigma^{k+1} \in S_{n+1}$,

we have the *elastic* constitutive equation with *initial* strains,

$$\Delta\Sigma = \mathscr{L} : [\Delta\mathscr{E}^{k+1} - \Delta\mathscr{E}^{\mathrm{p}}(\Delta\mathscr{E}^k)]; \qquad (11.14)$$

$k \to \infty \Rightarrow \Delta\mathbf{u}^k \to \Delta\mathbf{u}$.

Obviously, the iteration must be stopped at some moment. A good criterion for this is the evaluation of the distance separating the point $\Sigma_n + \Delta\Sigma^k$ in stress space from the elasticity domain C at *each* point in the structure since the plasticity condition is reached at convergence only.

The method outlined above resembles the *gradient* method in the search for a minimum of the functional

$$A(\Delta\mathbf{u}^*) = \tfrac{1}{2}\|\mathscr{L} : \Delta\mathscr{E}^* - \Delta\Sigma^{\mathrm{E}}\|^2 - \tfrac{1}{2}\|\mathscr{L}\Delta\mathscr{E}^{\mathrm{p}}(\Delta\mathbf{u}^*)\|^2. \qquad (11.15)$$

The computation is performed at *constant* rigidity as illustrated in the one-dimensional case in Fig. 11.2. A method with *variable* rigidity (Mercier, 1977) allows an acceleration in convergence. The method with constant rigidity is also illustrated in Fig. 11.3. This follows Nguyen Quoc Son (1977) where further developments concerning convergence may be found. We give now a worked out example.

Fig. 11.2. Iterative method with constant rigidity

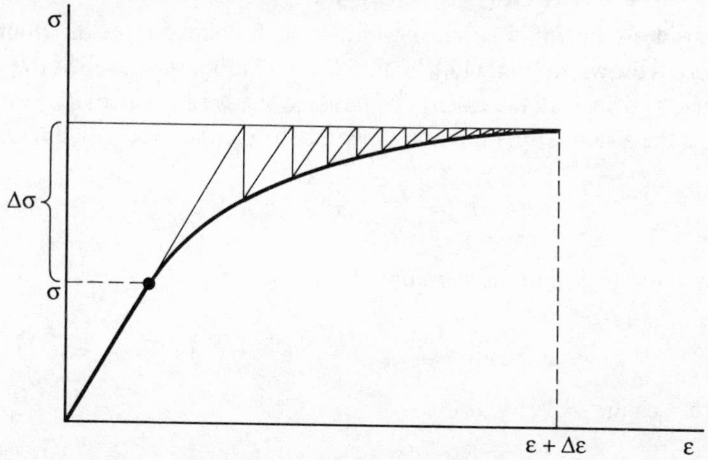

Fig. 11.3. Iterative method (Ilyushin) *Origin*: intersection of $\mathscr{E}_0$, S_0 and P_0. *Computation*: at $\Sigma_n \in S_n$ do $\Sigma_n + \mathscr{L}(\Delta\mathscr{E})$. (After Nguyen Quoc Son, 1977)

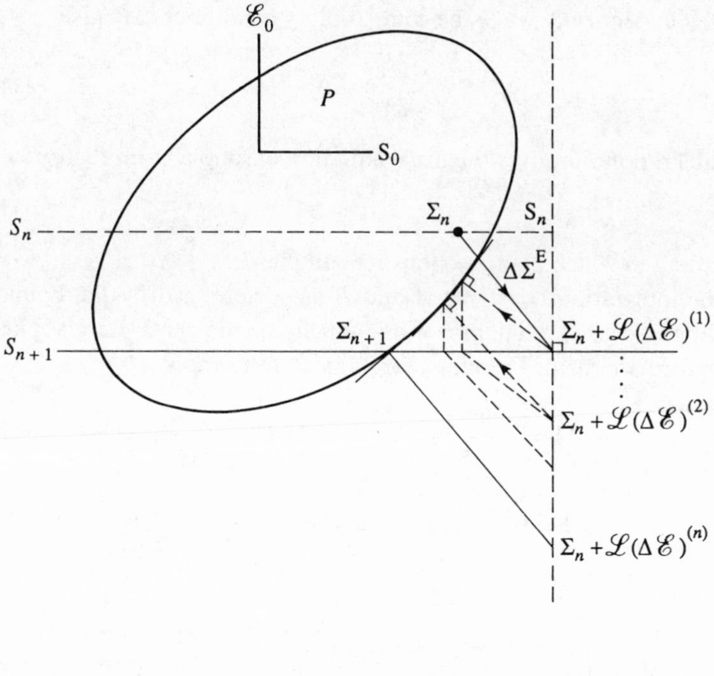

11.3.5 An elastoplastic thin flat plate with a thermal loading

We consider a thin infinite strip of width $2w$ with a temperature distribution $\theta(y)$ across the width (Fig. 11.4). With E Young's modulus and α the thermal expansion coefficient, the essentially nonzero stress component $\sigma_{xx} = \sigma_x(y)$ satisfies the strain–stress relation (linear isotropic thermoelasticity)

$$\varepsilon_{xx} = \frac{\sigma_{xx}}{E} + \alpha\theta + \varepsilon^{\mathrm{p}}_{xx}. \tag{11.16}$$

This is re-written in nondimensional form as

$$\varepsilon_x = \frac{S}{H} + T + \varepsilon^{\mathrm{p}} \tag{11.17}$$

with the scaling

$$S = \frac{\sigma_{xx}}{\sigma_0}, \quad \varepsilon_x = \frac{\varepsilon_{xx}}{\varepsilon_0}, \quad T = \frac{\alpha\theta}{\varepsilon_0}, \quad Y = \frac{y}{c}, \quad H = \frac{E}{E_0}.$$

The stress distribution $\sigma_x(y)$ or $S(Y)$ satisfies the equilibrium equation $\operatorname{div}\boldsymbol{\sigma} = \mathbf{0}$ identically while the compatibility conditions (2.6) give

$$\frac{\partial^2 \varepsilon_x}{\partial Y^2} = 0, \tag{11.18}$$

By integration this gives the usual equation of simple beam theory

$$\varepsilon_x = A + BY, \tag{11.19}$$

according to which plane sections remain plane.

The integration constants A and B are determined by the boundary conditions, e.g., an axial force P and a bending moment M are applied at the ends of the strip. Therefore, for a plate of thickness h,

$$\int_{-1}^{+1} S\,\mathrm{d}Y = \frac{p}{\sigma_0 wh} = p^*, \quad \int_{-1}^{+1} SY\,\mathrm{d}Y = \frac{M}{\sigma_0 w^2 h} \equiv M^* \tag{11.20}$$

Fig. 11.4. Infinite thin strip with thermal loading

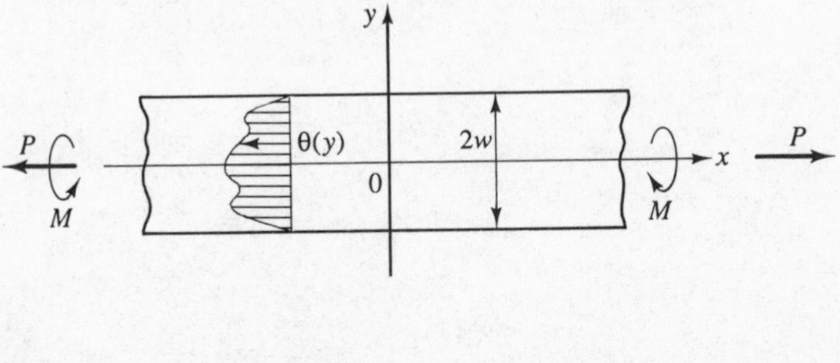

One obtains thus, assuming that E, hence H, can still depend on temperature,

$$\left.\begin{aligned} A &= A_1 \int_{-1}^{+1} H(T+\varepsilon^{\mathrm{p}})\,\mathrm{d}Y - A_2 \int_{-1}^{+1} H(T+\varepsilon^{\mathrm{p}})Y\,\mathrm{d}Y + A_1 P^* - A_2 M^*, \\ B &= -A_2 \int_{-1}^{+1} H(T+\varepsilon^{\mathrm{p}})\,\mathrm{d}Y - A_3 \int_{-1}^{+1} H(T+\varepsilon^{\mathrm{p}})Y\,\mathrm{d}Y + A_2 P^* - A_3 M^*, \end{aligned}\right\} \tag{11.21}$$

and

$$\begin{aligned} \varepsilon_x = (A_1 - A_2 Y) \times \left[\int_{-1}^{+1} H(T + \varepsilon^{\mathrm{p}})\,\mathrm{d}Y + P^*\right] \\ - (A_2 - A_3 Y) \times \left[\int_{-1}^{1} H(T + \varepsilon^{\mathrm{p}})Y\mathrm{d}Y + M^*\right] \end{aligned} \tag{11.22}$$

where

$$\left.\begin{aligned} A_1 &= \frac{\int_{-1}^{+1} HY^2\,\mathrm{d}Y}{\int_{-1}^{+1} H\mathrm{d}Y \int_{-1}^{+1} HY^2\,\mathrm{d}Y - (\int_{-1}^{+1} HY\mathrm{d}Y)^2}, \\ A_2 &= \frac{\int_{-1}^{+1} HY\,\mathrm{d}Y}{\int_{-1}^{+1} H\mathrm{d}Y \int_{-1}^{+1} HY^2\,\mathrm{d}Y - (\int_{-1}^{+1} HY\mathrm{d}Y)^2}, \\ A_3 &= \frac{\int_{-1}^{+1} H\,\mathrm{d}Y}{\int_{-1}^{+1} H\mathrm{d}Y \int_{-1}^{+1} HY^2\,\mathrm{d}Y - (\int_{-1}^{+1} HY\mathrm{d}Y)^2}. \end{aligned}\right\} \tag{11.23}$$

In the simple case then E is a constant *independent* of temperature ($E = E_0$),

$$H = 1, \quad A_1 = \tfrac{1}{2}, \quad A_2 = 0, \quad A_3 = \tfrac{3}{2}, \tag{11.24}$$

and the *mechanical strain* $\varepsilon = \varepsilon_x - T$ is simply given by

$$\begin{aligned} \varepsilon = -T + \frac{1}{2}\left(\int_{-1}^{+1} T\mathrm{d}Y + P^*\right) + \frac{3}{2}Y\left(\int_{-1}^{+1} TY\mathrm{d}Y + M^*\right) \\ + \frac{1}{2}\int_{-1}^{+1} \varepsilon^{\mathrm{p}}\,\mathrm{d}Y + \frac{3}{2}Y\int_{-1}^{+1} \varepsilon^{\mathrm{p}} Y\,\mathrm{d}Y. \end{aligned} \tag{11.25}$$

This further simplifies if the temperature distribution is an even function of Y (i.e., $\theta(y) = \theta(-y)$ so that $\int_{-1}^{+1} TY\,\mathrm{d}Y = 0$) and when there are no externally applied P and M. Then (11.25) reduces to

$$\varepsilon = T + \int_0^1 T\,\mathrm{d}Y + \int_0^1 \varepsilon^{\mathrm{p}}\,\mathrm{d}Y. \tag{11.26}$$

The iterative procedure starts at this point as ε^{p} will, in general, be a *non-*

linear function of ε, the two being related through the stress–strain curve of the material. Then eqns (11.22)–(11.26) represent *integral equations* for the solution of the *elastoplastic strains* during loading of a thermally stressed body made of a work-hardening material with temperature-dependent properties. For a *nonlinear strain-hardening*, the above-stated equations are solved by the Ilyushin iterative method. The scheme of solution is as follows.

(1) Assume $\varepsilon^p = 0$ everywhere.

(2) Solve for ε from eqn (11.25) or (11.26) obtaining then an elastic solution.

(3) For each value of ε read off the value ε^p from the stress–strain curve (it may be convenient to plot the curve $\varepsilon^p(\varepsilon)$).

(4) Using eqn (11.25) or (11.26) compute a better approximation to ε. The integrals have to be evaluated numerically.

(5) Repeat steps (3) and (4) until convergence is obtained. Half a dozen iterations are generally sufficient for practical purposes.

If the material properties are temperature dependent a different curve is used at step (3) at every station. Fig. 11.5 gives the strain and stress distribution (after Mendelson and Spero, 1962: $P = M = 0$; stress–strain curve in Fig. 11.6 with the data

$$w = 1, \quad \alpha = 9.5 \times 10^{-6}, \quad \sigma_0 = 28000, \quad E = 28 \times 10^6,$$

$$\theta = 600(y^2 - \tfrac{1}{3})).$$

For *linear* strain-hardening the above-presented problem can be solved in closed form but it is also instructive to apply the iterative method (Problem 11.1).

The reader will find other examples of application of this method in Mendelson (1968): thin circular shell with an axial temperature gradient, long solid cylinder with a radial temperature distribution, rotating disk with a temperature gradient, circular hole in a uniformly stressed infinite plate. For structures with a more complex geometry, however, each step has to be solved spatially by the FEM.

11.4 Application of the finite-element method

Globally, in quasi-statics, the solution of an elastoplasticity problem in *discretized* form consists in solving a *matricial* system through the implicit scheme for the time evolution and using the element method in space. We

Fig. 11.5. Strains (a) and stress (b) distribution in an elastoplastic infinite strip obtained by the method of successive elastic solutions (after Mendelson and Spero, 1962)

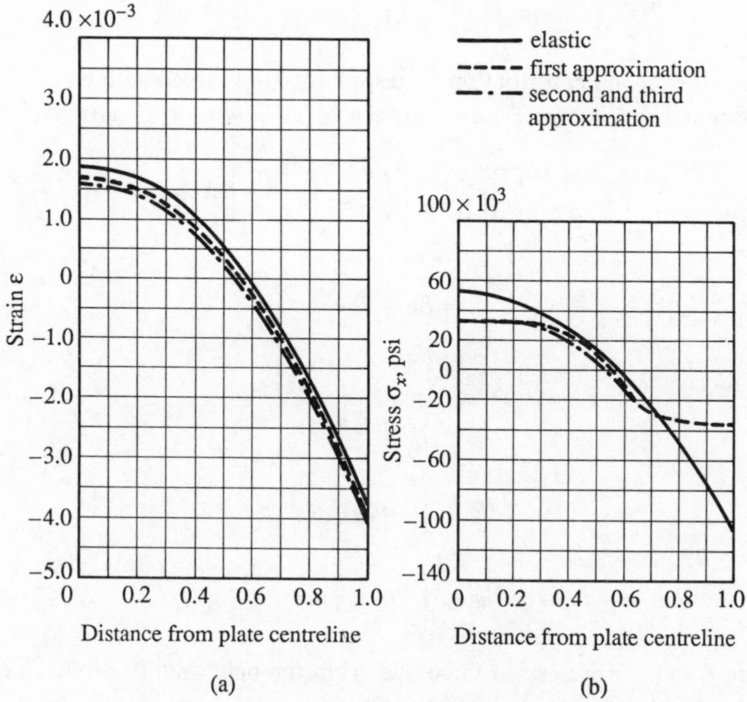

Fig. 11.6. Strain-hardening stress–strain relation used to obtain Fig. 11.5 (after Mendelson and Spero, 1962)

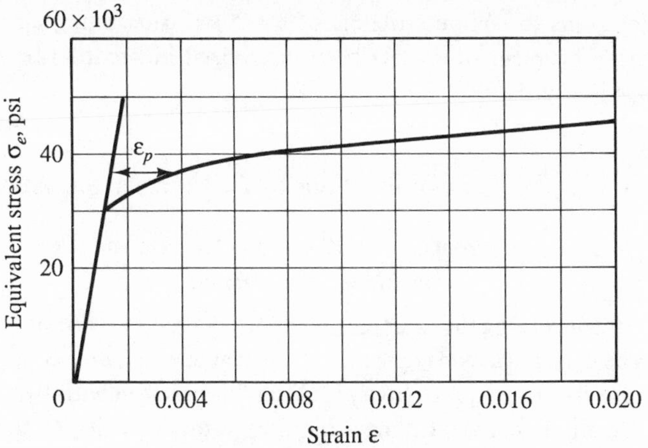

use the classical notation for the latter method (e.g. Zienkiewicz, 1971). Classically, if N and $\bar{N}$ are appropriately chosen basic functions, the displacement is expressed in terms of the displacement parameters q and $\bar{q}$ by

$$U = Nq + \bar{N}\bar{q}, \tag{11.27}$$

where the second contribution corresponds to prescribed data on $\partial\Omega_u$.

For an *elastic system* the discretized problem is written in matricial form as

$$\begin{bmatrix} K & \bar{K} \\ \bar{K}^{\mathrm{T}} & \bar{\bar{K}} \end{bmatrix}\begin{bmatrix} q \\ \bar{q} \end{bmatrix} = \begin{bmatrix} \bar{R} \\ R \end{bmatrix} - \int_\Omega \begin{bmatrix} B^{\mathrm{T}} \\ \bar{B}^{\mathrm{T}} \end{bmatrix} \boldsymbol{\sigma}^{\mathrm{T}}\, \mathrm{d}\Omega, \tag{11.28}$$

where

$$\left.\begin{aligned} \varepsilon &= Bq + \bar{B}\bar{q}, \\ \sigma &= D\varepsilon, \\ K &= \int_\Omega B^{\mathrm{T}} D B\, \mathrm{d}\Omega, \\ \bar{K} &= \int_\Omega B^{\mathrm{T}} D \bar{B}\, \mathrm{d}\Omega, \\ \bar{\bar{K}} &= \int_\Omega \bar{B}^{\mathrm{T}} D \bar{B}\, \mathrm{d}\Omega. \end{aligned}\right\} \tag{11.29}$$

Here $\bar{R}$ are the prescribed forces (i.e. $\mathbf{g}^{\mathrm{d}}$ in the bulk and $\mathbf{T}^{\mathrm{d}}$ on $\partial\Omega_T$). On account of this the first of eqns (11.12) yields

$$Kq = \bar{R} - \bar{K}\bar{q} - \int_\Omega B^{\mathrm{T}} \boldsymbol{\sigma}^{\mathrm{T}}\, \mathrm{d}\Omega; \tag{11.30}$$

This allows one to compute the increment Δq at each time step t_n of *the elastoplastic* problem which has been discretized in accordance with the flow chart in Fig. 11.7.

11.5 Examples of computations by FEM in elastoplasticity

11.5.1 Elastoplastic torsion of a cylindrical rod with a multiconnected section

The results concerning the example are reported in Glowinski and Lanchon (1973). The cylindrical rods considered have either a square section with a centred or off-centre square cavity, or a circular section with two circular cavities. In the first case, calling l the area ratio of cavity to the whole cross-section, using a criterion of von Mises type and stress functions (see Appendix 3.2), we obtain the results of Fig. 11.8 for the influence of the relative size of the centred cavity, keeping the tension angle α fixed and those in Fig. 11.9 for the evolution of the torsion torque as a function of α for various values of l. Fig. 11.10 exhibits the plasticized zones and those that are still in the elastic regime for different values of α (case of off-centre cavity). Fig. 11.11 is based on a minimization of another functional, but it shows pretty well the plasticization and the level curves for the stress

Fig. 11.7. Finite-element computation: flowchart (Nguyen Quoc Son, 1981a)

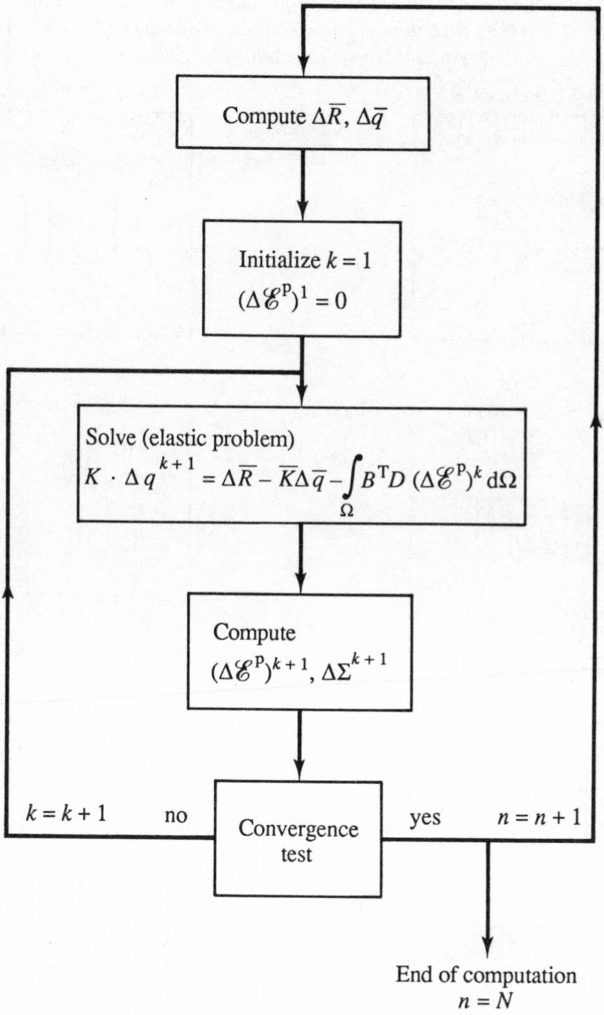

function (see Appendix 3.2). Fig. 11.12 reports on the plasticization for the circular section containing two circular cavities. The interested reader can find in Nádai (1950, p. 501) photographs showing the growth of the plastic zone in a rod of square section subjected to torsion, in perfect agreement with the first case in Fig. 11.8 The case of a multiconnected cross-section is also illustrated in that book. In particular, Nádai's Figure 35.22 (Nádai, 1950, p. 506) compares to the present Fig. 11.11(b). Note that Nádai often uses membrane models which exploit the Prandtl analogy of the 'sand hill' (see Appendix 3.2).

Fig. 11.8. Influence of the size of cavity (rod of square cross-section with centred square cavity) for a fixed torsion angle α (after Glowinski and Lanchon, 1973). Plastic zones hatched; elastic zones dotted.

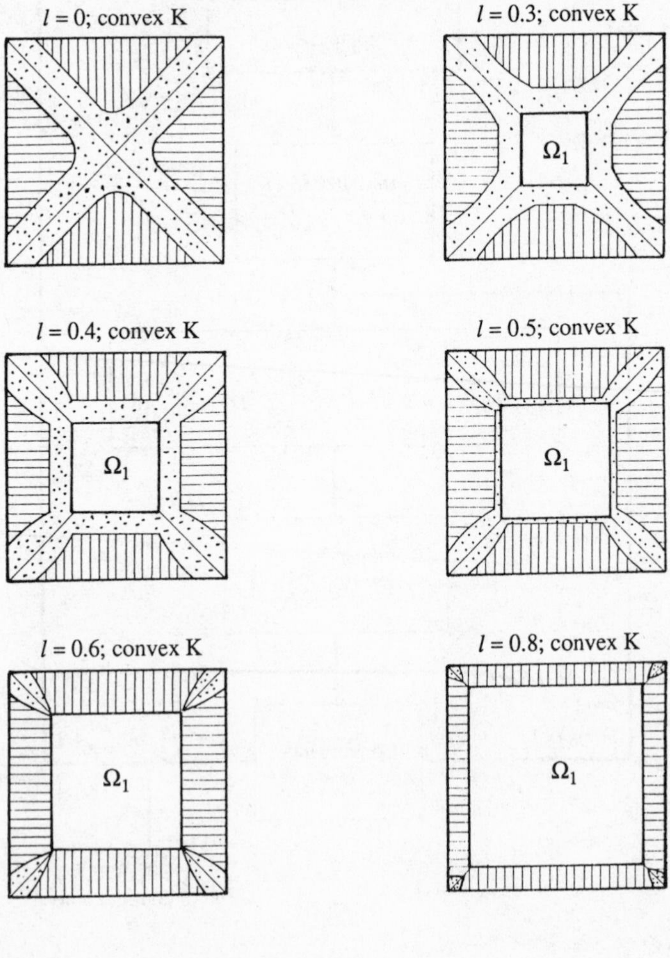

11.5.2 Traction of a cracked rectangular plate

This solution is due to Nguyen Quoc Son (1977). The material is elastic–perfectly-plastic obeying Mises' criterion. The discretization with 390 knots (and linear interpolation) is shown in Fig. 11.3. The iterative scheme of Fig. 11.3 is implemented in this example. If Young's modulus is $E = 5000$ K with an accuracy $\varepsilon_{ac} = 0.01$, the number of required iterations as a function of traction P varies according to the following table:

P	K	$1.1\,K$	$1.2\,K$	$1.3\,K$	$1.4\,K$
No. of iterations	7	9	12	22	40

Fig. 11.9. Variations of the torsion torque $\mathcal{M}^*$ vs torsion angle α for various sizes l of cavity (same problem as in Fig. 11.5; von Mises criterion; after Glowinski and Lanchon, 1973)

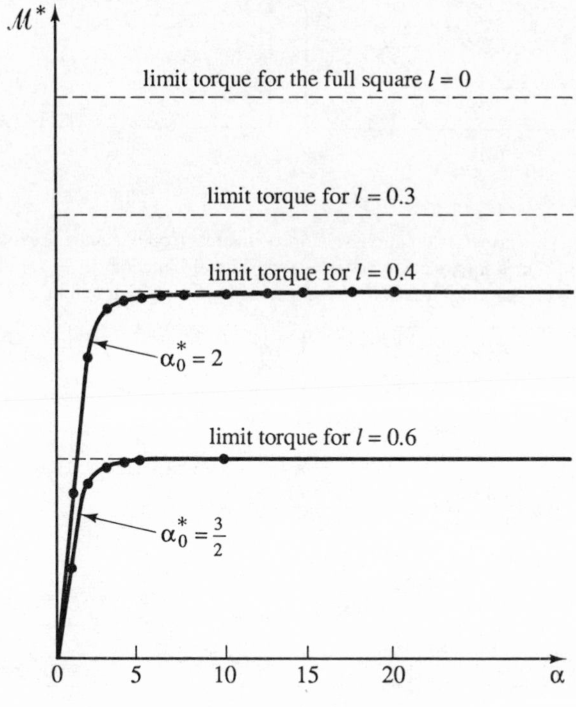

In agreement with Fig. 11.3, the convergence rate depends on the relative position of P and S_{n+1}. In particular, the iterative procedure diverges if $P \cap S_{n+1} = \emptyset$.

Understanding the contents of Chapter 11 is a matter of practice and working out examples while implementing numerical methods, in particular the finite-element method (FEM). The following problems aim at introducing the reader to variational formulations and FEM for (nonlinear)

Fig. 11.10. Plasticization as a function of torsion angle α (for a square cross-section with an off-centre square cavity) (after Glowinski and Lanchon, 1973). (a) $\alpha = 1.5$. (b) $\alpha = 2.0$.

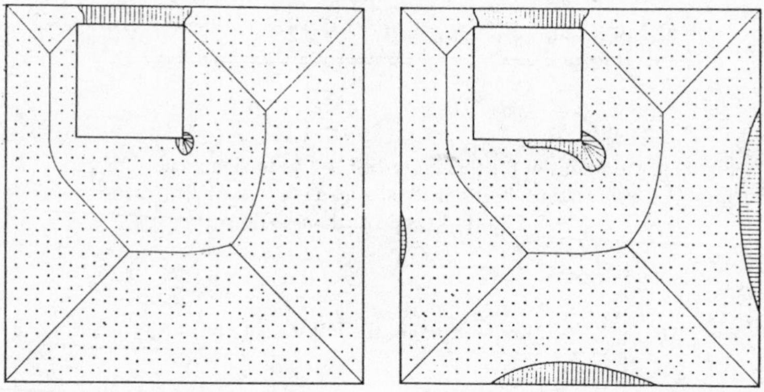

(a) (b)

Fig. 11.11. Plasticization and level curves in stress (rod of square cross-section with off-centre square cavity; after Glowinski and Lanchon, 1973). (a) Plasticization for $\alpha = 5$. (b) Level curves for the stress function.

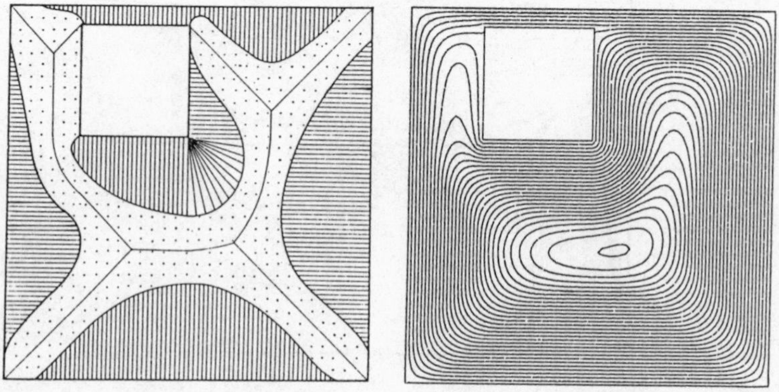

(a) (b)

elasticity and elastoplasticity in their incremental form with application to simple types of structural members.

Problems for Chapter 11

11.1. Ilyushin's iterative method

Consider the elastoplastic problem of the infinite plane strip in Subsection 11.3.5 but with *linear* strain-hardening (stress–strain curve in Fig. 6.4; same

Fig. 11.12. Plasticization as a function of α (rod of circular cross-section with two circular cavities; Mises criterion; after Glowinski and Lanchon, 1973). (a) $\alpha = 1.5$. (b) $\alpha = 5.0$.

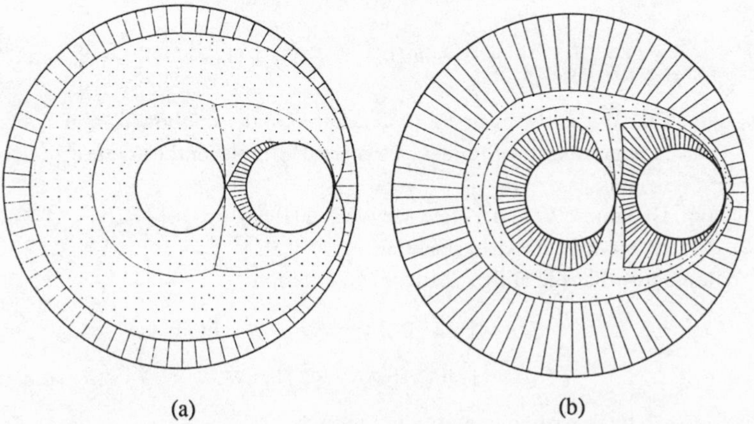

Fig. 11.13. Discretization of problem of a cracked rectangular plate subjected to a traction P (Material: $E = 5000\ K$, $\nu = 0.3$. After Nguyen Quoc Son, 1977.)

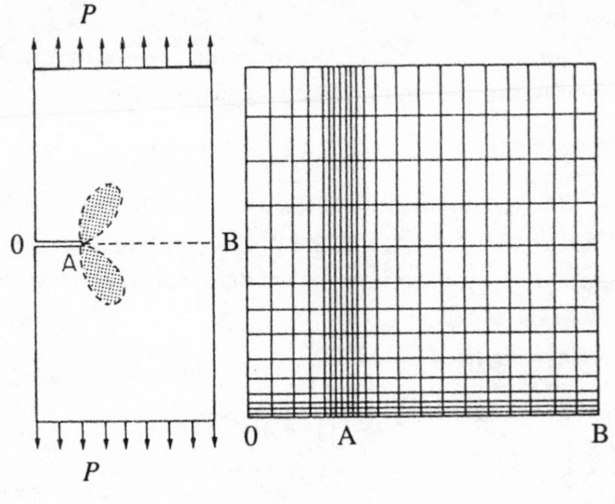

notation) and temperature loading $T(Y) = T_0(Y^2 - \frac{1}{3})$, all material properties being constant and assuming $M = N = 0$.

(i) Express ε^p in terms of ε [Answer: $\varepsilon^p = (1 - m)(\varepsilon \mp 1)$].

(ii) Show that ε satisfies a linear Fredholm integral equation.

(iii) Assuming that a plastic zone of the *tensile* type exists for $0 < Y < Y_1$, and another plastic zone of the *compressive* type exists for $Y_2 \leqslant Y < 1$, write the special form of the integral equation of (ii) [*Answer*:

$$\varepsilon = -T_0\left(Y^2 - \frac{1}{3}\right) + (1 - m)(1 - Y_2) + (1 - m)\int_{Y_2}^{1} \varepsilon \, \mathrm{d}Y,$$

$$T_0 \leqslant T_{oc} \equiv \frac{3}{2}$$

as yielding occurs at $T_0 = T_{oc}$. For $T_0 > T_{oc}$,

$$\varepsilon = -T_0\left(Y^2 \frac{1}{3}\right) + (1 - m)(1 - Y_2 - Y_1) + (1 - m)\left[\int_0^{Y_1} + \int_{Y_2}^{1} \varepsilon \, \mathrm{d}Y\right].]$$

(iv) Find explicitly the first *four* iterates $\varepsilon^{(k)}$ of the solution (Y_1 and Y_2 being assumed as known) and then a general formula for the iterate $\varepsilon^{(n+1)}$. Find its limit as $n \to \infty$.

(v) Determine Y_1 and Y_2 (see Mendelson (1968), pp. 181–2).

11.2. Matrix notation for yield condition

(a) Show that the Huber–Mises yield criterion

$$f = \tfrac{1}{2}[(\sigma_x - \sigma_y)^2 + (\sigma_x - \sigma_z)^2 + (\sigma_y - \sigma_z)^2 + 3(\sigma_{xy}^2 + \sigma_{xz}^2 + \sigma_{yz}^2)] - \sigma_0^2 \tag{1}$$

can be re-written in matrix notation as

$$f = \tfrac{1}{2}\boldsymbol{\sigma}^{\mathrm{T}}\mathbf{M}\boldsymbol{\sigma} - \sigma_0^2 = 0 \tag{2}$$

where $\boldsymbol{\sigma} = (\sigma_x, \sigma_y, \sigma_z, \sigma_{xy}, \sigma_{xz}, \sigma_{yz})$. What is the expression for the 6×6 matrix $\mathbf{M}$?

(b) Show that a piecewise linear yield surface can be described by means of a system of linear equations whose matrix form is

$$f = \mathcal{N}^{\mathrm{T}}\boldsymbol{\sigma} - \mathbf{k} = \mathbf{0}, \tag{3}$$

where $\mathcal{N}$ is the matrix of yield-surface gradients such that $\mathcal{N} = \nabla\mathbf{f} = \{\delta\mathbf{f}^{\alpha}/\delta\boldsymbol{\sigma}\}$, $\alpha = 1, 2, \ldots$, and $\mathbf{k}$ are plastic moduli. In particular, show that Tresca's yield condition can be written in such a form with

$$\left.\begin{aligned} \boldsymbol{\sigma} = (\sigma_1, \sigma_2), \qquad \mathcal{N} &= \begin{bmatrix} 0 & 1 & 1 & 0 & -1 & -1 \\ 1 & 0 & -1 & -1 & 0 & 1 \end{bmatrix}, \\ \mathbf{k} &= \{1\ 1\ 1\ 1\ 1\ 1\}\sigma_0. \end{aligned}\right\} \tag{4}$$

11.3. Matrix notation for yield condition with strain-hardening

(a) For a piecewise linear form of the initial yield surface show that the matrix formula

$$\mathbf{f} = \mathcal{N}^{\mathrm{T}}\boldsymbol{\sigma} - \mathcal{H}\boldsymbol{\lambda} - \mathbf{k} = \mathbf{0}, \tag{1}$$

where $\boldsymbol{\lambda}$ is a matrix of plastic multipliers and $\mathcal{H}$ is the hardening matrix, accounts for both isotropic and kinematic hardening.

(b) In particular, one gets

$$\mathcal{H} = h\mathbf{k}\mathbf{k}^{\mathrm{T}} \qquad \text{and} \qquad \mathcal{H} = h\mathcal{N}^{\mathrm{T}}\mathcal{N} \tag{2}$$

for isotropic hardening and Prager's kinematic hardening, where h is the hardening modulus.

11.4. Matrix notation for elastoplasticity

(a) Show that a possible matrix notation for elastoplasticity in the spirit of 11.2 and 11.3 is

$$\left.\begin{array}{c} \dot{\boldsymbol{\varepsilon}} = \dot{\boldsymbol{\varepsilon}}^{\mathrm{e}} + \dot{\boldsymbol{\varepsilon}}^{\mathrm{p}}, \quad \dot{\boldsymbol{\varepsilon}}^{\mathrm{e}} = \mathbf{S}\dot{\boldsymbol{\sigma}}, \quad \dot{\boldsymbol{\varepsilon}} = \mathcal{N}\dot{\boldsymbol{\lambda}}, \quad \dot{\lambda} \geqslant 0, \quad \mathcal{N} = \partial f/\partial\boldsymbol{\sigma}, \\ f \leqslant 0, \quad \dot{\lambda}^{\mathrm{T}} f = 0, \quad \dot{f}_{\mathrm{a}} = 0, \quad \dot{\lambda}^{\mathrm{T}}\dot{f}_{\mathrm{a}} = 0, \end{array}\right\} \tag{1}$$

where f_{a} denotes the potentially active planes of $f = 0$.

(b) As local unloading seldom occurs in the study of loading increasing in proportion to a single parameter, show that the above equations can be *simplified* by being re-written in terms of strains and stresses (*no* time rates; these are Hencky's ideas):

$$\boldsymbol{\varepsilon} = \boldsymbol{\varepsilon}^{\mathrm{e}} + \boldsymbol{\varepsilon}^{\mathrm{p}}, \quad \boldsymbol{\varepsilon}^{\mathrm{e}} = \mathbf{S}\boldsymbol{\sigma}, \quad \boldsymbol{\varepsilon}^{\mathrm{p}} = \mathcal{N}\boldsymbol{\lambda}, \quad \lambda \geqslant 0, \quad f \leqslant 0, \quad \boldsymbol{\lambda}^{\mathrm{T}} f = 0. \tag{2}$$

11.5. Matrix notation for energies and dissipation

For linear elasticity the stress and strain energies can be written as

$$W_\sigma = \frac{1}{2}\int_V \boldsymbol{\sigma}^{\mathrm{T}}\mathbf{S}\boldsymbol{\sigma}\,\mathrm{d}v, \qquad W_\varepsilon = \frac{1}{2}\int_V \boldsymbol{\varepsilon}^{\mathrm{T}}\mathbf{E}\boldsymbol{\varepsilon}\,\mathrm{d}v \tag{1}$$

with $W_\sigma = W_\varepsilon$. In an elastic–plastic body the strain energy is the sum of the elastic and plastic contributions, i.e., $W_\varepsilon = W_\varepsilon^{\mathrm{e}} + W_\varepsilon^{\mathrm{p}}$. By integrating over time show that the rate of energy dissipation is given by

$$\dot{W}_\varepsilon^{\mathrm{p}} = \int_V \boldsymbol{\sigma}^{\mathrm{T}}\dot{\boldsymbol{\varepsilon}}^{\mathrm{p}}\,\mathrm{d}v = \frac{1}{2}\int_V \dot{\boldsymbol{\lambda}}^{\mathrm{T}}\mathcal{H}\dot{\boldsymbol{\lambda}}\,\mathrm{d}v + \int_V \mathbf{k}^{\mathrm{T}}\dot{\boldsymbol{\lambda}}\,\mathrm{d}v. \tag{2}$$

For a perfectly plastic body, one obtains the following simplified expression:

$$\dot{W}_\varepsilon^{\mathrm{p}} = \int_V \mathbf{k}^{\mathrm{T}}\dot{\boldsymbol{\lambda}}\,\mathrm{d}v \geqslant 0. \tag{3}$$

11.6. Skeleton structures in elastoplasticity

(a) Let w, p, q and s denote the matrices of generalized displacements, loads, strains and stresses. Then show that one can write

$$p^{\mathrm{T}}w = \int_A \mathbf{f}^{\mathrm{T}}\mathbf{u}\,\mathrm{d}A + \int_R \mathbf{t}^{\mathrm{T}}\mathbf{u}\,\mathrm{d}R, \qquad s^{\mathrm{T}}q = \int_A \boldsymbol{\varepsilon}^{\mathrm{T}}\boldsymbol{\sigma}\,\mathrm{d}A \qquad (1)$$

(where A denotes the cross-sectional area and R the circumference) for the external and internal *virtual works* in the cross-section of a beam. Write down these expressions for a *Timoshenko beam* (so-called $M + N + T$ model; $s = \{N, T, M\}$) and the *Bernoulli beam* ($M + N$ model; $s = \{N, M\}$). This applies to many 'skeleton' structures (beams, grids, trusses, frames).

(b) Show that for a cross-section of elastic–plastic skeleton structures the equations of 11.4 take on the following matrix form:

$$\left.\begin{aligned} \mathbf{q} = \dot{\mathbf{q}}_{\mathrm{e}} + \dot{\mathbf{q}}_{\mathrm{e}}, \quad \dot{\mathbf{q}}_{\mathrm{e}} = \mathbf{S}\dot{\mathbf{s}}, \quad \dot{\mathbf{q}}_{\mathrm{p}} = \mathbf{N}\dot{\boldsymbol{\lambda}}, \quad \dot{\boldsymbol{\lambda}} \geqslant 0, \\ f \leqslant 0, \quad \dot{\lambda}^{\mathrm{T}}f = 0, \quad \dot{f}_{\mathrm{a}} \leqslant 0, \quad \dot{\lambda}_{\mathrm{a}}^{\mathrm{T}}\dot{f}_{\mathrm{a}} = 0, \end{aligned}\right\} \qquad (2)$$

with, e.g.,

$$f = N^{\mathrm{T}}s - H\lambda - k = 0$$

for a piecewise linear yield surface.

11.7. Energy dissipation and potential energy (P11.6 continued)

(c) Show that the rate of energy dissipation in a plastic beam (i) is given by

$$\dot{W}_{\mathrm{p}}^{(i)} = \int_{L_i} s^{\mathrm{T}}\dot{q}_{\mathrm{p}}\,\mathrm{d}L_i = \frac{1}{2}\int_{L_i} \dot{\lambda}^{\mathrm{T}}H\dot{\lambda}\,\mathrm{d}L + \int_{L_i} k^{\mathrm{T}}\dot{\lambda}\,\mathrm{d}L_i, \qquad (1)$$

while the total *potential energy* of an elastic member (i) is given by

$$P^{(i)} = W_q^{(i)} - 2U^{(i)} \qquad (2)$$

with

$$\left.\begin{aligned} W_q^{(i)} &= \int_0^t \dot{W}_q^{(i)}\,\mathrm{d}t, \qquad U^{(i)} = \int_0^t \dot{U}^{(i)}\,\mathrm{d}t, \\ \dot{U}^{(i)} &= \int_{L_i} p^{\mathrm{T}}W\,\mathrm{d}L_i + (\text{boundary terms}). \end{aligned}\right\} \qquad (3)$$

11.8. Discretization of elastoplasticity problems for a beam

Here discrete quantities are represented by symbols with overbars. Discretization does *not* change the value of virtual work, e.g.,

$$\overline{w}^{\mathrm{T}}\overline{p} = \int_0^L w^{\mathrm{T}}p\,\mathrm{d}\overline{x}, \qquad \overline{q}^{\mathrm{T}}\overline{s} = \int_0^L q^{\mathrm{T}}s\,\mathrm{d}\overline{x}, \qquad (1)$$

where $\overline{w}, \overline{p}, \overline{q}$ and $\overline{s}$ are algebraic column matrices while w and p are function matrices for one coordinate $\overline{x}$. Duality of compatibility equations in the discrete system reads

$$\overline{q} = \overline{C}\overline{w}, \qquad \overline{p} = \overline{C}^{\mathrm{T}}\overline{s}, \qquad (2)$$

where $\overline{C}$ is the so-called *compatibility matrix*.

Shape function matrices. The parametric form of the generalized displacement is written as

$$w = \Phi^w \bar{w}, \tag{3}$$

where Φ^w is the shape function matrix. In like manner, $q = \Phi^q \bar{q}$, where the shape functions Φ^q are related in such a way that the elements of $\bar{q}$ are the generalized strain measures at certain *points*, called sampling points. Examples of Φ^w for a beam are Hermite polynomials in a kinematic approach and Lagrange polynomials in a static approach. For a mixed approach, show that writing the principle of virtual work for the beam allows one to obtain an explicit formula for the compatibility matrix $\bar{C}$.
[*Hint*: start from

$$\int_0^L s^{\mathrm{T}} q \, \mathrm{d}\bar{x} = \int_0^L p^{\mathrm{T}} w \, \mathrm{d}\bar{x}$$

and pass to $\bar{s}^{\mathrm{T}} \bar{C} \bar{w} = \bar{p}^{\mathrm{T}} \bar{w}$.]

11.9. Introduction of the plastic multiplier

In elastoplasticity we also have a *discrete* representation of the plastic multiplier by

$$\lambda = \Phi^\lambda \bar{\lambda} \tag{1}$$

and the three shape function matrices introduced (noting that $q_e = \Phi^q \bar{q}_e$, $\bar{q}_p = \Phi^q \bar{q}_p$) must satisfy the conditions

$$\left.\begin{aligned} \Phi^q \bar{q} &= C \Phi^w \bar{w}, \\ \Phi^q \bar{N} &= N \Phi^\lambda. \end{aligned}\right\} \tag{2}$$

Prove these conditions. This allows one to find the matrix $\bar{N}$. Finally, show that

$$\left.\begin{aligned} \dot{U} = \bar{p}^{\mathrm{T}} \dot{\bar{w}}, \qquad \dot{\bar{W}} &= \bar{s}^{\mathrm{T}} \dot{\bar{q}}, \\ \dot{\bar{W}}_{\mathrm{p}} = \bar{s}^{\mathrm{T}} \dot{\bar{q}}_{\mathrm{p}} &= \tfrac{1}{2} \dot{\bar{\lambda}} \bar{H} \bar{\lambda}^{\mathrm{T}} + \bar{k}^{\mathrm{T}} \bar{\lambda}. \end{aligned}\right\} \tag{3}$$

11.10. FEM inelastic analysis of skeleton structures: minimum principles

For a structure made of an assembly of skeleton-structural members one introduces the *global* matrices **p, q, s, λ, k, C, E, H, N** instead of the local ones (with superimosed bar). In a purely elastic analysis, if $\mathbf{p}_*$ denotes the given load, then show that

$$\mathbf{K}\mathbf{w} - \mathbf{p}_* = \mathbf{0}, \tag{1}$$

where $\mathbf{K} = \mathbf{C}^{\mathrm{T}}\mathbf{E}\mathbf{C}$ is the so-called *stiffness matrix*. [*Hint*: use equilibrium and compatibility equations.] The above equation is the *canonical* set of equations of the displacement method in the elastic analysis. Prove the following *dual minimum principles.*

(a) Of all *kinematically compatible* displacement and strain states the actual state is the one that ensures the minimum of the *total potential energy* of the elastic system, i.e.,

$$\mathrm{Min}\{\varphi = \tfrac{1}{2}\mathbf{q}^{\mathrm{T}}\mathbf{E}\mathbf{q} - \mathbf{w}^{\mathrm{T}}\mathbf{p}^*\}.$$

(b) Of all stress states compatible with the given load, the actual state is the one that ensures the minimum of the *complementary energy* of the elastic system, i.e.,

$$\mathrm{Min}\{\varphi^* = \tfrac{1}{2}\mathbf{s}^{\mathrm{T}}\mathbf{S}\mathbf{s}\},$$

subject to $\mathbf{C}^{\mathrm{T}}\mathbf{s} - \mathbf{p}_* = \mathbf{0}$, $\mathbf{S} = \mathbf{E}^{-1}$.

11.11. FEM in elastic–plastic analysis: minimum principle

In the case of loading–unloading show that the complete set of equations for an elastic–plastic structure is given by (Δ = increment)

$$\left.\begin{aligned} C\Delta w - \Delta q = 0, &\qquad C^{\mathrm{T}}\Delta s - \Delta p_* = 0,\\ \Delta q = \Delta q^{\mathrm{e}} + \Delta q^{\mathrm{p}}, &\qquad \Delta q^{\mathrm{e}} = S\Delta s,\\ \Delta q^{\mathrm{p}} = N_{\mathrm{a}}\Delta\lambda_{\mathrm{a}}, &\qquad \Delta\lambda_{\mathrm{a}} \geqslant 0,\\ \Delta f_{\mathrm{a}} = N_{\mathrm{a}}^{\mathrm{T}}\Delta s - H_{\mathrm{a}}\Delta\lambda_{\mathrm{a}} \leqslant 0, &\qquad \Delta\lambda_{\mathrm{a}}^{\mathrm{T}}\Delta f_{\mathrm{a}} = 0, \end{aligned}\right\} \tag{1}$$

where Δp_* is the *known* load increment. Then show that the minimum principles of Chapters 4 and 5 read

Greenberg: $\quad I(\mathbf{u}) = \mathrm{Min}[\tfrac{1}{2}\Delta\lambda_{\mathrm{a}}^{\mathrm{T}}H\Delta\lambda_{\mathrm{a}} + \tfrac{1}{2}\Delta q_{\mathrm{e}}^{\mathrm{T}}E\Delta q_{\mathrm{e}} - \Delta w^{\mathrm{T}}\Delta p_*]$

subject to $C\Delta w - N_{\mathrm{a}}\Delta\lambda_{\mathrm{a}} - \Delta q_{\mathrm{e}} = 0, \quad \Delta\lambda_{\mathrm{a}} \geqslant 0,$

Hodge-Prager: $\quad J(\mathbf{s}) = \mathrm{Min}[\tfrac{1}{2}\Delta k_{\mathrm{a}}^{\mathrm{T}}H_{\mathrm{a}}^{-1}\Delta k_{\mathrm{a}} + \tfrac{1}{2}\Delta s^{\mathrm{T}}S\Delta s]$

subject to $\Delta k_{\mathrm{a}} - N_{\mathrm{a}}^{\mathrm{T}}\Delta s \geqslant 0, \quad C^{\mathrm{T}}\Delta s - \Delta p_* = 0,$

where

$$\Delta k_{\mathrm{a}} = H_{\mathrm{a}}\Delta\lambda_{\mathrm{a}}, \qquad \Delta\lambda_{\mathrm{a}} = H_{\mathrm{a}}^{-1}\Delta k_{\mathrm{a}}$$

(assuming that the submatrix H_{a} is not singular).

11.12. FEM in elastic–plastic analysis (unloading neglected)

When unloading is assumed not to take place, fields themselves instead of increments can be used. Show that the minimum principles of 11.11 take on the following form.

Kinematic principle: $\quad \mathrm{Min}[\tfrac{1}{2}\lambda^{\mathrm{T}}H\lambda + \tfrac{1}{2}q_{\mathrm{e}}^{\mathrm{T}}Eq_{\mathrm{e}} + \lambda^{\mathrm{T}}k_* - w^{\mathrm{T}}p_*]$

subject to $Cw - N\lambda - q_{\mathrm{e}} = 0, \quad \lambda \geqslant 0.$

Static principle: $\quad \mathrm{Min}[\tfrac{1}{2}\lambda^{\mathrm{T}}H\lambda + \tfrac{1}{2}s^{\mathrm{T}}Ss]$

subject to $H\lambda - N^{\mathrm{T}}s + k_* \geqslant 0, \quad C^{\mathrm{T}}s - p_* = 0.$

Here k_* are plastic moduli.

In the absence of hardening show that the above reduce to

$$\operatorname{Min}(\tfrac{1}{2}q_e^T E q_e + \lambda^T k_* - w^T p_*)$$

$$\text{subject to } Cw - N\lambda - q_e = 0, \lambda \geqslant 0,$$

and

$$\operatorname{Min}(\tfrac{1}{2}s^T S s)$$

$$\text{subject to } -N^T s + k_* \geqslant 0, \quad C^T s - p_* = 0.$$

11.13. Hu–Washizu variational principle for increments (3D problems)
Implementation of elastic–plastic computations by the finite-element method implies the use of variational principles. Several fields can be varied in the variation. For instance, consider the following functional:

$$J_{HW}(\Delta\mathbf{u}, \Delta\boldsymbol{\varepsilon}, \Delta\boldsymbol{\sigma}, \Delta\mathbf{T}) = \int_\Omega \Big\{ W(\Delta\boldsymbol{\varepsilon}) + \frac{1}{2}\sigma_{ij}\Delta u_{k,i}\Delta u_{k,j} - \rho\Delta\mathbf{f}\cdot\Delta\mathbf{u}$$

$$- \Delta\sigma_{kl}\Big[\Delta\varepsilon_{kl} - \frac{1}{2}(\Delta u_{k,l} + \Delta u_{l,k} + u_{i,k}\Delta u_{i,l}$$

$$+ \Delta u_{i,k}u_{i,l} + \Delta u_{i,k}\Delta u_{i,l})\Big]\Big\}\,\mathrm{d}\Omega$$

$$- \int_{\partial\Omega} [\Delta\mathbf{T}^d\cdot\Delta\mathbf{u} + \Delta\mathbf{T}\cdot(\Delta\mathbf{u} - \Delta\mathbf{u}^d)]\,\mathrm{d}s, \qquad (1)$$

where $W(\Delta\boldsymbol{\varepsilon}) = \frac{1}{2}E_{ijkl}\Delta\varepsilon_{ij}\Delta\varepsilon_{kl}$.
By effecting the variation δJ_{HW} and imposing stationariness with *no* subsidiary conditions, show that there follow the following equations:

$$\operatorname{div}(\Delta\boldsymbol{\sigma}) + \Delta\mathbf{f} = 0, \qquad \text{in } \Omega,$$

$$\mathbf{n}\cdot\Delta\boldsymbol{\sigma} = \Delta\mathbf{T}^d \qquad \text{on } \partial\Omega_T,$$

$$\mathbf{n}\cdot\Delta\boldsymbol{\sigma} = \Delta\mathbf{T} \qquad \text{on } \partial\Omega,$$

$$\Delta\mathbf{u} = \Delta\mathbf{u}^d \qquad \text{on } \partial\Omega_u,$$

$$\Delta\varepsilon_{kl} = \tfrac{1}{2}(\Delta u_{k,l} + \Delta u_{l,k} + u_{i,k}\Delta u_{i,l} + \Delta u_{i,k}u_{i,l} + \Delta u_{i,k}\Delta u_{i,l})$$

in Ω, where the last is the *incremental geometric* relation (accounting for second-order effects).

11.14. Hellinger–Reissner variational principle (Two-field variational principle – after Hellinger, 1914, and Reissner, 1953)
Introduce the *incremental complementary energy* (with $\mathbf{S} = \mathbf{E}^{-1}$, $\Delta\boldsymbol{\varepsilon} = \mathbf{S}:\Delta\boldsymbol{\sigma}$) by

$$W^*(\Delta\boldsymbol{\sigma}) = \Delta\boldsymbol{\sigma}:\Delta\boldsymbol{\varepsilon} - W, \qquad \text{e.g. } \tfrac{1}{2}S_{klmn}\Delta\sigma_{kl}\Delta\sigma_{mn}. \qquad (1)$$

Then consider the functional

$$J_{\mathrm{HR}}(\Delta\mathbf{u},\Delta\boldsymbol{\sigma}) = -\int_{\Omega} W^*\,\mathrm{d}\Omega + \int_{\Omega}\left\{\frac{1}{2}\Delta\sigma_{kl}[\Delta u_{k,l}\Delta u_{l,k} + u_{i,k}\Delta u_{i,l} + \Delta u_{i,k}u_{i,l} + \Delta u_{i,k}\Delta u_{i,l}] + \frac{1}{2}\sigma_{kl}\Delta u_{i,k}\Delta u_{i,l} - \rho\Delta\mathbf{f}\cdot\Delta\mathbf{u}\right\}\mathrm{d}\Omega - \int_{\partial\Omega}[\Delta\mathbf{T}^{\mathrm{d}}\cdot\Delta\mathbf{u} - \mathbf{n}\cdot\Delta\boldsymbol{\sigma}\cdot(\Delta\mathbf{u}-\Delta\mathbf{u}^{\mathrm{d}})]\,\mathrm{d}s. \qquad (2)$$

By considering the stationariness condition $\delta J_{\mathrm{HR}} = 0$ with *no* subsidiary condition, show that there result the equations

$$\left.\begin{aligned} \operatorname{div}(\Delta\boldsymbol{\sigma}) + \Delta\mathbf{f} = \mathbf{0} \quad & \text{in } \Omega,\\ \mathbf{n}\cdot\Delta\boldsymbol{\sigma} = \Delta\mathbf{T}^{\mathrm{d}} \quad & \text{on } \partial\Omega_T,\\ \Delta\mathbf{u} = \Delta\mathbf{u}^{\mathrm{d}} \quad & \text{on } \partial\Omega_u, \end{aligned}\right\} \qquad (3)$$

and the condition

$$S_{klmn}\Delta\sigma_{mn} = \tfrac{1}{2}(\Delta u_{k,l} + \Delta u_{l,k} + u_{i,k}\Delta u_{i,l} + \Delta u_{i,k}u_{i,l} + \Delta u_{i,k}\Delta u_{i,l}). \qquad (4)$$

11.15. Variational principles with subsidiary conditions

(a) Consider the functional

$$J_{\mathrm{C}}(\Delta\boldsymbol{\sigma}) = \int_{\Omega} W^*(\Delta\boldsymbol{\sigma})\,\mathrm{d}\Omega - \int_{\partial\Omega}\mathbf{n}\cdot\Delta\boldsymbol{\sigma}\cdot\Delta\mathbf{u}^{\mathrm{d}}\,\mathrm{d}s. \qquad (1)$$

Show that the stationariness condition of the variational principle $\delta J_{\mathrm{C}} = 0$ subjected to the subsidiary conditions $\operatorname{div}(\Delta\boldsymbol{\sigma}) + \Delta\mathbf{f} = \mathbf{0}$ in Ω and $\mathbf{n}\cdot\Delta\boldsymbol{\sigma} = \Delta\mathbf{T}^{\mathrm{d}}$ on $\partial\Omega_T$ provides the correct solution of the problem.

(b) Consider the functional

$$J_{\mathrm{p}}(\Delta\mathbf{u}) = \int_{\Omega}\left[W(\Delta\boldsymbol{\varepsilon}) + \frac{1}{2}\sigma_{ij}\Delta u_{k,i}\Delta u_{k,j} - \rho\Delta\mathbf{f}\cdot\Delta\mathbf{u}\right]\mathrm{d}\Omega \qquad (2)$$

with $\Delta\boldsymbol{\varepsilon} = \Delta\boldsymbol{\varepsilon}(\Delta\mathbf{u})$. Show that the stationariness condition $\delta J_{\mathrm{p}} = 0$ subject to the subsidiary condition $\Delta\mathbf{u} = \Delta\mathbf{u}^{\mathrm{d}}$ on $\partial\Omega$ provides a correct solution of the problem (the constitutive equation and the compatibility of reaction forces – $\mathbf{n}\cdot\Delta\boldsymbol{\sigma} = \Delta\mathbf{T}$ – on the boundary are assumed to hold *a priori*).

11.16. Incorporating the yield condition in the variational principle

The incremental forms of elastoplasticity equations are (Huber–Mises material with hardening)

$$\Delta\boldsymbol{\varepsilon} = [\nabla(\Delta\mathbf{u})]_s, \quad \Delta\boldsymbol{\varepsilon} = \Delta\boldsymbol{\varepsilon}^{\mathrm{e}} + \Delta\boldsymbol{\varepsilon}^{\mathrm{p}}, \qquad (1)$$

$$\operatorname{div}(\Delta\boldsymbol{\sigma}) = \mathbf{0} \text{ in } \Omega, \quad \mathbf{n}\cdot\Delta\boldsymbol{\sigma} = \Delta\mathbf{T}^{\mathrm{d}} \text{ on } \partial\Omega_T, \quad \Delta\mathbf{u} = \Delta\mathbf{u}^{\mathrm{d}} \text{ on } \partial\Omega_u, \qquad (2)$$

$$\Delta\boldsymbol{\sigma} = \mathbf{E}^{\mathrm{e}} : \Delta\boldsymbol{\varepsilon}^{\mathrm{p}}, \quad \Delta\boldsymbol{\varepsilon}^{\mathrm{p}} = \Delta\lambda\frac{\partial f}{\partial\boldsymbol{\sigma}}, \quad f = \left(\frac{3}{2}\sigma^{\mathrm{d}}_{ij}\sigma^{\mathrm{d}}_{ij}\right)^{1/2} - \sigma_Y = 0, \qquad (3)$$

with

$$\Delta\boldsymbol{\sigma} = \mathbf{E}^{\mathrm{ep}} : \Delta\boldsymbol{\varepsilon}, \quad \frac{\partial f}{\partial \boldsymbol{\sigma}} : \Delta\boldsymbol{\sigma} = \Delta\sigma_Y = \zeta\Delta\lambda. \tag{4}$$

Consider the functional

$$J^*(\Delta\mathbf{u}, \Delta\boldsymbol{\sigma}, \Delta\lambda) = \int_\Omega \left[W^*(\Delta\boldsymbol{\sigma}) + \alpha\frac{\partial f}{\partial \boldsymbol{\sigma}} : \Delta\boldsymbol{\sigma}\Delta\lambda - \frac{1}{2}\alpha\zeta(\Delta\lambda)^2 - \frac{1}{2}\Delta\sigma_{ij}(\Delta u_{i,j} + \Delta u_{j,i}) \right] \mathrm{d}\Omega - \int_{\partial\Omega} [\Delta\mathbf{T}^{\mathrm{d}}\cdot\Delta\mathbf{u} - \mathbf{n}\cdot\Delta\boldsymbol{\sigma}\cdot(\Delta\mathbf{u}^{\mathrm{d}} - \Delta\mathbf{u})]\,\mathrm{d}s. \tag{5}$$

Check that the stationariness condition $\delta J^* = 0$, $\forall \delta\Delta\mathbf{u}, \delta\Delta\boldsymbol{\sigma}, \delta\Delta\lambda$, provides the compatibility conditions, the last of eqns (4) above, and the equilibrium condition.

11.17. FEM formulation of variational principles

Let $\Omega = \bigcup_{\alpha=1}^{E} \bar{\Omega}_\alpha$, $\Omega_\alpha \cap \Omega_\beta = \varnothing$ for $\alpha \neq \beta$, the partition of Ω into finite elements (subregions), $\partial\Omega_{\alpha\beta}$ the common boundary between two adjacent subregions α and β, and $\partial\Omega_{\bar{\alpha}}$ the part of the αth element that belongs to the boundary $\partial\Omega$ of Ω.

(a) Re-write the incremental form of the principle of stationary potential energy (J_{p} in P11.15) in the finite-element form (so-called compatible-displacement model).

(b) Now consider the interpolation

$$\Delta u_k(\mathbf{x}) = \sum_{\xi=1}^{N_\alpha} \varphi_{k\xi}(\mathbf{x})\Delta u_\xi^{(n)}, \tag{1}$$

where $\varphi_{k\xi}$ are *shape functions*, $\Delta\mathbf{u}^{(n)}$ is the vector of incremental nodal generalized displacements and N_α is the number of all degrees of freedom assumed in the αth element considered. Compute the corresponding incremental displacement *gradient*, linear (in $\Delta\mathbf{u}^{(n)}$) and quadratic (in $\Delta\mathbf{u}^{(n)}$) incremental strains, and show that (for the αth element; the standard summation convention being used)

$$J_{\mathrm{p}}^{(\alpha)}[\Delta u_\xi^{(n)}] = \tfrac{1}{2}k_{\xi\zeta}^{(2)}\Delta u_\xi^{(n)}\Delta u_\zeta^{(n)} + \tfrac{1}{3}k_{\xi\zeta\mu}^{(3)}\Delta u_\xi^{(n)}\Delta u_\zeta^{(n)}\Delta u_\mu^{(n)} + \tfrac{1}{4}k_{\xi\zeta\mu\nu}^{(4)}\Delta u_\xi^{(n)}\Delta u_\zeta^{(n)}\Delta u_\mu^{(n)}\Delta u_\nu^{(n)} - \Delta U_\xi^{(n)}\Delta u_\xi^{(n)}, \tag{2}$$

where $\mathbf{k}^{(k)}$ are *stiffness matrices* of the kth order and $\Delta\mathbf{u}^{(n)}$ accounts for incremental *forces* ($\Delta\mathbf{f}$ and $\Delta\mathbf{T}^{\mathrm{d}}$) acting upon the element through its nodes. Give the expression of these quantities in terms of the shape functions and their gradients. Show that the first-order stiffness consists of three contributions accounting respectively for constitutive properties, initial stresses, and initial displacement.

By introducing the ordered vector set of generalized displacements and the vector of external *nodal* load components $\Delta\mathbf{R}$ show that the *linearized* version of the stationarity condition of J_p yields

$$\mathbf{K}\cdot\Delta\mathbf{r} = \Delta\mathbf{R}. \tag{3}$$

11.18. FEM equilibrium model

Find the FEM formulation corresponding to the stationariness of J_C (built with the complementary energy; so-called equilibrium model).

11.19. FEM mixed model (term paper)

Search the technical literature and establish the FEM formulation for mixed models using the Hellinger–Reissner principle ($\delta J_{HR} = 0, \delta\Delta\mathbf{u}, \delta\Delta\boldsymbol{\sigma}$).

11.20. Equilibrium imbalance and compatiblity mismatch effects (term paper)

Incremental approaches are not in general exact. In particular, they may *not* satisfy exactly the equilibrium condition on stress and the compatibility of strains. The problem is of importance as errors may accumulate during the incremental procedure. Study the technical literature and write a short report on the various ways to correct this defect through appropriate algorithms (see Kleiber, 1989).

11.21. Tangent-stiffness and initial-load methods (term paper)

Write a short term paper comparing the *tangent-stiffness* method and the *initial-load method* for elastic–plastic computations in loading (in the first case progressive yielding brings a change in the matrix coefficients and the system is solved for each increment step; in the second method an *iterative process* is used in which the matrix coefficients remain constant while non-linearity of the problem is taken into account by changing the force terms in successive iterations. Illustrate this by an example concerning a truss). (See Borkowski, 1988.)

12

Experimental study using infrared thermography

The object of the chapter In previous chapters elastoplasticity and fracture were treated in the absence of thermal effects. Obviously, in general, these effects are all coupled and a thorough analysis of the heat propagation equation shows that an experimental study based on thermal measurements provides very precious information concerning *intrinsic dissipation.* This possibility has led to the development of the method of infrared thermography, in particular in relation to the study of singularities in mechanical fields.

12.1 Heat equation in a deformable solid

We remind the reader that in an *elastoplastic* medium described by means of internal variables $\boldsymbol{\alpha}$ and an (anisotropic) Fourier law of heat conduction, the heat equation reads (see eqn (2.61))

$$\mathscr{C}\dot{\theta} - \left(\phi_{\text{intr}} + \theta\frac{\partial \boldsymbol{\sigma}^{\text{e}}}{\partial \theta} : \dot{\boldsymbol{\varepsilon}}^{\text{e}} - \theta\frac{\partial \mathbf{A}}{\partial \theta} \cdot \dot{\boldsymbol{\alpha}}\right) = -\nabla \cdot \mathbf{q}, \tag{12.1}$$

i.e.

$$\mathscr{C}\dot{\theta} + \theta(\boldsymbol{\tau} : \dot{\boldsymbol{\varepsilon}}^{\text{e}} + \mathbf{l} \cdot \dot{\boldsymbol{\alpha}}) = -\nabla \cdot \mathbf{q} + \phi_{\text{intr}} \tag{12.2}$$

with

$$\phi_{\text{intr}} = \boldsymbol{\sigma}^{\text{v}} : \dot{\boldsymbol{\varepsilon}}^{\text{e}} + \boldsymbol{\sigma} : \dot{\boldsymbol{\varepsilon}}^{\text{p}} + \mathbf{A} \cdot \dot{\boldsymbol{\alpha}}, \tag{12.3}$$

$$\boldsymbol{\sigma}^{\text{v}} = \boldsymbol{\sigma} - \boldsymbol{\sigma}^{\text{e}}, \qquad \boldsymbol{\varepsilon} = \boldsymbol{\varepsilon}^{\text{e}} + \boldsymbol{\varepsilon}^{\text{p}}, \tag{12.4}$$

$$\boldsymbol{\sigma}^{\text{e}} = \frac{\partial W}{\partial \boldsymbol{\varepsilon}^{\text{e}}}, \qquad \mathbf{A} = -\frac{\partial W}{\partial \boldsymbol{\alpha}}, \qquad \mathscr{C} = -\theta\frac{\partial^2 W}{\partial \theta^2} \geqslant 0, \tag{12.5}$$

$$W = W(\boldsymbol{\varepsilon}^{\text{e}}, \boldsymbol{\alpha}, \theta) \tag{12.6}$$

and

$$q_i = -k_{ij}\theta_{,ij}, \quad k_{ij}V_iV_j \geqslant kV_iV_i \geqslant 0, \quad \forall \mathbf{V}, \quad k \geqslant 0, \tag{12.7}$$

$$\boldsymbol{\tau} = -\frac{\partial \boldsymbol{\sigma}^{\mathrm{e}}}{\partial \theta}, \qquad \mathbf{l} = \frac{\partial \mathbf{A}}{\partial \theta}. \tag{12.8}$$

Here, W is Helmholtz's free energy. This is a function concave in θ and convex in $\boldsymbol{\varepsilon}^{\mathrm{e}}$ *and* $\boldsymbol{\alpha}$. In an analogous way, $\mathscr{E}$ will be the internal energy per unit volume, a function that is convex in all of its arguments. As an example of W, for anisotropic thermoelasticity we have

$$\begin{aligned} W = {} & \frac{1}{2}\varepsilon^{\mathrm{e}}_{ij}E_{ijkl}\varepsilon^{\mathrm{e}}_{kl} - \frac{1}{2\theta_0}\mathscr{C}(\theta-\theta_0)^2 \\ & -(\theta-\theta_0)\tau_{ij}\varepsilon^{\mathrm{e}}_{ij} + W^{\alpha}(\boldsymbol{\alpha},\theta), \end{aligned} \tag{12.9}$$

where θ is the absolute temperature, which only slightly differs from the uniform reference temperature θ_0. We have thus

$$\sigma^{\mathrm{e}}_{ij} = E_{ijkl}\varepsilon^{\mathrm{e}}_{kl} - \tau_{ij}(\theta-\theta_0), \qquad \mathbf{l} = -\frac{\partial^2 W^{\alpha}}{\partial \boldsymbol{\alpha}\partial\theta}. \tag{12.10}$$

We aim at analysing in some detail eqn (12.1) or (12.2), where the contribution within parentheses may be viewed as a *source term* in the otherwise rather traditional expression of the heat equation. With a view to linearizing this general equation we need to define a *natural* state which is free of any stress *and* internal forces $\mathbf{A}$. We denote by

$$\mathscr{S}_n = \{\bar{\boldsymbol{\varepsilon}}(\theta), \bar{\boldsymbol{\alpha}}(\theta), \theta \,|\, \boldsymbol{\varepsilon} = \bar{\boldsymbol{\varepsilon}}^{\mathrm{e}}(\theta)\} \tag{12.11}$$

this natural state which is the solution of the equations

$$\boldsymbol{\sigma}^{\mathrm{e}} = \mathbf{0}, \qquad \mathbf{A} = \mathbf{0},$$

that is

$$\frac{\partial W}{\partial \boldsymbol{\varepsilon}}(\bar{\boldsymbol{\varepsilon}}, \bar{\boldsymbol{\alpha}}, \theta) = \mathbf{0}, \qquad \frac{\partial W}{\partial \boldsymbol{\alpha}}(\bar{\boldsymbol{\varepsilon}}, \bar{\boldsymbol{\alpha}}, \theta) = 0. \tag{12.12}$$

The *existence of at least* one such natural state is guaranteed by the continuity of the derivatives $\partial W/\partial\boldsymbol{\varepsilon}$ and $\delta W/\delta\boldsymbol{\alpha}$ and *the convexity of* W. Note that this is *not* always the case (in particular in elastoplasticity).

The *natural specific heat* $\mathscr{C}_{\mathrm{n}}$ is defined in the following manner. Let us assume that the body may go through a succession of natural states while temperature evolves. The *thermal dilatation coefficients* are defined by

$$\bar{\boldsymbol{\varepsilon}}' \stackrel{\mathrm{def}}{=} \frac{\mathrm{d}\bar{\boldsymbol{\varepsilon}}}{\mathrm{d}\theta}, \qquad \bar{\boldsymbol{\alpha}}' \stackrel{\mathrm{def}}{=} \frac{\mathrm{d}\bar{\boldsymbol{\alpha}}}{\mathrm{d}\theta}. \tag{12.13}$$

In the succession of natural states we have

$$\phi_{\text{intr}} = 0, \qquad \mathbf{A}\cdot\bar{\boldsymbol{\alpha}}'\dot{\theta} = 0, \tag{12.14}$$

so that (12.1) is reduced to

$$\mathscr{C}_{\text{n}}\dot{\theta} + \boldsymbol{\nabla}\cdot\mathbf{q} = 0; \qquad \mathbf{q} = -\mathbf{k}\cdot\boldsymbol{\nabla}\theta, \tag{12.15}$$

when $\mathscr{C}_{\text{n}}$ has been defined by

$$\mathscr{C}_{\text{n}} = \mathscr{C} + (\bar{\boldsymbol{\varepsilon}}', \bar{\boldsymbol{\alpha}}')\cdot\mathbf{H}_{\varepsilon}\cdot(\bar{\boldsymbol{\varepsilon}}', \bar{\boldsymbol{\alpha}}')^T\theta \tag{12.16}$$

with (compare eqns (5.72_2 and (5.73))

$$\mathbf{H}_{\varepsilon} = \begin{pmatrix} \mathbf{E} & \mathbf{N} \\ \mathbf{N}^{\text{T}} & \mathbf{M} \end{pmatrix}, \tag{12.17}$$

where

$$\mathbf{E} = \frac{\partial^2 W}{\partial\boldsymbol{\varepsilon}\partial\boldsymbol{\varepsilon}}, \qquad \mathbf{N} = \frac{\partial^2 W}{\partial\boldsymbol{\varepsilon}\partial\boldsymbol{\alpha}}, \qquad \mathbf{M} = \frac{\partial^2 W}{\partial\boldsymbol{\alpha}\partial\boldsymbol{\alpha}}. \tag{12.18}$$

Clearly

$$\mathscr{C}_{\text{n}} > \mathscr{C} \tag{12.19}$$

since W is convex and thus the second contribution in (12.16) is positive. Then (12.1) can be rewritten as

$$\mathscr{C}_{\text{n}}\dot{\theta} + \theta\bar{\boldsymbol{\varepsilon}}' : \dot{\boldsymbol{\sigma}} = \phi_{\text{intr}} - \boldsymbol{\nabla}\cdot\mathbf{q}. \tag{12.20}$$

Proof (See Bouc and Nayroles, 1985.) $\bar{\boldsymbol{\varepsilon}}'$ and $\bar{\boldsymbol{\alpha}}'$ are solutions of

$$\mathbf{H}_{\varepsilon}\cdot(\bar{\boldsymbol{\varepsilon}}', \bar{\boldsymbol{\alpha}}')^{\text{T}} = (\boldsymbol{\tau}, \mathbf{l})^{\text{T}} \tag{12.21}$$

obtained by differentiation of (12.12) and use of the definitions (12.17) and (12.8). We can write

$$\dot{\boldsymbol{\sigma}} = \mathbf{E} : \dot{\boldsymbol{\varepsilon}} + \mathbf{N}\cdot\dot{\boldsymbol{\alpha}} - \boldsymbol{\tau}\dot{\theta}. \tag{12.22}$$

By using (12.21) to express $\boldsymbol{\tau}$ and $\mathbf{l}$ in terms of a particular solution $(\bar{\boldsymbol{\varepsilon}}'_1, \bar{\boldsymbol{\alpha}}'_1)$ we have

$$\begin{aligned} &\boldsymbol{\tau} : \dot{\boldsymbol{\varepsilon}} + \mathbf{l}\cdot\dot{\boldsymbol{\alpha}} - (\bar{\boldsymbol{\varepsilon}}'_1 : \dot{\boldsymbol{\sigma}} + \bar{\boldsymbol{\varepsilon}}'_1 : \mathbf{E} : \bar{\boldsymbol{\varepsilon}}'_1\dot{\theta}) \\ &\quad = (\bar{\boldsymbol{\varepsilon}}' - \bar{\boldsymbol{\varepsilon}}'_1, \bar{\boldsymbol{\alpha}}')\mathbf{H}_{\varepsilon}\left\{\begin{pmatrix} \dot{\boldsymbol{\varepsilon}} \\ \dot{\boldsymbol{\alpha}} \end{pmatrix} + \dot{\theta}\begin{pmatrix} \bar{\boldsymbol{\varepsilon}}'_1 \\ 0 \end{pmatrix}\right\}. \end{aligned} \tag{12.23}$$

This quantity vanishes for any $(\dot{\boldsymbol{\varepsilon}}, \dot{\boldsymbol{\alpha}}, \dot{\boldsymbol{\theta}})$ if and only if $(\bar{\boldsymbol{\varepsilon}}' - \bar{\boldsymbol{\varepsilon}}'_1, \bar{\boldsymbol{\alpha}}')$ belongs to the kernel of $\mathbf{H}_{\varepsilon}$, i.e., if $(\bar{\boldsymbol{\varepsilon}}'_1, 0)$ is a solution of (12.21). It follows that

$$(\boldsymbol{\tau} : \dot{\boldsymbol{\varepsilon}} + \mathbf{l}\cdot\dot{\boldsymbol{\alpha}}) = (\bar{\boldsymbol{\varepsilon}}'_1 : \dot{\boldsymbol{\sigma}} + \bar{\boldsymbol{\varepsilon}}'_1 : \mathbf{E} : \bar{\boldsymbol{\varepsilon}}'_1\dot{\theta}) \tag{12.24}$$

if and only if $(\bar{\boldsymbol{\varepsilon}}'_1, 0)$ is the solution of (12.21). It is verified that (12.2) can be rewritten as (12.20) on account of (12.16). QED.

The interest for the result (12.20) stems from the fact that $\mathscr{C}_n$, $\bar{\boldsymbol{\varepsilon}}'$ (the natural thermal dilatation coefficients) and $\mathbf{E}$ (the elasticity coefficients) are directly accessible to measurement. It is also remarkable that the contribution of internal variables has altogether disappeared from eqn (12.20) and that this also applies to the definition of $\mathscr{C}_n$ which is reduced to

$$\mathscr{C}_n = \mathscr{C} + \theta \bar{\boldsymbol{\varepsilon}}' : \mathbf{E} : \bar{\boldsymbol{\varepsilon}}'. \tag{12.25}$$

We can state the following: If natural thermal dilatation can take place without alteration in the internal state of the body (i.e. $\dot{\boldsymbol{\alpha}} = \mathbf{0}$), then eqn (12.10) is the local thermal balance for the body. This follows Bouc and Nayroles (1985).

It remains to specify the evolution equation of $\boldsymbol{\varepsilon}^p$ and $\boldsymbol{\alpha}$ to be able to write down the expression of ϕ_{intr}.

12.2 Linearization about a natural reference state

Let

$$\mathscr{S}_0 = (\boldsymbol{\varepsilon}_0, \boldsymbol{\alpha}_0, \theta_0) \tag{12.26}$$

be a certain state at point $\mathbf{x}$ in the deformable body. We set

$$\left.\begin{aligned} \tilde{\theta} = \theta - \theta_0, \qquad \tilde{S} = S - S_0, \\ \tilde{W}(\tilde{\theta}) = W(\theta_0 + \tilde{\theta}) + S_0\tilde{\theta} - W_0, \\ \tilde{\mathscr{E}}(\tilde{S}) = \mathscr{E}(S_0 + \tilde{S}) - \theta_0\tilde{S} - \mathscr{E}_0. \end{aligned}\right\} \tag{12.27}$$

It is immediately verified that $-\tilde{W}$ and $\tilde{\mathscr{E}}$ are the Legendre–Fenchel transforms (i.e., conjugate functions; see Appendix 2) of one another. That is

$$\left.\begin{aligned} \tilde{\mathscr{E}}(\tilde{S}) &= \operatorname*{Sup}_{\tilde{\theta} \geqslant -\theta_0} [\tilde{\theta}\tilde{S} + \tilde{W}(\tilde{\theta})], \\ -\tilde{W}(\tilde{\theta}) &= \operatorname*{Sup}_{\tilde{S} \geqslant -S_0} [\tilde{\theta}\tilde{S} - \tilde{\mathscr{E}}(\tilde{S})]. \end{aligned}\right\} \tag{12.28}$$

In other terms, $\tilde{W}$ is deduced from W by substracting its tangent affine function at θ_0. The same holds true for $\tilde{\mathscr{E}}$ with respect to $\mathscr{E}$. By way of example we give the *proof* of $(12.28)_1$. This exercise in convex analysis goes as follows. We verify that

$$\begin{aligned}\operatorname*{Sup}_{\tilde{\theta}\geq-\theta_0}\,[\tilde{\theta}\tilde{S}+\tilde{W}(\tilde{\theta})] &= \operatorname*{Sup}_{\theta\geq 0}\,[\tilde{S}(\theta-\theta_0)+W(\theta)+S_0(\theta-\theta_0)-W_0]\\ &= -(W_0+S_0\theta_0)-\tilde{S}\theta_0+\operatorname*{Sup}_{\theta\geq 0}\,[(\tilde{S}+S_0)\theta+W(\theta)]\\ &= -(W_0+S_0\theta_0)-\theta_0\tilde{S}+\mathscr{E}(S_0+\tilde{S})\\ &= \mathscr{E}(S_0+\tilde{S})-\theta_0\tilde{S}-\mathscr{E}_0\\ &= \tilde{\mathscr{E}}(\tilde{S}).\end{aligned}$$

The situation is analogous for $-W$ and $\mathscr{E}$.

Considering the state (12.26) as a *natural* one, we have

$$\left(\frac{\partial W}{\partial \boldsymbol{\varepsilon}}\right)_0 = \left(\frac{\partial \mathscr{E}}{\partial \boldsymbol{\varepsilon}}\right)_0 = \mathbf{0}, \qquad \left(\frac{\partial W}{\partial \boldsymbol{\alpha}}\right)_0 = \left(\frac{\partial \mathscr{E}}{\partial \boldsymbol{\alpha}}\right)_0 = \mathbf{0}. \tag{12.29}$$

Linearizing about this state consists mathematically in replacing $\tilde{W}$ by W_1, the osculator quadratic form of $\tilde{W}$ at $(\bar{\boldsymbol{\varepsilon}}_0, \bar{\boldsymbol{\alpha}}_0, 0)$. Therefore, we can directly write $\overline{W_1}, (\boldsymbol{\varepsilon}-\boldsymbol{\varepsilon}_0, \boldsymbol{\alpha}-\boldsymbol{\alpha}_0, \tilde{\theta})$ as a quadratic form in its three variables. The same applies to $\mathscr{E}_1$ as the function $\mathscr{E}_1(\boldsymbol{\varepsilon}-\boldsymbol{\varepsilon}_0, \boldsymbol{\alpha}-\boldsymbol{\alpha}_0, \tilde{S})$ except that most coefficients are modified in terms of θ_0. The natural dilatation coefficient remains constant in this linearization procedure. Finally, we note that if a natural state is associated with the temperature $\tilde{\theta}$ by

$$\tilde{\boldsymbol{\varepsilon}}(\tilde{\theta}) = \boldsymbol{\varepsilon}_0 + \bar{\boldsymbol{\varepsilon}}'\tilde{\theta}, \qquad \tilde{\boldsymbol{\alpha}}(\tilde{\theta}) = \boldsymbol{\alpha}_0 + \bar{\boldsymbol{\alpha}}'\tilde{\theta}, \tag{12.30}$$

then we obtain the following expression for the quadratic free energy W:

$$W_1 = \frac{1}{2}(\boldsymbol{\varepsilon}-\tilde{\boldsymbol{\varepsilon}}(\tilde{\theta}), \boldsymbol{\alpha}-\tilde{\boldsymbol{\alpha}}(\tilde{\theta}))\mathbf{H}_\varepsilon(\boldsymbol{\varepsilon}-\tilde{\boldsymbol{\varepsilon}}(\tilde{\theta}), \boldsymbol{\alpha}-\tilde{\boldsymbol{\alpha}}(\tilde{\theta}))^{\mathrm{T}} - \frac{1}{2\theta_0}\mathscr{C}_{\mathrm{n}}\tilde{\theta}^2, \tag{12.31}$$

which contains both a mechanical and a thermal contribution. This can be illustrated on the example of a *rheological model* of the type of the one previously examined in eqn (5.53) *et passim* (Fig. 5.6).

Example: rheological model with elements in parallel We consider the model of Fig. 12.1 with elements in parallel, except for the 'thermal dilatation' element $(\bar{\boldsymbol{\varepsilon}}', \tilde{\theta})$; $E_0, E_1, \ldots, E_n$ are linear springs and $V_1, \ldots, V_n$ are dissipative elements with deformation α_i considered as internal variables. The reference natural state is $\mathscr{S}_0 = (\mathbf{0}, \mathbf{0}, \theta_0)$. The natural dilatation $\tilde{\boldsymbol{\varepsilon}} = \bar{\boldsymbol{\varepsilon}}'\tilde{\theta}$ is in series. In agreement with (12.31) we can write

$$W_1 = \frac{1}{2}E_0(\varepsilon - \bar{\varepsilon}'\tilde{\theta})^2 + \frac{1}{2}\sum_{i=1}^{n} E_i(\varepsilon - \alpha_i - \bar{\varepsilon}'\tilde{\theta})^2 - \frac{1}{2\theta_0}\mathscr{C}_{\mathrm{n}}\tilde{\theta}^2, \tag{12.32}$$

an expression that compares with (5.53). In what follows we admit (12.20) and suppose that the above linearization has already been performed.

12.3 Method of infrared thermography

This method consists in observing and measuring the thermal field in a dissipative deformable body by means of a camera that scans the two-dimensional optical field. With an appropriate signal processing (demodulation, improvement of the signal-to-noise ratio) it is then possible to extract a rather clean temperature field and this, in turn, allows us to evaluate the right-hand side, i.e., the *heat* sources, in the equation (see eqn (12.20))

$$\dot{\tilde{\theta}} + \mathscr{C}_{\mathrm{n}}^{-1}\nabla\cdot\mathbf{q} = \mathscr{C}_{\mathrm{n}}^{-1}(\phi_{\mathrm{intr}} - \theta_0\bar{\boldsymbol{\varepsilon}}' : \dot{\boldsymbol{\sigma}}), \tag{12.33}$$

through a numerical evaluation on the left-hand side from the recorded data. So, for an isotropic Fourier conduction we have

$$\dot{\tilde{\theta}} - d\Delta\theta = \mathscr{C}_{\mathrm{n}}^{-1}(\phi_{\mathrm{intr}} - \theta_0\bar{\boldsymbol{\varepsilon}}' : \dot{\boldsymbol{\sigma}}) \tag{12.34}$$

where $d = k/\mathscr{C}_n$ is the coefficient of thermal diffusivity. The Laplacian $\Delta\tilde{\theta}$ must be evaluated (for instance in two dimensions for a thin plane body) from the spatial distribution of $\tilde{\theta}$ which is recorded by the scanning camera.

For example, we may consider the case of a plane test body in which a crack grows during a fatigue test (PVC test sample, frequency of 20 Hz). For such a frequency, the evolution is practically quasi-static. The material considered (PVC) has a low conductivity and behaves almost as a linear

Fig. 12.1. Rheological model

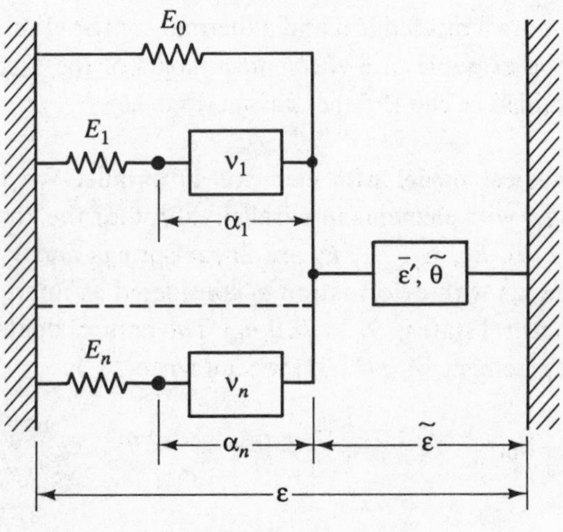

viscoelastic body outside the immediate vicinity of the crack tip. Eqn (12.34) is integrated throughout the thickness of the sample and there results

$$\dot{\tilde{\theta}} - \bar{d}\Delta_2\theta + \frac{\tilde{\theta}}{\tau} = \mathscr{C}_n^{-1}(\phi_{\text{intr}} - \theta_0\bar{\boldsymbol{\varepsilon}}' : \dot{\boldsymbol{\sigma}}), \tag{12.35}$$

where τ is a time characteristic of losses through the plane faces (for instance, $\tau \approx 1640$ s for a thickness of one centimetre so that the relaxation $\tilde{\theta}/\tau$ is negligible), $\bar{d}$ is the diffusivity in m^2/s and ϕ_{intr} is the *dissipation* averaged through the thickness (hence it is in W/m^3) and Δ_2 is the two-dimensional Laplacian. For PVC, $\bar{d} \approx 0.73 \times 10^{-7}$ m^2/s, $\theta_0 \approx 300$K (room temperature) $\bar{\varepsilon}' \approx 7 \times 10^{-5}$ K^{-1}. In the fatigue test the evolution $\tilde{\theta}(t)$ of temperature deviation from the reference, at a given point $\mathbf{x}$ in the body, has the characteristic time dependence in Fig. 12.2, where $\tilde{\theta} = \bar{\theta} + \hat{\theta}$, where $\bar{\theta}$ is a *slow* heating term due to the dissipation ϕ_{intr} (more precisely, its mean value $\overline{\phi_{\text{intr}}}$ over a cycle), and $\hat{\theta}$ is a term synchronous with the oscillation resulting essentially from the thermoelastic coupling term $\bar{\boldsymbol{\varepsilon}}' : \dot{\boldsymbol{\sigma}}$, and accessorily from the oscillating part $\hat{\phi}_{\text{intr}}$ of the dissipation.

Example of numerical values

$$\mathscr{C}_n = 2 \times 10^6 \text{ J/m}^3\text{/K};$$

$$\overline{\phi_{\text{intr}}} = \begin{cases} 0.25 \times 10^{-5} \text{ W/mm}^3 & \text{in a circle } |r| \leqslant R = 0.65 \text{ mm} \\ & \text{that supports the heat source,} \\ 0 & \text{elsewhere } (|r| > R), \end{cases}$$

with a profile given in Fig. 12.3 (x variable, $y = 0$).

Fig. 12.2. Fatigue test of a PVC sample $\tilde{\theta}(t)$

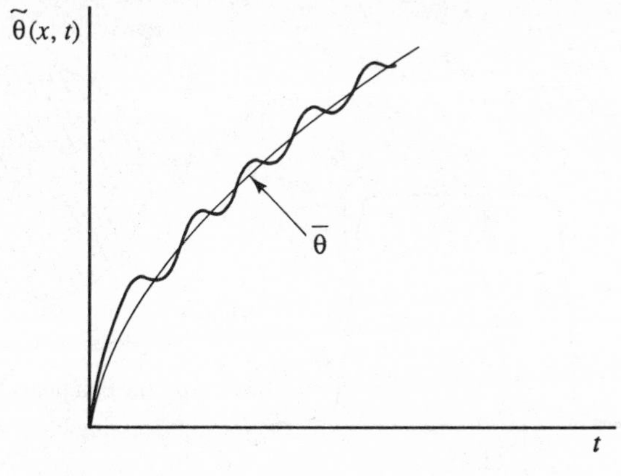

The expression in the right-hand side of (12.34) shows that, in some sense, the method allows us to measure the *trace* of the stress tensor (since in the isotropic case $\overline{\boldsymbol{\varepsilon}}' : \dot{\boldsymbol{\sigma}} = \overline{\varepsilon}'(\mathrm{tr}\,\dot{\boldsymbol{\sigma}})$), for the inphase (with excitation) component $\hat{\theta}(t)$ is practically exclusively linked to this thermoelastic term. On account of this a correspondence of the type $1°\mathrm{C} \Leftrightarrow 10^8$ Pa (or $3.2 \times 10^{-3}\,°\mathrm{C} \Leftrightarrow 0.32$ daN/m^2) can be established between temperature and stress scales in PVC (Nayroles *et al.*, 1981). The accuracy or 'resolution power' of this experiment is of the order 3.2×10^{-3} °C, or 0.03 daN/m^2 in stresses or 1/150 of the elasticity limit of PVC. The results obtained through different methods (these measurements, computations in plane elasticity, measurement in photoelasticity) are very close to each other. For instance, the stress intensity factor K_{I} obtained by interpolation (with a singularity $1/\sqrt{r}$ in stress, see Chapter 7) yields the following values:

in thermography, $K_{\mathrm{I}} = 0.93$ daN $\times$ mm$^{1/2}$
in photoelasticity, $K_{\mathrm{I}} = 0.96$ daN $\times$ mm$^{1/2}$
elasticity computation (by FEM). $K_{\mathrm{I}} = 1.07$ daN $\times$ mm$^{1/2}$

This example brings us to the study of the relation between the singularities in mechanical and thermal fields.

12.4 Temperature distribution in fracture

12.4.1 Consequences of thermodynamic laws

We consider anew some of the analyses of Chapter 7, but this time taking also into account the thermal effect. The situation considered is that in Fig.

Fig. 12.3. Example of measurement of ϕ_{intr} in PVC (after Bouc and Nayroles, 1985)

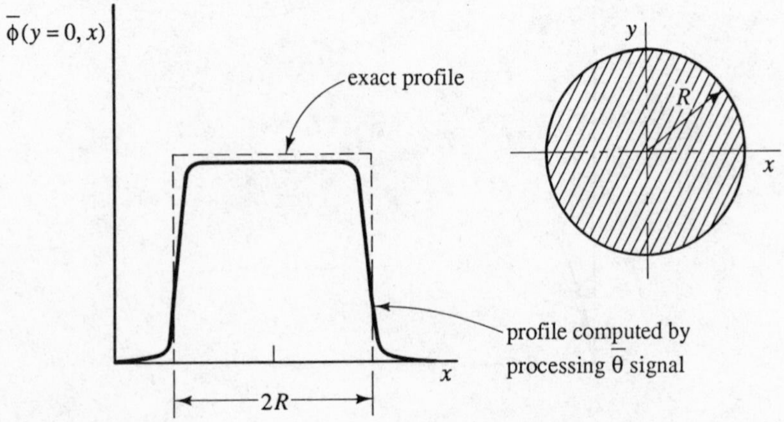

12.4 concerning the growth of a rectilinear crack. We follow Nguyen Quoc Son (1980, 1981b, 1982) and Bui *et al*, (1979). The origin of this type of approach is to be found in Cherepanov (1967, 1968).

Consequence of the first law of thermodynamics Following eqn (2.28) we have the global statement

$$\dot{K} + \dot{E} = \mathscr{P}_{(\mathrm{ext})} + \dot{Q}. \qquad (12.36)$$

The problem here is to compute the derivatives of integrals in the left-hand side since the *singular* point A (crack tip) is moving. We have

$$K + E = \int_{\Omega_\Gamma} \rho\left(e + \frac{1}{2}\mathbf{v}^2\right)\mathrm{d}\Omega + \int_{V_\Gamma} \rho\left(e + \frac{1}{2}\mathbf{v}^2\right)\mathrm{d}\Omega \qquad (12.37)$$

and

$$\frac{\mathrm{d}}{\mathrm{d}t}\int_{V_\Gamma} \rho\left(e + \frac{1}{2}\mathbf{v}^2\right)\mathrm{d}\Omega = \int_{V_\Gamma} \rho\left(e + \frac{1}{2}\mathbf{v}^2\right)^* \mathrm{d}\Omega, \qquad (12.38)$$

where

$$g^* \equiv \frac{\partial g}{\partial t}(x + l(t), y, t) = \dot{g} + \dot{l}\frac{\partial g}{\partial x}. \qquad (12.39)$$

It can be shown that the integrand in the right-hand side of (12.38) is an *integrable* function so that

Fig. 12.4. Rectilinear crack: notation

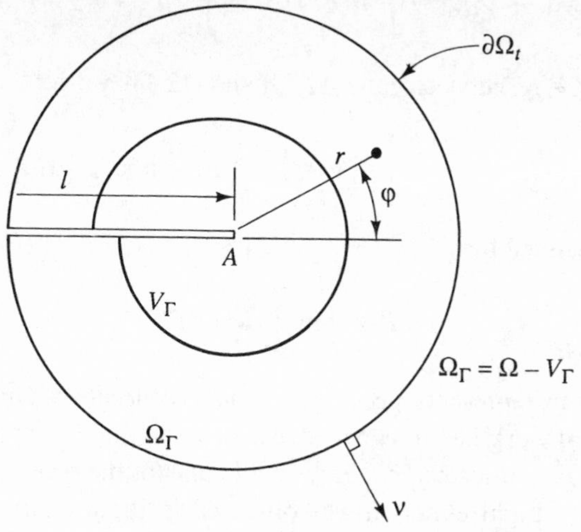

$$\lim_{\Gamma \to 0} \int_{V_\Gamma} \rho \left(e + \frac{1}{2} \mathbf{v}^2 \right)^* \mathrm{d}\Omega = 0. \tag{12.40}$$

Therefore, there remains

$$\dot{K} + \dot{E} = \lim_{\Gamma \to 0} \left[\frac{\mathrm{d}}{\mathrm{d}t} \int_{\Omega_\Gamma} \rho \left(e + \frac{1}{2} \mathbf{v}^2 \right) \mathrm{d}\Omega \right]. \tag{12.41}$$

The evaluation of the derivative in the right-hand side of this equation presents no problems. We obtain

$$\begin{aligned} \frac{\mathrm{d}}{\mathrm{d}t} \int_{\Omega_\Gamma} \rho \left(e + \frac{1}{2} \mathbf{v}^2 \right) \mathrm{d}\Omega = \int_{\Omega_\Gamma} \rho \frac{\mathrm{d}}{\mathrm{d}t} \left(e + \frac{1}{2} \mathbf{v}^2 \right) \mathrm{d}\Omega \\ - \int_{\Gamma} \rho \left(e + \frac{1}{2} \mathbf{v}^2 \right) \dot{l} n_1 \, \mathrm{d}\Gamma, \end{aligned} \tag{12.42}$$

where we use the notation introduced in Chapter 7.

Assuming now that the system Ω is thermally isolated so that $\dot{Q} = 0$ and accounting for (12.42) we can rewrite the first law (12.36) in the following form

$$\lim_{\Gamma \to 0} \left\{ \mathscr{P}_{(\mathrm{ext})} - \int_{\Omega_\Gamma} \rho \frac{\mathrm{d}}{\mathrm{d}t} \left(e + \frac{1}{2} \mathbf{v}^2 \right) \mathrm{d}\Omega + \int_{\Gamma} \rho \left(e + \frac{1}{2} \mathbf{v}^2 \right) \dot{l} n_1 \, \mathrm{d}\Gamma \right\} = 0. \tag{12.43}$$

We can also apply the first law to the system Ω_Γ only. This system, however, is *not* thermally isolated. For any Γ, we have

$$\int_{\Gamma} \mathbf{q} \cdot \mathbf{n} \, \mathrm{d}s + \mathscr{P}_{(\mathrm{ext})} - \int_{\Gamma} \mathbf{n} \cdot \boldsymbol{\sigma} \cdot \mathbf{v} \, \mathrm{d}\Gamma = \int_{\Omega_\Gamma} \rho \frac{\mathrm{d}}{\mathrm{d}t} \left(e + \frac{1}{2} \mathbf{v}^2 \right) \mathrm{d}\Omega. \tag{12.44}$$

Eliminating $\mathscr{P}_{(\mathrm{ext})}$ between eqns (12.43) and (12.44), we get

$$H = \lim_{\Gamma \to 0} \int_{\Gamma} \left[\rho \left(e + \frac{1}{2} \mathbf{v}^2 \right) \dot{l} n_1 + \mathbf{n} \cdot \boldsymbol{\sigma} \cdot \mathbf{v} \right] \mathrm{d}\Gamma, \tag{12.45}$$

when H is defined by

$$H = \lim_{\Gamma \to 0} \int_{\Gamma} \mathbf{q} \cdot \mathbf{n} \, \mathrm{d}\Gamma. \tag{12.46}$$

Physically, this represents a *heat source that is concentrated at the crack tip.* Experimental work has shown evidence of a very high temperature in fast fracture of glass and steel in a very small zone (of the order of 20Å) about the crack tip. Light emission was observed in the fracture of glass, and infrared thermography shows also evidence for this heat source localization in a steel plate (Bui, 1981; film by Bui, 1983; see Section 12.5).

Consequence of the second law of thermodynamics For the *whole* of Ω which is thermally isolated we can write (see eqn (2.10))

$$\frac{\mathrm{d}}{\mathrm{d}t}\int_{\Omega} \rho\eta \, \mathrm{d}\Omega \geqslant 0. \tag{12.47}$$

Therefore, an evaluation of the same type as in eqn (12.42) yields

$$\lim_{\Gamma \to 0}\left(\int_{\Omega_\Gamma} \rho\dot{\eta}\, \mathrm{d}\Omega - \int_{\Gamma} \rho\eta \dot{l} n_1 \, \mathrm{d}\Gamma\right) \geqslant 0. \tag{12.48}$$

For the subsystem Ω_Γ which is *not* thermally isolated, (2.33) reads

$$\int_{\Omega_\Gamma} \rho\dot{\eta}\, \mathrm{d}\Omega - \int_{\Gamma} \frac{1}{\theta}\mathbf{q}\cdot\mathbf{n}\, \mathrm{d}\Gamma \geqslant 0. \tag{12.49}$$

Let $\theta_A(\neq 0)$ be the temperature at the crack tip. It follows that

$$\lim_{\Gamma \to 0}\left\{\int_{\Omega_\Gamma} \rho\dot{\eta}\, \mathrm{d}\Omega\right\} - \frac{H}{\theta_A} \geqslant 0, \tag{12.50}$$

where H is the quantity *defined* by (12.46). It can also be shown that

$$\frac{\mathrm{d}}{\mathrm{d}t}\int_{\Omega} \rho\eta\, \mathrm{d}\Omega = \int_{\Omega} \rho\dot{\eta}\, \mathrm{d}\Omega = \lim_{\Omega_\Gamma \to 0}\int_{\Omega_\Gamma} \rho\dot{\eta}\, \mathrm{d}\Omega \geqslant 0 \tag{12.51}$$

and

$$\lim_{\Gamma \to 0}\int_{\Gamma} \rho\eta\, \dot{l} n_1 \, \mathrm{d}\Gamma = 0. \tag{12.52}$$

Now we write the expression of the second law for the subsystem V_Γ (which, obviously, is *not* thermally isolated either) at time t, i.e.

$$\int_{V_\Gamma} \rho\dot{\eta}\, \mathrm{d}\Omega + \int_{\Gamma} \frac{1}{\theta}\mathbf{q}\cdot\mathbf{n}\, \mathrm{d}\Gamma \geqslant 0. \tag{12.53}$$

Taking the limit as $\Gamma \to 0$, we obtain

$$\lim_{\Gamma \to 0}\int_{V_\Gamma} \rho\dot{\eta}\, \mathrm{d}\Omega = 0, \qquad \lim_{\Gamma \to 0}\int_{\Gamma} \frac{1}{\theta}\mathbf{q}\cdot\mathbf{n}\, \mathrm{d}\Gamma = \frac{H}{\theta_A}, \tag{12.54}$$

whence the inequality

$$H/\theta_A \geqslant 0, \qquad \text{i.e., } H \geqslant 0. \tag{12.55}$$

That is, the heat source at the crack tip necessarily is a *warm* source. The quantity may also be interpreted as the *dissipated intrinsic power* at the crack tip. Therefore, we conclude that the whole of the intrinsic power concentrated at the crack tip is transformed into heat.

Remarks (i) The result (12.45) can be written in the following equivalent form:

$$H = \lim_{\Gamma \to 0} \int_{\Gamma} \left[\rho \left(\psi + \frac{1}{2} \mathbf{v}^2 \right) \dot{l} n_1 + \mathbf{n} \cdot \boldsymbol{\sigma} \cdot \mathbf{v} \right] \mathrm{s}\Gamma, \tag{12.56}$$

where ψ is the free energy per unit mass, because

$$\lim_{\Gamma \to 0} \int_{\Gamma} \rho \theta \eta \, \dot{l} n_1 \, \mathrm{d}\Gamma = 0. \tag{12.57}$$

(ii) We can also write down the enlightening form

$$H = G\dot{l}, \tag{12.58}$$

where G is the thermodynamic force at the crack tip defined by ($\mathbf{d}$ is the displacement)

$$G = \lim_{\Gamma \to 0} \int_{\Gamma} \left[\rho \left(e + \frac{1}{2} \mathbf{v}^2 \right) n_1 - \mathbf{n} \cdot \boldsymbol{\sigma} \cdot \mathbf{d}_{,1} \right] \mathrm{d}\Gamma, \tag{12.59}$$

Consequently, a *propagation criterion* can be proposed in the following form: there exists G_c such that $G = G_c$ when propagation occurs and $G \leqslant G_c$ otherwise.

12.4.2 Singularity of the temperature distribution

We have the following 'heat propagation' equation (see eqn (2.50)):

$$\rho \theta \dot{\eta} + \nabla \cdot \mathbf{q} - \phi_{\text{intr}} = 0, \quad \forall \mathbf{x} \in \Omega_t, \tag{12.60}$$

along with the condition of concentrated heat source at the crack tip:

$$\lim_{\Gamma \to 0} \int_{\Gamma} \mathbf{q} \cdot \mathbf{n} \, \mathrm{d}s = H. \tag{12.61}$$

Bui *et al.* (1979) have shown that, if $H \neq 0$, then the asymptotic distribution of temperature depends *only* upon H and on the adopted conduction law. As a matter of fact, admitting that various physical quantities, at the crack tip, admit an asymptotic expansion of the form $K(t)r^{\alpha}$ $(\mathrm{Log}\, r)^{\beta}$, eqn (12.61) implies that $|q| \sim 1/r$ and $\nabla \cdot \mathbf{q} \sim 1/r^2$. One can thus state the following.

Proposition (i) *The temperature distribution is singular and its singularity is logarithmic when the linear Fourier conduction law is adopted, in the presence of a concentrated heat source* $H \neq 0$.

(ii) *If* $H = 0$, *then the temperature remains finite at the crack tip.*

Bui *et al.* (1979) – also Nguyen Quoc Son (1981b) – have obtained the singularities at the crack tip for a *dynamical* thermoelastic response of the material. For the present, the asymptotic analysis of the solution remains an open problem in the case of *elastoplasticity with thermal coupling.* However, in mode III crack (see Chapter 7), the mechanical and thermal problems uncouple. Actually, in that case, $G = 0$, $H = 0$ and the temperature remains finite at the crack tip in agreement with part (ii) of the above-stated proposition. A *numerical* evaluation shows that the temperature remains finite also in mode I and thus $H = 0$ (Nguyen Quoc Son, 1982) but there are no known asymptotic solutions. Some numerical results are reproduced for illustrative purposes (Fig. 12.5). They concern mode I in which one has to determine

the displacement $\mathbf{d}(X, Y)$,

the stress $\boldsymbol{\sigma}(X, Y)$,

the temperature $\theta(X, Y)$,

in such a way that, in stationary evolution, we have to satisfy

the mechanical equations

$$\left.\begin{aligned} 2\varepsilon_{ij} &= d_{i,j} + d_{j,i}, \\ \sigma_{ij} &= \partial W/\partial\varepsilon_{ij}, \\ -\dot{l}\varepsilon^{\mathrm{p}}_{ij,x} &= \dot{\lambda}\partial f/\partial\sigma_{ij}, \qquad \dot{\lambda} \geqslant 0 \text{ if } f(\boldsymbol{\sigma}) = 0, \\ \sigma_{ij,j} &= 0, \end{aligned}\right\} \tag{12.62}$$

the thermal equation (this is eqn (12.34) with $\tilde{\theta} \approx 0$)

$$k\Delta\tilde{\theta} + \mathscr{C}_n\dot{l}\tilde{\theta}_{,x} + \bar{\varepsilon}'\dot{l}\sigma_{kk,x} - \dot{l}\boldsymbol{\sigma} : \boldsymbol{\varepsilon}^{\mathrm{p}}_{,x} = 0, \tag{12.63}$$

to which must be adjoined appropriate boundary conditions. Here both the third of eqns (12.62) and (12.63) have been written taking account of the fact that $\dot{A} \approx -\dot{l}A_{,x}$ (lemma in Section 7.3). The interested reader will compare Fig. 12.5 to Figures 10.3.2 to 10.3.5 in Mendelson (1968), who was mostly concerned with computational difficulties (influence for grid size), the choice between plane strains and plane stresses, or the influence of strain hardening, but *not* with the coupling with the temperature field. These differences highlight the progress made in the field of *theoretical and computational elastoplasticity* within an interval of twenty years.

12.5 Illustrative examples

The best illustrations are provided by videofilms that reproduce experimental recordings by infrared thermography. We recommend the following films (available at *Imagiciel*, 91128 Palaiseau, France).

'Thermographic analysis of a fine sand in large distortion'.
Author: M.P. Luong, Length 5 mm (1982).

'Fracture observed by means of infrared thermography': (experimental study of thermomechanical couplings during initiation and progression at low speed, of a crack in a steel plate).
Author: H.D. Bui. Length: 5 mm (1983).

'Infrared vibrothermography of concrete'.
Author: M.P. Luong. Length: 6 mm (1984).

The following problems deal essentially with singularities in temperature and mechanical fields for various mechanical behaviours of bodies containing a running crack.

Problems for Chapter 12

12.1. Reduced form of the heat equation

Prove the results (12.16), (12.20), and (12.25) concerning the heat equation in a mechanically dissipative material.

12.2. Reduced form of the free energy

By linearizing about a natural state defined by eqns (12.26) and (12.29) establish the quadratic form (12.31) of the free energy.

12.3. Heat source at a crack tip

Redo all the steps of the proof of eqn (12.45). In a like manner prove the inequality (12.55).

12.4. Heat equation in a cracked body

Consider the balance of energy for a system involving possible reversible surface energy R, i.e.,

$$\dot{K} + \dot{E} + \dot{R} = \mathscr{P}_{(\mathrm{ext})} + \dot{Q}, \tag{1}$$

where K and E have the same significance as in the text. Consider the special case where $R = 2\gamma l(t)$ in which $l(t)$ is the length of a rectilinear crack (compare Problem 7.9).

(i) Taking account of the nonnecessary integrability of $\rho\dot{e}$ and $\rho\ddot{\mathbf{u}}\cdot\dot{\mathbf{u}}$ and using the transport condition of the singularity (lemma in Section 7.3), show that (1) implies that

$$(G - 2\gamma)\dot{l} = H, \tag{2}$$

where

$$\begin{aligned} H &\equiv \lim_{\Gamma\to 0}\int_\Gamma \mathbf{q}\cdot\mathbf{n}\,\mathrm{d}\Gamma, \\ G &= \lim_{\Gamma\to 0}\int_\Gamma \left[\rho\left(e + \frac{\dot{\mathbf{u}}^2}{2}\right) + (\mathbf{n}\cdot\boldsymbol{\sigma}\cdot\dot{\mathbf{u}}/\dot{l})\right]\mathrm{d}\Gamma. \end{aligned} \tag{3}$$

(ii) Show that the total intrinsic dissipation has the expression

$$\Phi = \int_\Omega \phi_{\text{intr}}\,\mathrm{d}\Omega + (G - 2\gamma)\dot{l}. \tag{4}$$

(iii) Show that the heat conduction equation can be written in the sense of distribution theory as

$$\theta\dot{S} + \nabla\cdot\mathbf{q} = \phi_{\text{intr}} + (G - 2\gamma)\dot{l}\delta(M), \tag{5}$$

where $\delta(M)$ is Dirac's delta generalized-function of which the support is the crack tip M. This result is due to Bui *et al.* (1979).

12.5. Singularity in temperature at the tip of a running crack

Consider the special case where $\mathbf{q} = -k\nabla\theta$ (Fourier's law). Then prove, from the assumption that ϕ_{intr}, $\theta\dot{S}$ and $S\dot{\theta}$ are Lebesgue-integrable in $\Omega(t)$ for almost all times t, the lemma concerning the transport condition for the singularity, and eqn (4) of 12.4, that the temperature field has the following asymptotic behaviour with distance from the crack tip:

$$\theta = -\frac{(G - 2\gamma)\dot{l}}{2\pi k}\operatorname{Log} r + (\text{more regular terms}). \tag{1}$$

[*Note*: The result in elasticity is due to Bui *et al.* (1979). However, this result holds good for a large class of materials satisfying the integrability condition on ϕ_{intr}.]

12.6. Singularities and energy release rate in Kelvin–Voigt viscoelasticity

Taking into account the scheme introduced in Chapter 2 (no internal variable), $\boldsymbol{\sigma}^{\text{v}} = \boldsymbol{\sigma} - \boldsymbol{\sigma}^{\text{e}}$, $\boldsymbol{\sigma}^{\text{e}} = \mathbf{E}:\boldsymbol{\varepsilon}$, $\phi_{\text{intr}} = \boldsymbol{\sigma}^{\text{v}}:\dot{\boldsymbol{\varepsilon}}$, show that $\boldsymbol{\sigma}^{\text{v}} \sim 1/\sqrt{r}$ while $\boldsymbol{\varepsilon} \sim \boldsymbol{\sigma} \sim \sqrt{r}$. Then deduce from the expression for G in which $\dot{\mathbf{u}} \sim \dot{l}\mathbf{u}_{,1}$ that $G = 0$ in this case.

12.7. Singularity and energy release rate in Maxwell viscoelasticity

Taking into account the scheme introduced in Chapter 2, where $\phi_{\text{intr}} = \boldsymbol{\sigma}:\dot{\boldsymbol{\varepsilon}}^{\text{v}}$, $\boldsymbol{\sigma} = \mathbf{E}:(\boldsymbol{\varepsilon} - \boldsymbol{\varepsilon}^{\text{v}})$, show that $\boldsymbol{\sigma} \sim \boldsymbol{\varepsilon} \sim 1/\sqrt{r}$ and $G \neq 0$. Assuming isotropy for the state law (with Young's modulus E and Poisson's ratio ν), show that

$$G = \frac{1 - \nu^2}{E}(K_{\mathrm{I}}^2 + K_{\mathrm{II}}^2)$$

in *plane strains*, K_{I} and K_{II} being stress intensity factors.

12.8. Singularities and energy-release rate

(i) Suppose that the variable $\boldsymbol{\varepsilon}^{\mathrm{v}}$ of 12.7 has an evolution equation of the Norton–Hoff (creep) type:

$$\dot{\boldsymbol{\varepsilon}}^{\mathrm{v}} = C(|\boldsymbol{\sigma}|/\sigma_0)^{m-1}\boldsymbol{\sigma}/\sigma_0, \qquad |\boldsymbol{\sigma}|^2 = \sigma_{ij}\sigma_{ji}, \tag{1}$$

Fig. 12.5. Thermomechanical coupling at the crack tip (after Nguyen Quoc Son, 1982)

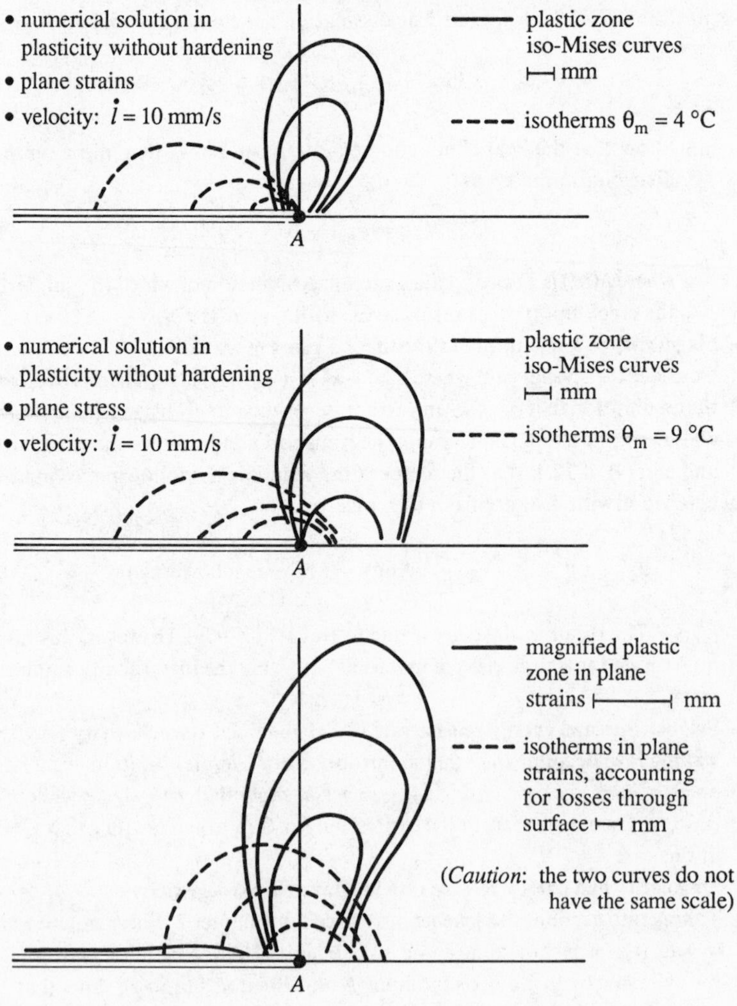

where C, σ_0, and m are material parameters. Then show that the integrability of $\phi_{intr} = \boldsymbol{\sigma} : \dot{\boldsymbol{\varepsilon}}^v$ implies that $(m+1)\beta > -2$, where β is the singularity order of the temperature, i.e., we write

$$\theta = H(t)r^{\alpha}(\mathrm{Log}\, r)^{\beta}\hat{\theta}(\dot{l}, \text{angle distribution}).$$

(ii) If $\boldsymbol{\varepsilon}^v$ is *more* regular than $\boldsymbol{\varepsilon}^e$ (i.e., $m\beta > \beta - 1$) show that $\beta = -1/2$ as in elasticity. In particular, if $m < 3$, the stress singularity is that of linear elasticity ($\boldsymbol{\sigma} \sim 1/\sqrt{r}$). Determine β for $m > 3$ by using the equation of motion $\rho\ddot{\mathbf{u}} = \operatorname{div} \boldsymbol{\sigma}$ in the asymptotic analysis. What is the value of G when $\beta = -(1/m - 1)$ and $m > 3$ (prove that $G = 0$)? What happens for $m = 3$?

12.9. Singularities and energy release rate in perfect plasticity
This is a problem of *ductile* fracture. Consider the scheme where $\boldsymbol{\sigma} = \mathbf{E} : \boldsymbol{\varepsilon}^e$, $\boldsymbol{\varepsilon} = \boldsymbol{\varepsilon}^e + \boldsymbol{\varepsilon}^p$, $\phi_{intr} = \boldsymbol{\sigma} : \dot{\boldsymbol{\varepsilon}}^p$ and $\dot{\boldsymbol{\varepsilon}}^p = \dot{\lambda}\partial f/\partial\boldsymbol{\sigma}$, in quasi-static and isothermal evolution. Show in mode I of cracking that the integrability conditions on ϕ_{intr} and $\theta\dot{S}$ of 12.5 are fulfilled. In particular, $\phi_{intr} \sim 1/r$ and G, defined in 12.4, vanishes. Conclude that the dissipation of the whole body is simply the volume dissipation, as there are no other dissipative terms, and that the point heat source carried by the crack tip does *not* exist in perfect plasticity (Nguyen Quoc Son, 1980, 1981b).

Appendix 1

Thermodynamics of continuous media

The purpose of the appendix Our aim here is, based upon a 'physical' basis, to account for the statement of principles (2.9) and (2.10) of the thermodynamics of continuous media and for the essential choice we operate among the thermodynamic variables, selecting those that allow us to give the definition of the mechanical and thermal state of a deformable body through the mediation of the constitutive equations called *laws of state* and *complementary laws.*

A1.1 General notions

A1.1.1 Thermodynamic systems

We call a *system* $\mathcal{S}$ a part of the material universe (open region of Euclidean three-dimensional space $\mathbb{E}^3$). The complement of $\mathcal{S}$ in $\mathbb{E}^3$ is called the exterior of $\mathcal{S}$, or $\mathcal{S}^{\text{ext}}$. A system is called closed when there is no exchange of matter between it and its exterior. We call a *thermodynamic system* a system whose energy exchange with the exterior is nothing but an exchange of *heat* and of *work done* by volume forces or by surface forces acting upon $\mathcal{S}$. A thermodynamic system in which there is no energy exchange with the exterior is said to be *isolated.*

The definitions just given bring into play three kinds of exchange: mass, work and heat. And this leads to the introduction of the notion of *membrane* and the specification of this notion for each kind of exchange.

Impermeable membrane (contrary: *permeable*): This is a membrane that does not permit mass exchange.

Rigid membrane (contrary: *flexible or deformable*): This is a membrane that does not permit work exchange.

Adiabatic membrane (contrary: *diathermal*): This is a membrane that does not permit heat exchange.

A1.1.2 Thermodynamic state variables

A thermodynamic state variable is a macroscopic quantity, which is characteristic of a system $\mathscr{S}$, and which can be a scalar, a tensor, an n-vector, etc., such as temperature or a stress tensor.

A state variable is said to be extensive if, within a homogeneous system $\mathscr{S}$, it is proportional to the mass of the system. If the variable does not depend upon the mass of the system, then it is said to be *intensive.* All thermodynamic state variables are either extensive or intensive. We accept that with each extensive variable a *specific* intensive variable (i.e., per unit mass) can be associated. Ultimately this is the fact that makes it possible to work with intensive variables alone.

A1.1.3 Thermodynamic state

The choice of the thermodynamic state variables is determined not only by the physical nature of the system $\mathscr{S}$ under study and its transformations, but also by the scheme adopted and the hoped-for precision of the description; so the number of thermodynamic state variables may vary from one system and theory to another; they may, e.g., depend upon the number of secondary effects and couplings taken into consideration.

The set of values of the thermodynamic state variables that characterize a system $\mathscr{S}$ at a certain moment constitutes the *thermodynamic state* $\mathscr{E}(\mathscr{S}, t)$ of the system at that given moment. We say that a system is in *thermodynamic equilibrium* if this system does not evolve with time. Still, in general, thermodynamic systems do evolve with time, under the action of external stimuli. The *transition* from one thermodynamic state to another is called a *thermodynamic process.* A thermodynamic process is said to be *reversible* if the inverse evolution of the system in time – i.e., the succession of thermodynamic states that the system has gone through – implies the reversal in time of the action of external stimuli. Otherwise the thermodynamic process is said to be *irreversible.*

Thermostatics is the science that compares systems in thermodynamic equilibrium. For example, it describes the transition from a state of equilibrium $\mathscr{E}_1(\mathscr{S})$ to another state of equilibrium $\mathscr{E}_2(\mathscr{S})$. Thermodynamics, in its main sense, is the study of phenomena outside a state of equilibrium, but actually not far outside this equilibrium. Everybody of course agrees that in the years 1890 to 1920 thermostatics was developed in an elegant mathematical form by Clausius, Gibbs, Duhem and Carathéodory, in harmony with the experiments. Unfortunately, we cannot say the same about thermodynamics outside equilibrium; schools strongly disagree with each other on this subject. Which means that thermostatics deserves some more careful study.

A1.2 Thermostatics

A1.2.1 Axioms of thermostatics

The exclusive object of thermostatics is systems $\mathcal{S}$ in thermodynamic equilibrium. An 'experimental' definition of equilibrium has just been given above; the exact mathematical definition is the following.

Definition The state $\mathscr{E}(\mathcal{S})$ of a system $\mathcal{S}$ in *thermodynamic equilibrium* is the set of the quantities proper to this equilibrium, whether of geometric, mechanical, or physicochemical, etc. nature, expressed by real numbers that remain invariable in time. The system $\mathcal{S}$ is said to be *finite* if $n+1$ of these characteristics, (denoted by $\chi_0, \chi_1, \ldots, \chi_n$, i.e., χ_α, $\alpha = 0, 1, \ldots, n$) constitute a system of independent variables, so that any other characteristic quantity becomes a well-defined function, called a *function of state* of the χ_α, this way forming a complete system of variables of state of the state $\mathscr{E}(\mathcal{S})$. In geometrical terms, the set of all possible states $\mathscr{E}$ *of a system forms a simply connected differentiable manifold* which is denoted by $\mathscr{V}(\mathcal{S}) = \mathscr{V}$.

This definition leads us to an *axiomatic formulation* of thermostatics as given by C. Carathéodory (1909, 1925), a mathematician well known for his work on the calculus of variations.

The *transformation*, denoted by $\mathscr{F}(\mathscr{E}_1, \mathscr{E}_2)$, is the transition of a system $\mathcal{S}$, from one state of equilibrium $\mathscr{E}_1$ to another state of equilibrium $\mathscr{E}_2$. Let $\{\mathscr{F}\}$ be the set of these transformations. During a transformation there is a possibility of exchange. According to the notion of membrane mentioned above, we can state that a system is *closed* if all membranes (external or internal) are impermeable. A system is said to be *simple* if there is no internal membrane, while all the parts of the system are homogeneous in any state of equilibrium $\mathscr{E}$.

When a *transformation* $\mathscr{F}(\mathscr{E}_1, \mathscr{E}_2)$ is such that $\mathscr{E}_2$ is identical to $\mathscr{E}_1$, it is then called a *cycle*. The obvious way to define the composition of transformations is (Fig. A1.1)

$$\mathscr{F}(\mathscr{E}_1, \mathscr{E}_3) = \mathscr{F}_2(\mathscr{E}_2, \mathscr{E}_3) \circ \mathscr{F}_1(\mathscr{E}_1, \mathscr{E}_2), \tag{A1.1}$$

so that the final state of the first transformation $\mathcal{F}_1$ becomes the initial state of the second $\mathcal{F}_2$. We can then state the following.

First principle in axiomatic form *To* any *transformation* $\mathcal{F}(\mathcal{E}_1, \mathcal{E}_2)$ *of a system* $\mathcal{S}$ *two numbers* $\mathcal{T}_{\mathcal{F}}$ *and* $\mathcal{Q}_{\mathcal{F}}$ *can be associated, called respectively* work *received and* heat *received by* $\mathcal{S}$ *during the transformation. The thermodynamic function called internal energy* $\mathbb{E}(\chi_\alpha)$, *defined up to an additive constant, is defined such that*

$$\mathbb{E}(\mathcal{E}_2) - \mathbb{E}(\mathcal{E}_1) = \mathcal{T}_{\mathcal{F}} + \mathcal{Q}_{\mathcal{F}}. \tag{A1.2}$$

Although the notion of work is the same as in mechanics, the notion of heat is not specified; which means that within an axiomatic system (A1.2) may also be considered as the definition of $\mathcal{Q}_{\mathcal{F}}$. If $\mathcal{F}$ is defined by a composition similar to (A1.1) then

$$\mathcal{T}_{\mathcal{F}} = \mathcal{T}_{\mathcal{F}_1} + \mathcal{T}_{\mathcal{F}_2}, \qquad \mathcal{Q}_{\mathcal{F}} = \mathcal{Q}_{\mathcal{F}_1} + \mathcal{Q}_{\mathcal{F}_2}. \tag{A1.3}$$

And this leads us to consider the existence of two particular classes of transformations.

Adiabatic transformations These are the transformations of the subset $\{\mathcal{A}\}$ of $\{\mathcal{F}\}$ that take place without heat exchange with the exterior (con-

Fig. A1.1. (a) Composition of general transformations between equilibrium states (Carathéodory's formulation). (b) Reversible transformation (the path $\mathcal{F}_{1',2'}$ is *entirely* in $\mathcal{V}$).

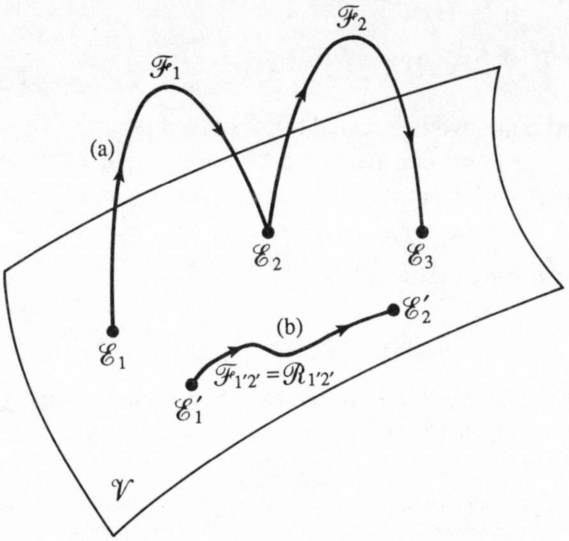

sequently within the limits of an adiabatic membrane); so

$$\mathcal{Q}_{\mathcal{F} \in \{\mathcal{A}\}} = 0. \tag{A1.4}$$

We have then the following axiom.

Axiom of adiabatic transformations *If $\mathscr{E}_0$ is a state of the system $\mathscr{S}$ and $\mathcal{N}(\mathscr{E}.)$ is the set of the states, in such a way that each one of them will be the final state of an adiabatic transformation of which $\mathscr{E}_0$ is the* initial *state, i.e.,*[†]

$$\mathcal{N}(\mathscr{E}_0) = \{\mathscr{E} \mid \exists \mathcal{A}(\mathscr{E}_0, \mathscr{E})\}, \tag{A1.5}$$

and $\mathcal{I}(\mathscr{E}_0)$ is the set of states where each state is the initial state of adiabatic transformation of which $\mathscr{E}_0$ is the final *state, i.e.*

$$\mathcal{I}(\mathscr{E}_0) = \{\mathscr{E} \mid \exists \mathcal{A}(\mathscr{E}, \mathscr{E}_0)\}, \tag{A1.6}$$

then

$$\mathcal{N}(\mathscr{E}_0) \cup \mathcal{I}(\mathscr{E}_0) \supset \mathcal{V}(\mathscr{S}). \tag{A1.7}$$

This means that two states of $\mathcal{V}$ can always be connected by an adiabatic transformation but, in general, we cannot arbitrarily select which one of these two states will be the initial state of the transformation.

Reversible transformations These are the transformations of the subset $\{\mathscr{R}\}$ of $\{\mathscr{F}\}$, where the path from $\mathscr{E}_1$ to $\mathscr{E}_2$ takes place *entirely* inside $\mathcal{V}$ (which is *not* the case for any $\mathscr{F}$!) If $\mathscr{L}$ is the transition path from $\mathscr{E}_1$ to $\mathscr{E}_2$ in $\mathcal{V}$, then we have

$$\mathcal{T}_{\mathscr{F} \in \{\mathscr{R}\}} = \int_{\mathscr{L}} \omega, \qquad \mathcal{Q}_{\mathscr{F} \in \{\mathscr{R}\}} = \int_{\mathscr{L}} \varphi, \tag{A1.8}$$

where ω and φ are two differential forms of the first order,[‡] not identically zero and not proportional, defined on $\mathcal{V}$. We say then that the system is in *reversible evolution* between the states $\mathscr{E}_1$ and $\mathscr{E}_2$ and that ω and φ are, respectively, the *elementary work received* and the *elementary heat received* in this reversible evolution. We can for example write

$$\omega = \sum_{\alpha=0}^{n} A_\alpha \, \mathrm{d}\chi_\alpha, \qquad \varphi = \sum_{\alpha=0}^{n} B_\alpha \, \mathrm{d}\chi_\alpha, \tag{A1.9}$$

and the first principle (A1.2), in the case of reversible transformations, is written here in an elementary form as

† '$\mathcal{N}$' refers to 'natural' evolutions and '$\mathcal{I}$' to 'impossible' evolutions

‡ See, for example, Y. Choquet-Bruhat (1968).

$$\mathrm{d}\mathbb{E} = \omega + \varphi. \tag{A1.10}$$

What is then the statement of the *second principle* of thermodynamics according to Carathéodory?

Statement of the second principle for a simple closed system (Carathéodory, 1925)

Given that $\mathscr{E}_0 \in \mathscr{V}$, *then* $\mathscr{E}_0 \notin \mathring{\mathscr{N}}(\mathscr{E}_0)$,
i.e., $\mathscr{E}_0$ *belongs to the boundary of* $\mathscr{N}(\mathscr{E}_0)$. (A1.11)

In a less esoteric language, this means that there are neighbouring states of $\mathscr{E}_0$ that cannot be reached through an adiabatic transformation starting from $\mathscr{E}_0$. We have then the following.

Carathéodory's Theorem *If the axiom (A1.11) is true, then there exist two functions of state* $\theta(\chi_\alpha)$ *and* $S(\chi_\alpha)$, *which are not unique, such that*

$$\varphi = \theta\,\mathrm{d}S. \tag{A1.12}$$

The proof is given in Carathéodory (1925), Buchdahl (1960), and Germain (1973).

A1.2.2 Scaling of temperature, Carnot's Theorem

The notion of temperature, the 'zero' principle *Two simple closed systems,* $\mathscr{S}_1$ *and* $\mathscr{S}_2$, *are considered to be in* thermal equilibrium *when, although they are separated by an adiabatic membrane, there is no change, even if this membrane is replaced by a diathermal membrane. It follows that if two systems are in thermal equilibrium with a third one, then they are in thermal equilibrium between them. This statement defines an* equivalence relation *and the classes of equivalence of* $\mathscr{V}$ *through this equivalence relation are isomorphic to* $\mathbb{R}$. *The states of equilibrium of a well-defined system in thermal equilibrium defines then a submanifold of n dimensions of* $\mathscr{V}$, *called* isothermal, *which we consider to be sufficiently regular. This manifold is a hypersurface of equation*

$$\tau = g(\chi_\alpha) = \text{const.}, \tag{A1.13}$$

so that τ *is a* function of state. *We say then that (A1.13) defines a* scaling *of temperature and that* τ *is the* temperature *of the system in the state* $\{\chi_\alpha\}$ *We can say then that a* complex *closed system* $\mathscr{S}$ *is thermally simple if all the simple subsystems* $\mathscr{S}_\beta$ *that compose it are at the same temperature. The second principle applies to thermally simple systems. We have then the following.*

Carnot's Theorem *There is a universal scaling of temperature θ, called* thermodynamic temperature *or* absolute temperature *and a function of state $S(\chi_\alpha)$ called* entropy *of the system, such that*

$$\varphi = \theta\,\mathrm{d}S, \quad S = S(\chi_\alpha), \quad \theta > 0, \quad \operatorname{Inf}\theta = 0, \tag{A1.14}$$

and the entropy of a combination of thermally simple systems is the sum of the entropy of each one of these systems. S, is defined up to an additive constant (often considered as equal to zero in the limit $\theta \to 0$).

A sketch proof of Carnot's theorem may be found in Germain (1973, pp. 344–5).

As a corollary of the previous theorem we have the following.

Corollary of Carnot's Theorem *We can always describe a thermodynamic system by the state variables called* normal, $\{\chi_\alpha\} = \{\chi_0 = S, \chi_1, \ldots, \chi_n\}$, *in such a way that*

$$\varphi = \theta\,\mathrm{d}S, \qquad \omega = \sum_{\beta=1}^{n} \tau_\beta\,\mathrm{d}\chi_\beta, \tag{A1.15}$$

where there is no dS *in ω.*

A1.2.3 Thermodynamic potentials

We should point out that $\mathbb{E}$ and S are extensive variables (i.e. proportional to the mass of the system). The χ_β's are also extensive variables. $\mathbb{E}$ is a positively homogeneous function of degree 1 with respect to the variables χ_α. But (A1.15) gives

$$\theta = \frac{\partial \mathbb{E}}{\partial S}, \qquad \tau_\beta = \frac{\partial \mathbb{E}}{\partial \chi_\beta}, \tag{A1.16}$$

in such a manner that θ and the χ_β's are positively homogeneous of degree zero. They are *intensive* variables. We say then that $\mathbb{E}(\chi_0 = S, \chi_1, \ldots, \chi_n)$ is a *thermodynamic potential.* Generally, speaking, we call thermodynamic potentials those variables from which we can deduce the variables characterizing *all* the thermodynamic properties of the system.

Associated thermodynamic potentials Let $Z = Z(\chi_\alpha) = Z(\chi_0, \chi_1, \cdots, \chi_n)$ be a given thermodynamic potential. We have then

$$\mathrm{d}Z = \sum_{\alpha=0}^{n} \mu_\alpha\,\mathrm{d}\chi_\alpha = \boldsymbol{\mu}\cdot\mathrm{d}\boldsymbol{\chi}, \qquad \mu_\alpha = \frac{\partial Z}{\partial \chi_\alpha}, \tag{A1.17}$$

where the dot indicates the scalar product of $(n+1)$-vectors. We define an *associated* thermodynamic potential $g(\mu_0, \mu_1, \ldots, \mu_p, \chi_{p+1}, \ldots, \chi_n)$ through the *partial Legendre transformation*† on the first p variables of state, by

$$\left.\begin{aligned} g(\boldsymbol{\mu}^{(1)}, \boldsymbol{\chi}^{(2)}) &= Z(\boldsymbol{\chi}) - \boldsymbol{\mu}^{(1)} \cdot \boldsymbol{\chi}^{(1)}, \\ \boldsymbol{\chi}^{(1)} = (\chi_0, \chi_1, \ldots, \chi_p), &\qquad \boldsymbol{\mu}^{(1)} = (\mu_0, \mu_1, \ldots, \mu_p)^{\mathrm{T}}, \\ \boldsymbol{\chi}^{(2)} = (\chi_{p+1}, \ldots, \chi_n), &\qquad \boldsymbol{\mu}^{(2)} = (\mu_{p+1}, \ldots, \mu_n)^{\mathrm{T}}, \end{aligned}\right\} \tag{A1.18}$$

in such a way that

$$\left.\begin{aligned} \chi_\alpha = -\frac{\partial g}{\partial \mu_\alpha}, &\qquad \alpha = 0, 1, \ldots, p, \\ \mu_\beta = \frac{\partial Z}{\partial \chi_\alpha}, &\qquad \beta = p+1, \ldots, n. \end{aligned}\right\} \tag{A1.19}$$

For example, let us consider the thermodynamic potentials that are associated to the internal energy. We have $\chi_0 = S$. The *Helmholtz free energy* F can be defined by

$$F = \mathbb{E} - \theta S, \tag{A1.20}$$

so that

$$F = F(\theta, \chi_1, \ldots, \chi_n), \qquad S = -\frac{\partial F}{\partial \theta}. \tag{A1.21}$$

If among the χ_α, $\alpha \geqslant 1$, we find the *volume* V, e.g. $\chi_1 = V$, then $\tau_1 = -p$, where p is the pressure, and we can define the *enthalpy* H by

$$H = \mathbb{E} + pV \tag{A1.22}$$

so that

$$H = H(S, p, \chi_2, \ldots, \chi_n), \qquad V = \frac{\partial H}{\partial p}. \tag{A1.23}$$

As to the *Gibbs* potential, this is obtained by combining (A1.20) and (A1.22), that is

$$\left.\begin{aligned} G = G(\theta, p, \chi_2, \ldots, \chi_n) &= \mathbb{E} - \theta S + pV, \\ S = -\frac{\partial G}{\partial \theta}, &\qquad V = \frac{\partial G}{\partial p}. \end{aligned}\right\} \tag{A1.24}$$

† Compare with the Legendre – Fenchel transformation of convex functions in Appendix 2

The potentials associated with S through Legendre transformations are called the *Massieu functions* of the system.

Different potentials present different properties of *convexity* with respect to the different choices of independent state variables. For example, $\mathbb{E}$ is convex in S and in the strain of an elastic solid. It follows that F is concave in θ and convex in strain. Actually, the properties of convexity are preserved by the Legendre–Fenchel transformation (see Appendix 2). To describe the transition from $\mathbb{E}$ to F the Legendre–Fenchel transformation is written (with the conventions of Appendix 2)

$$(-F) = \theta S - \mathbb{E}. \tag{A1.25}$$

Since $\mathbb{E}$ is convex in S, $(-F)$ is convex in the dual variable of S, that is θ; consequently, F is concave in θ. For a gas, S is concave in $\mathbb{E}$ and the volume V.† Let us also point out that $\mathbb{E}$ must be an *increasing* function of S, since θ is always positive (see (A1.14)).

A1.2.4 The evolution of real systems

What we want to do now is to apply the second principle to all transformations $\{\mathscr{F}\}$, including *irreversible* transformations. Starting from any state $\mathscr{E}_0$ we can show that the set $\mathscr{N}(\mathscr{E}_0)$ may be defined by the inequality (whose sign is prescribed by experiments)

$$S \geqslant S_0, \tag{A1.26}$$

the equality sign holding good for reversible transformations only. The inequality (A1.26) is a characteristic of the *entire set* of adiabatic transformations, whereas the reversible adiabatic transformations were used in order to introduce the universal concepts of temperature and entropy.

Isothermal transformations During such a transformation of the system the temperature remains constant. If this transformation is reversible, the evolution of the system between $\mathscr{E}_0$ and $\mathscr{E}$ is described by a succession of equilibrium states, and the quantity of heat received $\mathscr{Q}_r$ during this transformation is given by

$$S(\mathscr{E}) - S(\mathscr{E}_0) - \frac{\mathscr{Q}_r}{\theta} = 0. \tag{A1.27}$$

If the transformation is *irreversible* we may suppose that (A1.27) is replaced

† See Germain (1973, pp. 352–62) for the convexity of potentials. Convexity is in general necessary for questions of stability.

by

$$S(\mathscr{E}) - S(\mathscr{E}_0) - \frac{\mathscr{Q}}{\theta} \geqslant 0, \tag{A1.28}$$

in such a way that, for a given $\mathscr{Q}$, the final entropy will be greater than the one we would have observed in a reversible transformation. We can also say that the quantity of heat received during the transition from $\mathscr{E}_0$ to $\mathscr{E}$ is *smaller* in the case of irreversible transformations, and that the resulting *loss* of received energy is due to the intrinsic irreversibilities of the system.

According to (A1.28), these losses caused by *intrinsic irreversibilities* are defined by

$$\mathscr{P}(\mathscr{F}_{\text{isoth}}(\mathscr{E}_0, \mathscr{E})) = S(\mathscr{E}) - S(\mathscr{E}_0) - \frac{\mathscr{Q}(\mathscr{F}_{\text{isoth}}(\mathscr{E}_0, \mathscr{E}))}{\theta} \geqslant 0. \tag{A1.29}$$

We can define in the same manner the losses due to thermal *irreversibilities*

$$\mathscr{P}(\mathscr{F}_{\text{monoth}}(\mathscr{E}_0, \mathscr{E})) = S(\mathscr{E}) - S(\mathscr{E}_0) - \frac{\mathscr{Q}(\mathscr{F}_{\text{monoth}}(\mathscr{E}_0, \mathscr{E}))}{\theta_0} \geqslant 0, \tag{A1.30}$$

where $\mathscr{F}_{\text{monoth}}(\mathscr{E}_0, \mathscr{E})$ *is a monothermal transformation* which corresponds to an exchange of heat with a source at *constant* temperature.

For continuous media we shall use the form

$$S_2 - S_1 \geqslant \int_1^2 \frac{\mathrm{d}\mathscr{Q}}{\theta} \tag{A1.31}$$

of eqn (A1.28), which we will not justify here. For a material body occupying the volume Ω with a regular boundary of $\partial\Omega$ evolving between two instants t_1 and t_2, we shall write (A1.31) in the form

$$S(t_2) - S(t_1) \geqslant \int_{t_1}^{t_2} \left\{ \int_{\partial\Omega} \frac{q}{\theta} \mathrm{d}s + \int_\Omega \rho \frac{h}{\theta} \mathrm{d}\Omega \right\} \mathrm{d}t, \tag{A1.32}$$

where q and h are, respectively, a flux and a source (per unit mass) of heat. The inequality (A1.32) is known as the *inequality of Clausius* (who in 1854 postulated its existence for an isolated system). In fact, at moment t, this fundamental inequality can be written in a differential form as

$$\frac{\mathrm{d}S}{\mathrm{d}t} \geqslant \int_\Omega \rho \frac{h}{\theta} \mathrm{d}\Omega - \int_{\partial\Omega} \frac{\mathbf{q}\cdot\mathbf{n}}{\theta} \mathrm{d}s, \tag{A1.33}$$

where $\mathbf{q}$ is the *heat flux vector* (directed towards the interior of Ω, and $\mathbf{n}$ is the unit outward normal to $\partial\Omega$. Since entropy is an extensive quantity, at time t we can write,

$$S(\Omega) = \int_\Omega \rho\eta \,\mathrm{d}\Omega, \tag{A1.34}$$

where η is the *specific entropy* at $(\mathbf{x}, t)$, $\mathbf{x} \in \Omega$. In (A1.33) $\mathrm{d}/\mathrm{d}t$ is a *material derivative.*

The problem now is the following. The above statements and results apply to transformations linking between them states *in equilibrium.* What can we say about the states outside equilibrium, i.e., for *thermodynamics* in the full sense of the term?

A1.3 Thermodynamics

We know how to define the notion of equilibrium state of a system. The notions of *temperature* and *entropy*, in particular, are defined in *thermostatics* at *equilibrium.* We shall consider now three different presentations: the first is the classical one, leading to the *theory of irreversible processes* (which possesses a microscopic basis that we will not examine here); the second is the *axiomatic* one, the thermodynamics of Coleman and Noll; and the third, which, in a way, is somewhere in between, called *thermodynamics with internal variables.* This is the one used for elastoplasticity and viscoplasticity in this book.

A1.3.1 The theory of irreversible processes

The material points, or '*particles*', $\mathbf{X}$, of the deformable body under study, have an equation of motion

$$\mathbf{x} = \mathscr{H}(\mathbf{X}, t). \tag{A1.35}$$

It follows that generally speaking there is no equilibrium in a thermostatics sense. The fundamental axiom of the theory of irreversible processes is then the following.

Axiom of local state *For* each moment t, *we associate to the particle* $\mathbf{X}$ *a set of independent parameters* $(\chi_0, \chi_1, \ldots, \chi_n)$ *which are the variables of state of a certain thermostatic system in the sense defined above. These variables are considered to be* normal and extensive, *in the sense that we suppose that at a moment t there is a* specific internal energy $e(\chi_\alpha)$ *such that* ($\eta = \chi_0$ *being the specific entropy*)

$$\theta = \frac{\partial e}{\partial \eta}, \qquad \tau_\beta = \frac{\partial e}{\partial \chi_\beta}, \qquad \beta = 1, \ldots, n. \tag{A1.36}$$

A thermodynamic *process then is conceived here as a series of thermostatic equilibria, which permit the introduction of the notions of temperature θ and entropy η. According to our working hypothesis, i.e., that a material particle in motion should be in equilibrium at practically any moment, the response times which allow the thermostatic system to recover a new state of thermostatic equilibrium must be short, compared with the characteristic durations of the kinematic and dynamic evolutions of the medium. This will prove to be impracticable each time that the evolution of the system in motion is* too fast. *Obviously, this is a matter of relative appreciation (see Subsection A1.3.3, below). Through the use of material derivatives we may then write* locally *(A1.10), that is, the* first principle, *in the form*

$$\dot{e} = \theta\dot{\eta} + \varpi, \qquad \varpi = \sum_{\beta=1}^{n} \tau_\beta \dot{\chi}_\beta. \tag{A1.37}$$

We can also introduce the specific *Helmholtz free energy* by

$$\psi = e - \theta\eta$$

in such a manner that

$$\dot{\psi} = -\eta\dot{\theta} + \varpi, \qquad \varpi = \sum_{\beta=1}^{n} \tau_\beta \dot{\chi}_\beta, \tag{A1.38}$$

and

$$\eta = -\frac{\partial\psi}{\partial\theta}, \qquad \tau_\beta = \frac{\partial\psi}{\partial\chi_\beta}. \tag{A1.39}$$

The dissipation per unit volume ϕ is obtained by the combination of the local form of (A1.32), and (A1.37), as follows:

$$\rho\dot{\eta} \geqslant \rho\frac{h}{\theta} - \nabla\cdot(\mathbf{q}/\theta), \tag{A1.40}$$

which in turn gives

$$\phi = \rho\theta\dot{\eta} - \rho h + \theta\nabla\cdot\left(\frac{\mathbf{q}}{\theta}\right) = (\rho\theta\dot{\eta} + \nabla\cdot\mathbf{q}) - \rho h + \theta\mathbf{q}\cdot\nabla\left(\frac{1}{\theta}\right) \tag{A1.41}$$

and thus

$$\phi = \rho\left(\dot{e} - h + \frac{1}{\rho}\nabla\cdot\mathbf{q} - \varpi\right) + \theta\mathbf{q}\cdot\nabla\left(\frac{1}{\theta}\right) \geqslant 0. \tag{A1.42}$$

But, for a deformable continuous medium, the first principle of thermodynamics is evidently written as

$$\left.\begin{aligned}\frac{\mathrm{d}}{\mathrm{d}t}\int_{\Omega}\rho\left(e+\frac{1}{2}\mathbf{v}^2\right)\mathrm{d}\Omega &= \dot{\mathscr{T}}+\dot{Q},\\ \dot{\mathscr{T}}(\Omega) &= \int_{\Omega}\rho\mathbf{f}\cdot\mathbf{v}\,\mathrm{d}\Omega+\int_{\partial\Omega}\mathbf{T}\cdot\mathbf{v}\,\mathrm{d}s,\\ \dot{Q}(\Omega) &= \int_{\Omega}\rho h\,\mathrm{d}\Omega-\int_{\partial\Omega}\mathbf{q}\cdot\mathbf{n}\,\mathrm{d}s,\end{aligned}\right\} \tag{A1.43}$$

whereas the equations of motion and the accompanying boundary condition are written

$$\left.\begin{aligned}&\rho\dot{\mathbf{v}}=\operatorname{div}\boldsymbol{\sigma}+\rho\mathbf{f},\qquad \boldsymbol{\sigma}=\boldsymbol{\sigma}^{\mathrm{T}} \quad \text{in}\ \Omega,\\ &\boldsymbol{\sigma}\cdot\mathbf{n}=\mathbf{T} \qquad\qquad\qquad\qquad\quad \text{at}\ \partial\Omega.\end{aligned}\right\} \tag{A1.44}$$

so that (A1.43) leads locally to

$$\begin{gathered}\rho\dot{e}=\sigma_{ij}D_{ij}-\nabla\cdot\mathbf{q}+\rho h=\boldsymbol{\sigma}:\mathbf{D}-\nabla\cdot\mathbf{q}+\rho h\\ (D_{ij}=\tfrac{1}{2}(v_{i,j}+v_{j,i})=D_{ji}),\end{gathered} \tag{A1.45}$$

that is

$$\rho\left(\dot{e}-h+\frac{1}{\rho}\nabla\cdot\mathbf{q}\right)=\boldsymbol{\sigma}:\mathbf{D}. \tag{A1.46}$$

Given (A1.46), (A1.42) is also written as

$$\phi=\boldsymbol{\sigma}:\mathbf{D}-\rho\varpi+\theta\mathbf{q}\cdot\nabla\left(\frac{1}{\theta}\right)\geqslant 0. \tag{A1.47}$$

In concluding this point, eqns (A1.30) or (A1.39) provide the *state laws* in terms of derivatives of a thermodynamic potential and (A1.47) is the inequality of dissipation that the dissipative processes must satisfy. For example, we observe that for small strains we have $\mathbf{D}\approx\dot{\boldsymbol{\varepsilon}}$. If we consider $e=e(\eta,\boldsymbol{\varepsilon})$ – which is the case for elasticity in SPH for the nondissipative part – then the laws of state (A1.36) are written

$$\theta=\left.\frac{\partial e}{\partial\eta}\right|_{\boldsymbol{\varepsilon}=\text{const.}},\qquad \boldsymbol{\sigma}^{\mathrm{e}}=\rho\frac{\partial e}{\partial\boldsymbol{\varepsilon}}, \tag{A1.48}$$

and (A1.47) takes the form

$$\boldsymbol{\sigma}^{\mathrm{D}}:\dot{\boldsymbol{\varepsilon}}+\theta\mathbf{q}\cdot\nabla\left(\frac{1}{\theta}\right)\geqslant 0,\qquad \boldsymbol{\sigma}^{\mathrm{D}}\equiv\boldsymbol{\sigma}-\boldsymbol{\sigma}^{\mathrm{e}}. \tag{A1.49}$$

In the *theory of irreversible processes* (TIP) an inequality such as (A1.49) is re-written in a *bilinear form*

$$\phi = \mathbf{X} \cdot \mathbf{Y} = \sum_{\alpha} \mathbf{X}_\alpha \mathbf{Y}_\alpha \geqslant 0, \tag{A1.50}$$

where $\mathbf{Y}_\alpha$ are called *fluxes* and $\mathbf{X}_\alpha$ *forces*. In the *vicinity of equilibrium* (the only possibility that we might consider in view of the limited validity of the axiom of the local state) the *theory of irreversible processes* established by Onsager, Casimir, Meixner, de Donder, De Groot and Mazur, in the years 1940–60, considers *linear affine relations* between fluxes and forces, that is

$$\mathbf{Y} = \mathbf{L}(\mathbf{X}), \qquad \mathbf{L}(\mathbf{0}) = \mathbf{0} \tag{A1.51}$$

with (T = transpose)

$$\mathbf{L} = \mathbf{L}^{\mathrm{T}}, \tag{A1.52}$$

which are the celebrated *reciprocity relations of Onsager and Casimir*. We may notice that the existence of a *potential of dissipation quadratic* in $\mathbf{Y}$, in the manner of Rayleigh's potential for viscosity, $\mathscr{D}$ such that

$$\mathscr{D} = \mathscr{D}(\mathbf{X}), \qquad \mathbf{Y}_\alpha = \frac{\partial \mathscr{D}(\mathbf{X})}{\partial X_\alpha}, \tag{A1.53}$$

is equivalent to (A1.51)–(A1.52) as can easily be proven.

Finally, we may observe that we can re-write the expression $\mathrm{d}S/\mathrm{d}t$ as follows (exercise):

$$\frac{\mathrm{d}S(\Omega)}{\mathrm{d}t} = \mathscr{C}_{\mathrm{L}}(\Omega) + \mathscr{I}(\Omega), \tag{A1.54}$$

where

$$\mathscr{C}_{\mathrm{L}}(\Omega) \equiv \int_{\Omega} \frac{\rho h}{\theta} \mathrm{d}\Omega - \int_{\partial\Omega} \frac{\mathbf{q} \cdot \mathbf{n}}{\theta} \mathrm{d}s \tag{A1.55}$$

and

$$\mathscr{I}(\Omega) \equiv \int_{\Omega} \frac{\phi}{\theta} \mathrm{d}\Omega \tag{A1.56}$$

are, respectively, Clausius' term and Jouguet's dissipation term.

Conclusion The *complementary laws* that complete the thermodynamic description by the *laws of state* satisfy the inequality (A1.47) or (A1.50).

A1.3.2 The 'rational' theory of Coleman and Noll

In the so-called *rational* thermodynamics, following B.D. Coleman, W. Noll and C.A. Truesdell, the acquired experience of thermostatics is ignored. The hypothesis rather is that those notions that previously could only be defined at equilibrium exist *a priori* for any thermodynamic state whatever, *even outside equilibrium* (but then a certain measure – *the distance* – of the departure from equilibrium should be appropriately defined). This is the case for the notions of temperature and entropy so that the formal bases of this new thermodynamics (which developed from 1960 to 1970) are the *a priori* statements (A1.33) and (A1.43) as well as (A1.34). Dynamics enter through the fact that, in theory, all the past history must be taken into account (for example, the motion and the temperature field); in this type of thermodynamics we are then led to consider *the constitutive equations in functional form* (on the temporal interval), the *instantaneous behaviour* being actually formally the same as the one described by the laws of state of the previous paragraph. The *Clausius–Duhem inequality*, obtained by the combination of eqns (A1.40) and (A1.45) and the introduction of the free energy per unit mass $\psi = e - \eta\theta$, that is

$$-\rho(\dot{\psi} + \eta\dot{\theta}) + \boldsymbol{\sigma} : \mathbf{D} + \theta\mathbf{q} \cdot \nabla\left(\frac{1}{\theta}\right) \geqslant 0, \qquad \text{(A1.57)}$$

plays the role of a constraint that must be satisfied by the functional constitutive equations (for η, $\boldsymbol{\sigma}$, and $\mathbf{q}$) for any so-called *admissible* thermodynamic process. We shall not make use of this thermodynamic approach† which, in practice, can only be useful in the study of behaviours with memory (such as Volterra's hereditary media), such as we meet in certain types of viscoelastic media with a non-Newtonian viscous behaviour.

A1.3.3 Theory with internal variables

A1.3.3.1 General properties

The origins of *internal-variable thermodynamics* may first be traced in the kinetic description of physicochemical processes of evolution; but its spectacular development is related to the *rheological models* and the elastoviscoplasticity of deformable materials of the metallic type (alloys, polycrystals). This approach is adopting a somewhat intermediary line between the two thermodynamics already mentioned. It provides a new characterisation of

† See for example, Eringen (1967) and Maugin (1988), Sec. 2.10.

continuous media, which, in order to define the thermodynamic state of a system, introduces, in addition to the usual *observable* state variables, a certain number of *internal* variables, collectively denoted by $\boldsymbol{\alpha}$ and which are supposed to describe the internal structure (hidden to the eye of the external observer who can only see a *black box*). It follows that the value, at moment t, of the dependent variables (for example, the stress $\boldsymbol{\sigma}$) becomes simultaneously a *function* both of the values of the independent observable variables *and* of the internal variables. This state law $\boldsymbol{\sigma}(\boldsymbol{\chi}, \boldsymbol{\alpha})$, where $\boldsymbol{\chi}$ represents, as before, the controllable variables of state, must be complemented by an *evolution law* which describes the evolution of the variable $\boldsymbol{\alpha}$. So, we can write the following:

$$\boldsymbol{\sigma} = \tilde{\boldsymbol{\sigma}}(\boldsymbol{\chi}, \boldsymbol{\alpha}) \quad \textit{state law} \text{ (mechanical here);} \tag{A1.58}$$

$$\dot{\boldsymbol{\alpha}} = \mathscr{A}(\boldsymbol{\chi}, \boldsymbol{\alpha}) + \mathscr{B}(\boldsymbol{\chi}, \boldsymbol{\alpha})\dot{\boldsymbol{\chi}} \quad \textit{evolution law,} \tag{A1.59}$$

In fact, we may suppose that we have been able to choose the $\boldsymbol{\alpha}$ in such a way that $\mathscr{B}(\boldsymbol{\chi}, \boldsymbol{\alpha})$ might be identically zero and that an instantaneous variation of $\boldsymbol{\chi}$ does not cause any instantaneous variations in the $\boldsymbol{\alpha}$'s (if $\boldsymbol{\chi}$ is a strain, the hypothesis $\mathscr{B}(\boldsymbol{\chi}, \boldsymbol{\alpha}) = \mathbf{0}$ corresponds to the fact that instantaneous strains are elastic or zero). The following observations seem to be imperative.

(i) Nature and choice of internal variables The tensorial nature of the variable $\boldsymbol{\alpha}$ (scalar, vector, tensor, n-vector) as well as its 'physical' nature must in general be specified: does it represent the average of some microscopic effect or is it a local structural rearrangement? We must also point out that the notion of '*internal*' for a variable of state depends upon the level of observation: we can perfectly well think of a variable that might be considered as *internal* from the *macroscopic* observation point of view (usual macroscopic scale of continuous media, in which a strong non-locality is not taken into account), or as *observable* from a point of view of mesoscopic observation. So that, along with Mandel (1980), we can always say 'that a clever physicist will always manage to detect the "internal" variables and measure them'. Which means that, in practice, these variables are *measurable* but not controllable.

(ii) Internal variables and functional constitutive equations We understand that the elimination of $\boldsymbol{\alpha}$ between eqns (A1.58) and (A1.59) leads to a *time*-functional law for $\boldsymbol{\sigma}$ with respect to the observable variables alone. But, from a practical point of view, the 'approximation' (A1.58)–(A1.59) of a functional constitutive equation through the internal variables offers two main advantages: (a) it only requires a *finite number* of variables (the space of states is of finite dimension); the past history has been condensed into the present values of a finite number of internal variables, and this number either increases with the size of the sample (if the structural arrangements under study are local) or is relatively small (if the internal variables represent average effects); and (b) it enables us to use all the lessons of the irreversible processes of thermodynamics (see iii). In this manner we combine the two thermodynamics previously mentioned and at the same time we are led to a *differential* mathematical problem of the *evolution* type, which *a priori* seems quite desirable.

(iii) Internal variables and theory of irreversible processes Eqn (A1.59) shows that we are placed within a *dynamic* framework. What are we to do then with those notions, like heat and entropy, that are well defined only in thermostatics? It is obvious that we must formulate an *axiom of the local-state type* and that we must be careful that the characteristic times that eventually enter into (A1.59) (this is not always the case; see elasto-plasticity which has no time scale) remain relatively small as against the macroscopic evolution. If this is possible, we can then actually use the theory of irreversible processes, *enlarged so as to encompass the internal state variables*, and in this case the theory with internal variables remains a thermodynamic theory in the vicinity of equilibrium.

A1.3.3.2 Local accompanying state

At this point we follow essentially Kestin and Rice (1970), Kestin and Bataille (1975), Rice (1971) and Germain *et al.* (1983). In order to illustrate our argument we will take a concrete case that belongs to the area of *elastoviscoplasticity.* We wish to take into account a nonelastic behaviour, as a result, for example, of the microscale structural rearrangements of the constitutive elements of the given material. This kind of behaviour may well be the result of a plastic deformation due to a dislocation movement (slip of the crystallographic planes in a metal), to the 'twinning' of crystals, to the slip of the grain boundaries, or to a phase transition induced by strain, etc. In this case the internal variables are represented by an **n**-vector $\boldsymbol{\alpha}$, each component of which is characteristic of a *local* structural rearrangement (at a given point in the structure). It is obvious that the number of this kind of variables will increase according to the size of the sample under study. Another, more usual, type of internal variable is the '*average*' type. The internal variables then represent the *mean measures* of the structural rearrangements that take place at *several* points. In this case, $\mathbf{n}$ (which we hope will be small, so that the method will be really interesting) is independent of the size of the sample. For example, already in 1963, E. Kröner had suggested that the different statistical moments of the distribution of dislocations could be used as internal variables.

Let us suppose in theory (if not in practice) that the internal variables may well keep a prescribed value by the imposition of the appropriate stresses $\boldsymbol{\sigma}$; the sample then will tend towards a state of thermodynamic *equilibrium*, characterized by a stress $\boldsymbol{\sigma}$ a strain $\boldsymbol{\varepsilon}$ and a temperature θ. In that case, $\boldsymbol{\sigma}$ (or $\boldsymbol{\varepsilon}$), θ and $\boldsymbol{\alpha}$ are *thermodynamic variables of state*. Let us then, along with Rice and Kestin, suppose that *if* different *states of equilibrium are possible for the* same *values of internal variables, then the* neighbouring *states are connected by the laws of ordinary thermoelasticity* (which can be established in thermostatics).

We have then

$$\boldsymbol{\sigma} = \frac{\partial W}{\partial \boldsymbol{\varepsilon}}, \qquad W = \rho\psi = \tilde{W}(\boldsymbol{\varepsilon}, \theta, \boldsymbol{\alpha}) \tag{A1.60}$$

If W^* be the Legendre–Fenchel transform of W such that (see Appendix 2)

$$W^* = \operatorname*{Sup}_{\boldsymbol{\varepsilon}} (\boldsymbol{\sigma} : \boldsymbol{\varepsilon} - W), \tag{A1.61}$$

then

$$\boldsymbol{\varepsilon} = \frac{\partial W^*}{\partial \boldsymbol{\sigma}}, \qquad W^* = \tilde{W}^*(\boldsymbol{\sigma}, \theta, \boldsymbol{\alpha}). \tag{A1.62}$$

The *anelastic strain* will be the contribution resulting from a variation of internal variables, *when stresses and temperature are fixed*, that is

$$(\delta\boldsymbol{\varepsilon})^{\mathrm{p}} = \frac{\partial \boldsymbol{\varepsilon}}{\partial \boldsymbol{\alpha}} \cdot \delta\boldsymbol{\alpha}. \tag{A1.63}$$

But the thermodynamic force associated with $\boldsymbol{\alpha}$ is

$$A = -\partial W/\partial \alpha = -\partial W^*/\partial \alpha, \tag{A1.64}$$

whence *Maxwell's relation*

$$\frac{\partial \boldsymbol{\varepsilon}}{\partial \alpha} = \frac{\partial A}{\partial \boldsymbol{\sigma}} \left(= \frac{\partial^2 W^*}{\partial \alpha \partial \boldsymbol{\sigma}} \right). \tag{A1.65}$$

As to the *thermoelastic strain*, this is the result of the variation of $\boldsymbol{\sigma}$ and θ alone, so that

$$(\delta\boldsymbol{\varepsilon})^{e} = \underbrace{\left(\frac{\partial^2 W^*}{\partial\boldsymbol{\sigma}\partial\boldsymbol{\sigma}}\right)}_{\text{compliances}} : \delta\boldsymbol{\sigma} + \underbrace{\left(\frac{\partial^2 W^*}{\partial\boldsymbol{\sigma}\partial\theta}\right)}_{\substack{\text{thermal}\\ \text{dilatations}}} \delta\theta. \tag{A1.66}$$

We can then apply the thermodynamics of irreversible processes (TIP), since a process of homogeneous macroscopic strains may be approached through a sequence of constrained (controlled) states of equilibrium.

Certain writers talk of a *local accompanying* (thermostatic) *state.*

A1.3.3.3 Evolution laws for internal variables

What is left to do now is to introduce a kinetic relation for internal variables. Kestin and Rice suppose that '*The rate at which any structural rearrangement happens is entirely determined by the force associated with the rearrangement.*' So, we should write

$$\dot{\boldsymbol{\alpha}} = \tilde{\boldsymbol{\alpha}}(A, \theta, \boldsymbol{\alpha}) \tag{A1.67}$$

where A depends upon the stress, since W^* also depends upon it. Such a formalism is particularly well suited to the *viscoplasticity of metals.* In this case (A1.67) is translated by 'the force on a segment of dislocation line governs its movement'. Let us then prove that there exists a flow potential $\mathscr{D}^*$ such that

$$\dot{\boldsymbol{\varepsilon}}^{\mathrm{p}} = \frac{\partial \mathscr{D}^*(\boldsymbol{\sigma}, \theta, \boldsymbol{\alpha})}{\partial \boldsymbol{\sigma}}. \tag{A1.68}$$

Actually if we evaluate the integral with θ and $\boldsymbol{\alpha}$ *fixed*, we can write in an identical way

$$\dot{\boldsymbol{\alpha}} = \frac{\partial}{\partial A} \int_0^A \tilde{\boldsymbol{\alpha}}(A, \theta, \boldsymbol{\alpha})\,\mathrm{d}A. \tag{A1.69}$$

We set then

$$\mathscr{D}^*(\boldsymbol{\sigma}, \theta, \boldsymbol{\alpha}) = \int_0^{A(\boldsymbol{\sigma}, \theta, \boldsymbol{\alpha})} \tilde{\boldsymbol{\alpha}}(A, \theta, \boldsymbol{\alpha})\,\mathrm{d}A, \tag{A1.70}$$

where from

$$\frac{\partial \mathscr{D}^*}{\partial \boldsymbol{\sigma}} = \left(\tilde{\boldsymbol{\alpha}} \cdot \frac{\partial}{\partial \boldsymbol{\sigma}}\right) A = \frac{\partial \boldsymbol{\varepsilon}}{\partial \boldsymbol{\alpha}} \cdot \dot{\boldsymbol{\alpha}} = \dot{\boldsymbol{\varepsilon}}^{\mathrm{p}}, \tag{A1.71}$$

where we have used Maxwell's relation (A1.65) as well as (A1.63). Eqn (A1.68) expresses the fact that the increment $\dot{\boldsymbol{\varepsilon}}^{\mathrm{p}}$ is *normal* to a *constant* flow potential in the space of stresses. This situation is met with in the elasto-viscoplasticity of metals where a time scale (related to the phenomenon of viscosity) plays an important part. To illustrate (A1.68) we may mention the *continuous* model of slip in metal viscoplasticity (Zarka, 1972) where (A1.68) is written as

$$\dot{\gamma}^{(\alpha)} = \tilde{\gamma}^{(\alpha)}(\tau^{(\alpha)}, \theta, \gamma) = \frac{\partial \mathscr{D}^*}{\partial \tau^{(\alpha)}}, \tag{A1.72}$$

where $\gamma^{(\alpha)}$ is the slip in the $\boldsymbol{\alpha}$ simple slip system of the crystal, $\gamma^{(\alpha)}$ is a variable of the 'structural-rearrangement type', and $\tau^{(\alpha)} = \partial W^*/\partial \gamma^{(\alpha)}$ is the *resolved shear stress* related to the system $\boldsymbol{\alpha}$.

It should be noted that a behaviour that *does not depend upon time* may be considered as a limit case, if we accept that the flow potential $\mathscr{D}^*$ is represented by a *singular limit*. What we get then is *elastoplasticity* independent of time, and this is the object of the main part of this book.

We can sum up the *consequences of the axiom of the accompanying local state by* the following.

(i) The laws of state are deduced from

$$\mathrm{d}e = \theta\,\mathrm{d}\eta + \tilde{\varpi} = \theta\,\mathrm{d}\eta + \sum_{\beta} \frac{1}{\rho} \tilde{\tau}_\beta\,\mathrm{d}\chi_\beta - \sum_{\gamma} \frac{1}{\rho} \tilde{A}_\gamma\,\mathrm{d}\alpha_\gamma, \tag{A1.73}$$

or else, after a partial Legendre transformation,

$$\mathrm{d}\psi = -\eta\,\mathrm{d}\theta + \sum_{\beta} \tau_\beta\,\mathrm{d}\chi_\beta - \sum_{\gamma} A_\gamma\,\mathrm{d}\alpha_\gamma, \tag{A1.74}$$

whence

$$\eta = -\frac{\partial \psi}{\partial \theta}, \qquad \tilde{\tau}_\beta = \rho \frac{\partial \psi}{\partial \chi_\beta}, \qquad \tilde{A}_\gamma = -\rho \frac{\partial \psi}{\partial \alpha_\gamma}, \tag{A1.75}$$

where χ_β are the observable variables of state, other than θ, and α_γ are the internal variables. With $\chi_\beta = (\chi_1 = \boldsymbol{\varepsilon}; \chi_\beta = 0, \beta \geqslant 1)$ we get then

$$\eta = -\frac{\partial \psi}{\partial \theta}, \qquad \boldsymbol{\sigma}^{\mathrm{e}} = \rho \frac{\partial \psi}{\partial \boldsymbol{\varepsilon}}, \qquad \mathbf{A} = -\rho \frac{\partial \psi}{\partial \boldsymbol{\alpha}}, \qquad \boldsymbol{\sigma}^{\mathrm{D}} \equiv \boldsymbol{\sigma} - \boldsymbol{\sigma}^{\mathrm{e}}. \tag{A1.76}$$

(ii) The inequality of dissipation (A1.47) then takes the form (in SPH)

$$\phi = (\boldsymbol{\sigma}^{\mathrm{D}} : \dot{\boldsymbol{\varepsilon}} + \mathbf{A} \cdot \dot{\boldsymbol{\alpha}}) + \theta \mathbf{q} \cdot \nabla\left(\frac{1}{\theta}\right) \geqslant 0, \tag{A1.77}$$

while the primitive Clausius–Duhem inequality (A1.57) remains *unchanged* since it is only based on the general principles (A1.43), (A1.34), which remain unaltered by the introduction of internal variables.

The thermodynamical approach of internal variables owes a lot to the French school of thermomechanics of continuous media. It is applied not only to the *anelastic* behaviour that often constitutes our object in this course, but also to numerous areas of physics (for example in fluids, in solutions of polymers and polyelectrolytes; see Sidoroff, 1976, Maugin and Drouot, 1983, 1988).

Appendix 2
Convexity

The object of the appendix This is a reminder of the main definitions and properties related to convex functions and sets as they occur in many problems of elastoplasticity; of the notions of differential and gradient adapted to these; of the notion of continuity that is appropriate (lower semicontinuity) in the 'calculus' involving convex functions, and of the important notions of duality and conjugate functions through the Legendre–Fenchel transformation.

A2.1 Definitions

Convexity plays a fundamental role in thermodynamics and in many problems of mechanics, especially in plasticity. We only introduce the minimum prerequisites, frequently used in the technical literature.

A2.1.1 Convex function

Let V be a Banach space and φ a function of V in $\hat{\mathscr{A}} := \mathscr{A} \cup \{+\infty\}$. The function f is said to be convex if

$$\varphi(\lambda x + (1 - \lambda)y) \leqslant \lambda\varphi(x) + (1 - \lambda)\varphi(y), \qquad \forall x, y \in V, \quad \forall \lambda \in [0, 1]. \tag{A2.1}$$

φ is said to be *strictly* convex if the strict inequality sign holds in (A2.1) whenever $x \neq y$.

It can be checked that φ is convex, if and only if its *epigraph*, epi φ, is such that (Fig. A2.1(a))

$$\operatorname{epi}\varphi = \{(z, x) | z \geqslant \varphi(x)\} \tag{A2.2}$$

is *convex*.

A2.1.2 Indicator function of a convex set

Let K be a subset of V. The indicator function, I_K of K is defined by

$$I_K(x) = \begin{cases} 0 & \text{if } x \in K, \\ +\infty & \text{if } x \notin K, \end{cases} \tag{A2.3}$$

If K is a convex set, then I_K is a convex function. Indeed,

if x or $y \notin K$, the right-hand side of (A2.1) takes a value $+\infty$ and the inequality (A2.1) is obvious,

if x and $y \in K$, then $\lambda x + (1-\lambda)y \in K$ if $\lambda \in [0,1]$ and the two sides of (A2.1) vanish.

It is readily checked that

Fig. A2.1 (a) Epigraph. It is said that 'the curve is below the chord'. (b) Convexity and derivative. Geometrical characterization of convexity property in one dimension:

$$\forall x, y, \quad \varphi(x) + \varphi'(x)(y-x) \leqslant \varphi(y) \quad \text{(convexity)};$$

$$\varphi(x) + \varphi'(x)(y-x) < \varphi(y) \quad \text{(strict convexity)}, \; y \neq x.$$

It is said that 'the curve is above the tangent everywhere' (see subdifferential in (A2.4))

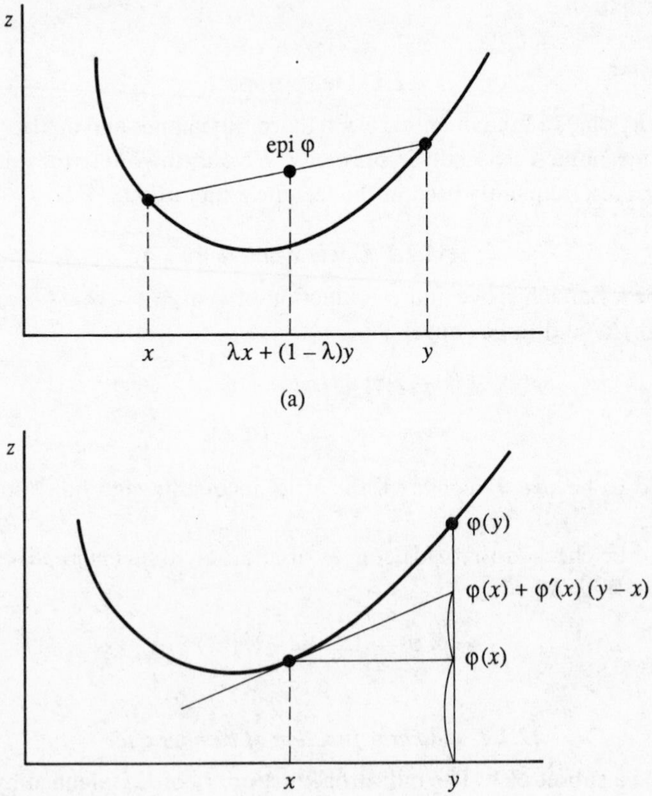

the sum of two convex functions is a convex function,

the upper hull $\varphi(x) = \text{Sup}[\varphi_1(x), \varphi_2(x)]$ of two convex functions $\varphi_1(x)$ and $\varphi_2(x)$ is convex (this can be generalized to a finite or nonfinite family of convex functions).

A2.2 Subdifferentials

Let V' be the dual vector space of V and $\langle \cdot, \cdot \rangle_{V,V}$ the inner product defined by this duality. Then it is said that $z' \in V'$ belongs to the *subdifferential* of the function φ at point x if and only if,

$$\forall y \in V, \varphi(y) - \varphi(x) \geqslant \langle z, y - x \rangle_{V',V}. \tag{A2.4}$$

We thus let

$$\partial\varphi = \{z | z \text{ has property (A2.4)}\}; \tag{A2.5}$$

the set $\partial\varphi$ is a subset of V' which may be empty or contain more than one point.

First example Let $\varphi(x) = |x|$ (Fig. A2.2). Then

$$\partial\varphi(x) = \begin{cases} -1 & \text{if } x < 0, \\ [-1, +1] & \text{if } x = 0, \\ 1 & \text{if } x > 0. \end{cases}$$

Second example If V is a Hilbert space and $\varphi = I_K$ is the indicator function of a convex K of V, then

Fig. A2.2. First example of subdifferential

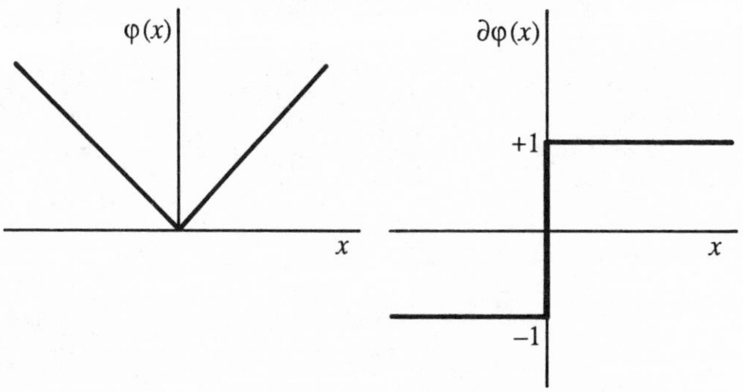

$$\partial\varphi(x) = \begin{cases} \varnothing \text{ if } x \notin K, \\ \text{(cone of outward normals at } x \text{ to } K) = \\ N_C(x) \text{ if } x \text{ is on the boundary of } K, \\ \{0\} \text{ if } x \text{ is inside } K. \end{cases} \tag{A2.6}$$

Proof If $x \in K$, then necessarily

$$0 \geqslant (z, y - x), \forall y \in K \quad \text{(convexity)}. \tag{A2.7}$$

If x is inside K, we can take $y - x = w$ where w is a vector with any 'small' norm. Replacing w by $-w$ and using (A2.7) we are led to $z = 0$, necessarily.

If x belongs to the boundary of K, we set

$$\Pi(x, y) = \{u | (u, y - x) \leqslant 0\}. \tag{A2.8}$$

This set is a half-space that contains the origin. Then (A2.7) allows us to show that

$$\partial\varphi(x) = \bigcap_{y \in K} \Pi(x, y). \tag{A2.9}$$

This is a *cone*, called the cone of outward normals to K at x. When K is not regular at x, this cone is not reducible to a half straight-line (Fig. A2.3).

Third example (Particular case of the previous one). Let $\varphi(x) = I_{[a,b]}$, i.e.,

Fig. A2.3. Second example of subdifferential. (a) K is regular. (b) K is not regular at x.

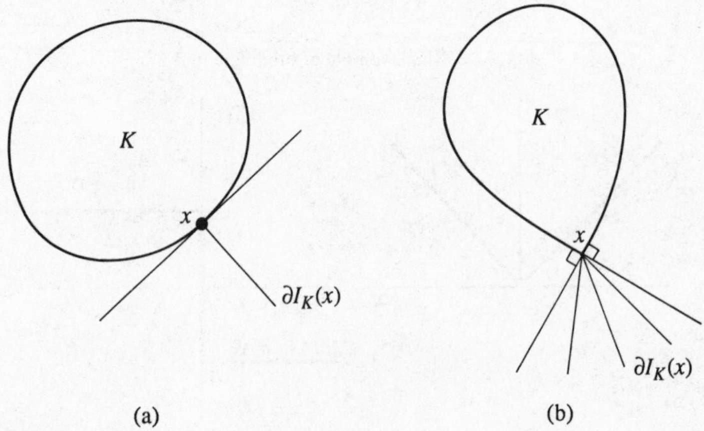

$$\varphi(x) = \begin{cases} +\infty & \text{if } x \notin [a,b], \\ 0 & \text{if } x \in [a,b]. \end{cases} \tag{A2.10}$$

It can be shown that (exercise; Fig. A2.4)

$$\partial\varphi(x) = \begin{cases} \varnothing & \text{if } x < a, \\ \mathbb{R}^- & \text{if } x = a, \\ \{0\} & \text{if } a < x < b, \\ \mathbb{R}^+ & \text{if } x = b, \\ \varnothing & \text{if } x > b. \end{cases} \tag{A2.11}$$

The notion of subdifferential is related to the classical notion of differential through the following result. *If φ is convex and differentiable at x, then*

$$\partial\varphi(x) = \{\mathrm{D}\varphi(x)\} \tag{A2.12}$$

where $\mathrm{D}\varphi$ *is the differential of* φ.

In particular, we know that a convex differentiable function admits a *minimum* at a point x inside its domain if and only if $\mathrm{D}\varphi(x) = 0$. This generalizes to the case of nondifferentiable functions for which we can state 'φ admits a minimum at x' is *equivalent to* $0 \in \partial\varphi(x)$.

In one dimension the *derivative of a convex function is always increasing.* This translates to the following: *the subdifferential of a convex function is a monotone operator*, i.e.,

$$(\partial\varphi(x) - \partial\varphi(x'), x - x') \geqslant 0, \qquad \forall x \text{ and } x' \in V, \tag{A2.13}$$

or else

$$(z - z', x - x') \geqslant 0, \qquad \forall z \in \partial\varphi(x), \qquad \forall z' \in \partial\varphi(x'). \tag{A2.14}$$

Fig. A2.4. Third example of subdifferential

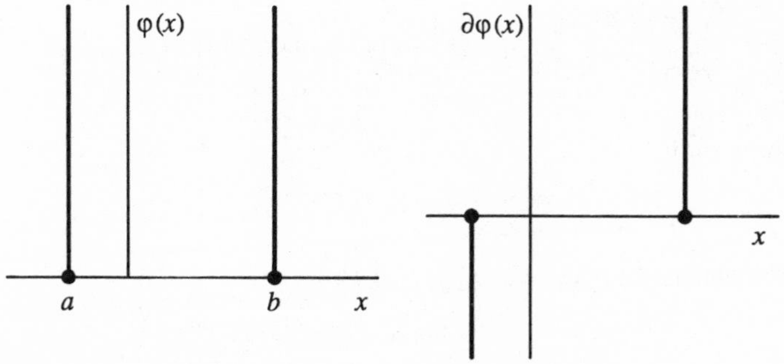

Indeed, on account of (A2.4),

$$\varphi(x) - \varphi(x') \geqslant (z', x - x'), \qquad \forall z' \in \partial\varphi(x'),$$

$$\varphi(x') - \varphi(x) \geqslant (z, x' - x), \qquad \forall z \in \partial\varphi(x).$$

Adding the two inequalities we obtain (A2.14) or (A2.13)

A2.3 Lower semicontinuity

We say that $\varphi : V \to \mathbb{R}$ is *lower semicontinuous* (or 'l.s.c') for the strong topology (respectively, the weak topology) of V, if, whenever $x_n \to \infty$ for the strong topology (respectively, the weak one) of V, we have

$$\varphi(x) \leqslant \liminf_{n\to\infty} \varphi(x_n). \tag{A2.15}$$

It is obvious that weak lower semicontinuity implies strong lower semicontinuity. The reciprocal statement is generally wrong, *except for convex functions*: if φ is a strongly lower semicontinuous convex function, then it is also weakly lower semicontinuous. For instance, if φ is the indicator function of a convex set K, φ is l.s.c. (strongly and weakly) if and only if K is a *closed* convex of V.

A2.4 Conjugate functions, Legendre–Fenchel transformation

Let φ be a proper (i.e., not identically equal to $+\infty$) l.s.c. convex function. Its conjugate or Legendre–Fenchel transform φ^* is defined by

$$\varphi^*(z) = \operatorname*{Sup}_{x\in V} \{(x, z)_{V,V'} - \varphi(x)\}, \qquad \forall z \in V'. \tag{A2.16}$$

It is easy to show (exercise) that φ^*, also, is an l.s.c. convex function, for it is the upper hull of l.s.c. convex functions.

First example

$$f(x) = \begin{cases} -(1 - x^2)^{1/2} & \text{if } |x| \leqslant 1, \\ +\infty & \text{if } |x| > 1; \end{cases}$$

then show that

$$f^*(z) = (1 + z^2)^{1/2}.$$

Second example Let $f(x) = a|x|$, $a > 0$; show that

$$f^*(z) = \begin{cases} +\infty & \text{if } |z| > a, \\ 0 & \text{if } |z| \leqslant a, \end{cases}$$

and thus

$$f^* = I_{[-a,+a]}.$$

An elementary graphical construction of the graph of the *conjugate* or *dual* φ^* of φ is given in Fig. A2.5.

Directly from the definition of φ^* there follows between φ and φ^* the so-called *Young inequality*,

$$\varphi(x) + \varphi^*(z) \geqslant (x, z)_{V,V'}, \quad \forall x \in V, \forall z \in V'. \tag{A2.17}$$

We also have the following result (proof by way of exercise). *The following three properties are equivalent*:

(i) $z \in \partial\varphi(x)$,
(ii) $x \in \partial\varphi^*(z)$;
(iii) $\varphi(x) + \varphi^*(z) = (x, z)_{V,V'}$.

The equivalence between (i) and (ii) shows that φ and φ^* are the *inverses* of one another:

$$\partial\varphi^* = (\partial\varphi)^{-1}. \tag{A2.18}$$

From this there follows that

$$(\varphi^*)^* = \varphi \tag{A2.19}$$

if φ is proper lower semicontinuous.

If $\varphi = I_K$ where K is a closed† convex set of V, then

Fig. A2.5. Drawing the graph of φ^* from φ. (1) Draw $\varphi(x)$. (2) Select y and draw the line yx of slope y. (3) Substract $\varphi(x)$ from that line yx. (4) Take the supremum of the result. (5) Carry the result in the (φ^*, y) plane.

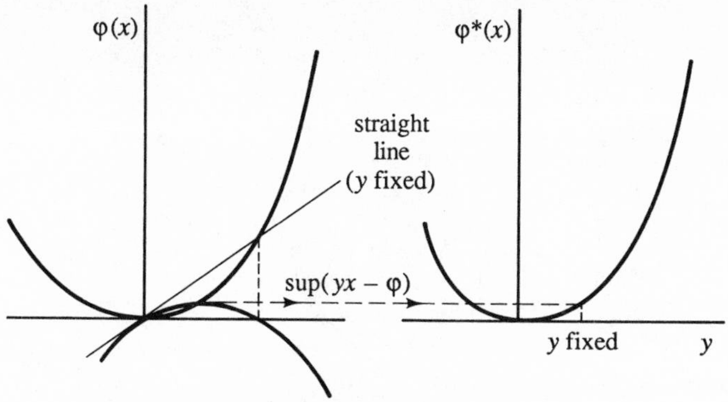

† A convex function is said to be closed if its epigraph is closed.

$$\varphi^*(z) = \underset{x \in V}{\operatorname{Sup}} [(x, z)_{V,V'} - \varphi(x)] = \underset{x \in K}{\operatorname{Sup}} (x, z)_{V,V'} \tag{A2.20}$$

and φ^* is the *support function* of the convex K at x (Fig. A2.6).

The conditions (i)–(iii) are none other than those of the classical Legendre transformation between two potentials (compare eqn (A1.25)).

Property *The function* φ^* *defined by* (*A2.20*) *is a* positively homogeneous function of degree 1, *i.e.*,

$$\varphi^*(\lambda z) = \lambda \varphi^*(z), \qquad \forall z \in V', \forall \lambda \geqslant 0. \tag{A2.21}$$

This property may be said to be *characteristic* in that if ϕ is an l.s.c. convex function on V, satisfying (A2.21), then there exists a closed convex set K such that

$$\phi = (I_K)^*. \tag{A2.22}$$

This property is of utmost importance in the formulation of the normality rules in relation to a dissipation function that is a positively homogeneous convex function of degree 1 (see Chapters 2 and 3). We shall give the proof of (A2.21).

Proof of (*A2.21*) We seek ϕ^* such that

$$\begin{aligned} \phi^*(x) &= \underset{z \in V'}{\operatorname{Sup}} \{(x, z) - \phi(z)\} \\ &= \underset{z \in V'}{\operatorname{Sup}} |z|_{V'} \left[\left(x, \frac{z}{|z|_{V'}} \right) - \phi\left(\frac{z}{|z|_{V'}} \right) \right] \end{aligned} \tag{A2.23}$$

thanks to homogeneity. Thus

Fig. A2.6. Support function of a convex. $\varphi^*(z) = (x, z)$.

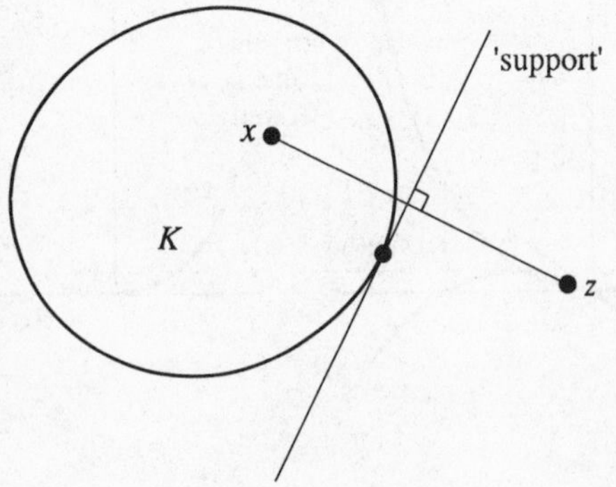

$$\phi^*(x) = \operatorname*{Sup}_{\substack{\lambda \geqslant 1 \\ |z|_{V'}=1}} \lambda[(x,z)_{V,V'} - \phi(z)]. \tag{A2.24}$$

There are two cases.

(a) If there exists $z_0 \in V'$ such that

$$(x, z_0)_{V,V'} - \phi(z_0) > 0, \qquad |z_0| = 1,$$

then

$$\phi^*(x) \geqslant \operatorname*{Sup}_{\lambda \geqslant 0} \lambda[(x, z_0) - \phi(z_0] = +\infty. \tag{A2.25}$$

(b) If for any z in V', $|z| = 1$, we have

$$(x, z) - \phi(z) \leqslant 0, \tag{A2.26}$$

then

$$\phi^*(x) = 0. \tag{A2.27}$$

But (A2.26) implies that x belongs to a closed convex set of V that is defined by

$$K \bigcap_{\substack{z \in V' \\ |z|_{V'}=1}} E_z \tag{A2.28}$$

where E_z is the closed half-space

$$E_z = \{y \in V | (y,z)_{V,V'} - \phi(z) \leqslant 0\}. \tag{A2.29}$$

Thus (A2.26) and (A2.28) show that $\phi^* = I_K$, whence (A2.22) through (A2.19). QED.

A2.5 Minimization of functions

We simply state the following result. *Let f be a function defined on V, with values in $\mathbb{R}$, and not identically equal to $+\infty$. We want to minimize f on a closed convex K of V. Then, if f is weakly l.s.c. on K and*

(a) *either K is bounded in V,*
(b) *or $\lim_{|x| \to +\infty, x \in K} f(x) = +\infty$,*

then

(i) *there exists $x_0 \in K$ such that*

$$f(x) \geqslant f(x_0), \forall x \in K, \tag{A2.30}$$

(ii) *if f is convex, the set of all vectors x_0 satisfying (A2.30) is a closed convex,*

(iii) *if f is strictly convex, the set of vectors x_0 satisfying (A2.30) is reduced to a point.*

Bibliographical guide In the preceding, we have mostly followed the very clear exposition of Suquet (1982). The mathematically inclined reader will deepen his approach to convexity by consulting the quite complete syntheses of Moreau (1966) and Tyrrell Rockafellar (1970), two authors to whom many rich developments are due, let alone Werner Fenchel, who introduced the general conjugacy correspondence for convex functions. The book of Ekeland and Temam (1974) is also useful. An exposition for specialists in continuum mechanics, with an emphasis on applications to thermodynamics, is to be found in Germain (1973, Reminder R2, pp. 317–33) and Sewell (1987). The last author gives many applications involving variational principles, including for elastic–plastic bodies.

Appendix 3

Analytic solutions of some problems in elastoplasticity

The object of the appendix It is clear that elastoplasticity problems are not easily amenable to analytical methods, but for a few exceptions as in the case of the spherical envelope in Chapter 6. In particular, the elastoplastic borderline separating the region where the material still behaves elastically and the already plasticized region is an unknown in such problems. For complex geometries then a numerical implementation seems necessary (Chapter 11). However, the few cases that admit analytical solutions are typical of a methodology of which any student and practitioner of elastoplasticity must be aware. We have thus selected four examples, the first in *plane strain* (the wedge problem), the second in *torsion*, the third exhibiting a *complex loading* and the fourth accounting for *anisotropy* in a composite material.

A3.1 Elastoplastic loading of a wedge

A.3.1.1 General equations

A wedge of angle $\beta < \pi/2$ is made of an isotropic elastoplastic material, satisfying Hooke's law in the elastic regime and Tresca's criterion without hardening at the yield limit. On its upper face it is subjected to a pressure p which increases with time (Fig. A3.1). We look first for the fully elastic solution and then for the elastoplastic solution in which the plasticized zone progresses until the whole wedge has become plastic. The solution of this problem in the elastoplastic framework is due to Naghdi (1957) – see also Murch and Naghdi (1958) and Calcotte (1968, pp. 158–64). The *rigid–perfectly-plastic* solution may be found in Prager and Hodge (1951, pp. 156–8). Thanks to the symmetry present, computations are performed in polar coordinates. The material is assumed *incompressible* and no volume forces are active. The general equations valid in both elastic and plastic regimes are as follows.

Strains in polar coordinates

$$\varepsilon_{rr} = \frac{\partial u_r}{\partial r}, \qquad \varepsilon_{\theta\theta} = \frac{u_r}{r} + \frac{1}{r}\frac{\partial u_\theta}{\partial \theta}, \qquad z\varepsilon_{r\theta} = \frac{1}{r}\frac{\partial u_r}{\partial \theta} + \frac{\partial u_\theta}{\partial r} - \frac{u_\theta}{r}. \tag{A3.1}$$

For *plane* strains $\varepsilon_{zz} = 0$ and the incompressibility condition $\theta = \operatorname{tr} \boldsymbol{\varepsilon} = 0$ requires that

$$\varepsilon_{rr} = -\varepsilon_{\theta\theta}. \tag{A3.2}$$

Equilibrium equations Let $\mathbf{s}$ and σ_{m} denote the deviator and mean stress associated with the stress tensor $\boldsymbol{\sigma}$. We have

$$\left.\begin{aligned} s_r &\equiv s_{rr} = \sigma_{rr} - \sigma_{\mathrm{m}}, \\ s_\theta &\equiv s_{\theta\theta} = \sigma_{\theta\theta} - \sigma_{\mathrm{m}}, \\ s_z &\equiv s_{zz} = \sigma_{zz} - \sigma_{\mathrm{m}} = -(s_r + s_\theta), \end{aligned}\right\} \tag{A3.3}$$

and the equilibrium equations read

$$\left.\begin{aligned} &\frac{\partial s_r}{\partial r} + \frac{1}{r}\frac{\partial \sigma_{r\theta}}{\partial \theta} + \frac{s_r - s_\theta}{r} + \frac{\partial \sigma_{\mathrm{m}}}{\partial r} = 0, \\ &\frac{\partial \sigma_{r\theta}}{\partial r} + \frac{1}{r}\frac{\partial s_\theta}{\partial \theta} + 2\frac{\sigma_{r\theta}}{r} + \frac{1}{r}\frac{\partial \sigma_{\mathrm{m}}}{\partial \theta} = 0. \end{aligned}\right\} \tag{A3.4}$$

A3.1.2 Elastic law of state

This is Hooke's law in which we set $2G = E/(1 + \nu)$ if E is Young's modulus and ν is Poisson's ratio.

Fig. A3.1. Elastoplastic wedge

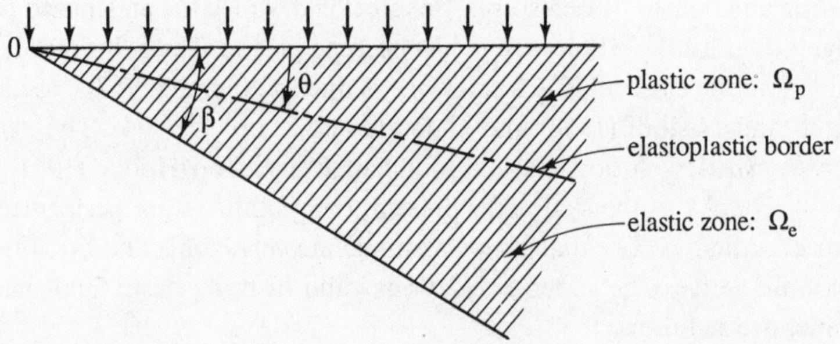

A3.1.3 Prandtl–Reuss equations

These are the equations

$$\dot{\boldsymbol{\varepsilon}} = \mathbf{S} : \dot{\boldsymbol{\sigma}} + \dot{\lambda}\frac{\partial f}{\partial \boldsymbol{\sigma}}. \tag{A3.5}$$

On account of Mises' criterion

$$s_r^2 + s_r s_\theta + s_\theta^2 + \sigma_{r\theta}^2 = k^2, \tag{A3.6}$$

eqns (A3.5) read

$$2G(\dot{\varepsilon}_{rr}, \dot{\varepsilon}_{\theta\theta}, \dot{\varepsilon}_{zz}, \dot{\varepsilon}_{r\theta}) = (\dot{s}_r, \dot{s}_\theta, \dot{s}_z, \dot{\sigma}_{r\theta}) + \dot{\lambda}(s_r, s_\theta, s_z, \sigma_{r\theta}). \tag{A3.7}$$

A3.1.4 The full elastic solution

Taking account of Hooke's equation and eqns (A3.1) we can write eqns (A3.4) and the condition (A3.2) in the following form:

$$\left.\begin{aligned} G\left[\Delta u_r - \frac{1}{r}\left(\frac{u_r}{r} + \frac{2}{r}\frac{\partial u_\theta}{\partial \theta}\right)\right] + \frac{\partial \sigma_{\mathrm{m}}}{\partial r} = 0, \\ G\left[\Delta u_\theta + \frac{1}{r}\left(\frac{2}{r}\frac{\partial u_r}{\partial \theta} - \frac{u_\theta}{r}\right)\right] + \frac{1}{r}\frac{\partial \sigma_{\mathrm{m}}}{\partial \theta} = 0, \end{aligned}\right\} \tag{A3.8}$$

and

$$\left(\frac{\partial}{\partial r} + \frac{1}{r}\right)u_r + \frac{1}{r}\frac{\partial u_\theta}{\partial \theta} = 0, \tag{A3.9}$$

where we have set

$$\Delta \equiv \frac{\partial^2}{\partial r^2} + \frac{1}{r}\frac{\partial}{\partial r} + \frac{1}{r^2}\frac{\partial^2}{\partial \theta^2}. \tag{A3.10}$$

The solution of eqns (A3.8)–(A3.9) is written as

$$\left.\begin{aligned} Gu_r &= -r(a\cos 2\theta + c\sin 2\theta), \\ Gu_\theta &= r(a\sin 2\theta - c\cos 2\theta) - \mathrm{d}r\log r, \\ \sigma_{\mathrm{m}} &= 2(b + d\theta) \end{aligned}\right\} \tag{A3.11}$$

where *a*, *b*, *c*, and *d* are constants of integration. The corresponding stresses are obtained as

$$\left.\begin{aligned} s_r &= -s_\theta = -2(a\cos 2\theta + c\sin 2\theta),\\ \sigma_{r\theta} &= -d + 2(a\sin 2\theta - c\cos 2\theta),\\ s_z &= 0. \end{aligned}\right\} \tag{A3.12}$$

The boundary conditions are

$$\boldsymbol{\sigma}\cdot\mathbf{n} = -p\mathbf{n} \quad \text{for } \theta = 0,$$

$$\boldsymbol{\sigma}\cdot\mathbf{n} = 0 \quad \text{for } \theta = \beta,$$

whence

$$\left.\begin{aligned} &\sigma_\theta = -p, \quad \sigma_r = \sigma_{r\theta} = 0 \quad &&\text{at } \theta = 0,\\ &\sigma_\theta = \sigma_r = \sigma_{r\theta} = 0 &&\text{at } \theta = \beta. \end{aligned}\right\} \tag{A3.13}$$

One obtains thus (Naghdi, 1957)

$$\left.\begin{aligned} &a = -p\frac{\tan\beta}{4\alpha}, \quad b = -p\left(\tfrac{1}{2} + \frac{\tan\beta}{4\alpha}\right),\\ &c = \frac{p}{4\alpha}, \quad d = -\frac{p}{2\alpha}, \quad \alpha \equiv \tan\beta - \beta. \end{aligned}\right\} \tag{A3.14}$$

A3.1.5 Elastoplastic border

With $s_r = -s_\theta$, eqns (A3.12) and (A3.4), the criterion (A3.6) can be rewritten in the form $s_\theta^2 + \sigma_{r\theta}^2 = k^2$, i.e.,

$$\left(\frac{p}{2\alpha}\right)^2 f(\theta) = k^2, \tag{A3.15}$$

where we have set

$$f(\theta) = \tan^2\beta + 2(1 - \cos 2\theta) - 2\tan\beta\sin 2\theta. \tag{A3.16}$$

As p and α are constants at a prescribed time t, plastic flow will occur whenever $f(\theta)$ reaches its maximum. $f(\theta)$, for $0 < \theta < \beta$, is minimum at $\theta = \beta/2$ and maximum for $\theta = 0$ and $\theta = \beta$ when pressure reaches the following critical value:

$$p^* = 2k\alpha/\tan\beta. \tag{A3.16α}$$

Obviously, eqn (A3.15) shows that the plastic flow condition does not depend on r so that when p is increased and goes over the value p^*, plastic regions Ω_p, are formed as indicated in Fig. A3.2.

A3.1.6 The elastoplastic solution

In the region $\Omega_{el} = \{\theta_1 < \theta < \theta_2\}$ the elastic solution is still of the form (A3.11)–(A3.12) but the boundary conditions at $\theta = \theta_1$ and $\theta = \theta_2$ are none other than the fulfilment of Mises's criterion, i.e.,

$$s_\theta^2 + \sigma_{r\theta}^2 = k^2 \qquad \text{at } \theta = \theta_1 \text{ and } \theta = \theta_2. \tag{A3.17}$$

Moreover, eqns (A3.11), and (A3.12) indicate that $\boldsymbol{\sigma}$ does not depend on r in the elastic region and this, by continuity at the elastoplastic border, extends also to Ω_p, so that the equilibrium equations, at *any* point in $\Omega = \Omega_{el} \cup \Omega_p$, read

$$\left.\begin{aligned} &\frac{\partial \sigma_{r\theta}}{\partial \theta} + (s_r - s_\theta) = 0, \\ &\frac{\partial s_\theta}{\partial \theta} + 2\sigma_{r\theta} + \frac{\partial \sigma_m}{\partial \theta} = 0. \end{aligned}\right\} \tag{A3.18}$$

Taking now the time derivative of eqns (A3.4), multiplying the result by $\dot{\lambda}$, combining this with (A3.18) and (A3.7), and finally using (A3.1) and (A3.5), we deduce an equation for $\dot{u}_r$ in Ω_p:

$$\frac{1}{r}\frac{\partial^2 \dot{u}_r}{\partial \theta^2} - r\frac{\partial^2 \dot{u}_r}{\partial r^2} + 3\frac{\partial \dot{u}_r}{\partial r} + \frac{\dot{u}_r}{r} = \sigma_{r\theta}\frac{\partial \dot{\lambda}}{\partial \theta}. \tag{A3.19}$$

As no radial slip can occur along the elastoplastic borderline, u_r must also be a linear function of r in Ω_p, i.e., we can write

$$u_r = r\varepsilon_{rr}(\theta, \phi). \tag{A3.20}$$

Fig. A3.2. Formation of plastic zones in the wedge,

$\Omega_e : \theta_1 < \theta < \theta_2,\ \Omega_p : \Omega_{p1} : 0 < \theta < \theta_1.\ \Omega_{p2} : \theta_2 < \theta < \beta$

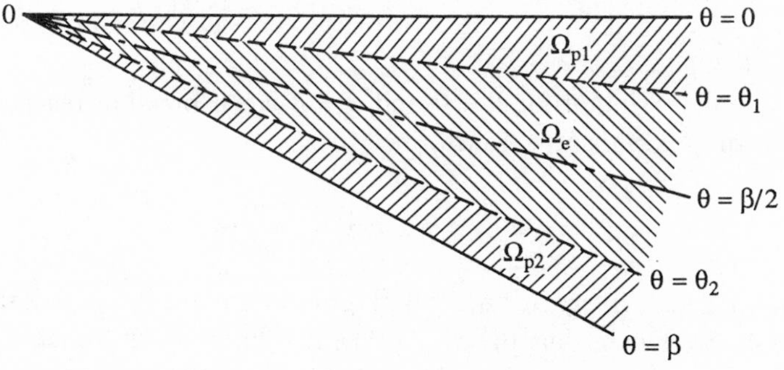

Substituting from this into eqn (A3.19) and integrating in time, we obtain

$$\frac{\partial^2 \varepsilon_{rr}}{\partial \theta^2} + 4\varepsilon_{rr} = \int \sigma_{r\theta} \frac{\partial \dot{\lambda}}{\partial \theta} \mathrm{d}t, \tag{A3.21}$$

where an arbitrary function of θ has been set equal to zero without loss in generality and $\dot{\lambda}$, which is proportional to the rate of performed work, is given by

$$\dot{\lambda} = \frac{1}{2k^2}(s_r \dot{\varepsilon}_{rr} + s_\theta \dot{\varepsilon}_{\theta\theta} + 2\sigma_{r\theta}\dot{\varepsilon}_{r\theta}). \tag{A3.22}$$

Invoking now the continuity of stresses at $\theta = \theta_1$ and $\theta = \theta_2$ we can show that $s_r = s_\theta$ at all points of Ω_p and the flow condition $s_\theta^2 + \sigma_{r\theta}^2 = k^2$ can be rewritten as

$$(s_r - s_\theta) = \pm 2(k^2 - \sigma_{r\theta}^2)^{1/2}. \tag{A3.23}$$

Using this in eqn (A3.18), and taking into account the boundary conditions

$$\left.\begin{aligned} &\sigma_{\theta\theta} = -p, \quad \sigma_{r\theta} = 0 \quad &&\text{for } \theta = 0, \theta = \theta_1, \\ &\sigma_{\theta\theta} = \sigma_{r\theta} = 0 &&\text{for } \theta = \beta, \theta = \theta_2, \end{aligned}\right\} \tag{A3.24}$$

we can establish the stress solution as

$$\left.\begin{aligned} s_r &= -s_\theta = k\cos 2\theta, \qquad s_2 = 0, \\ \sigma_{r\theta} &= -k \sin 2\theta, \\ \sigma_{\mathrm{m}} &= -p + k, \end{aligned}\right\} \tag{A3.25}$$

for $0 < \theta < \theta_1$, and

$$\left.\begin{aligned} s_r &= -s_\theta = -k\cos 2(\theta - \beta), \\ \sigma_{r\theta} &= k \sin 2(\theta - \beta), \\ \sigma_{\mathrm{m}} &= -k, \end{aligned}\right\} \tag{A3.26}$$

for $\theta_2 < \theta < \beta$.

Finally, eqn (A3.21) is now transformed to

$$\frac{\partial^2 \varepsilon_{rr}}{\partial \theta^2} + 2(\tan 2\theta)\frac{\partial \varepsilon_{rr}}{\partial \theta} + 4(\sec^2 2\theta)\varepsilon_{rr} = 0. \tag{A3.27}$$

In turn this can be integrated while taking account of the incompressibility condition, producing thus the displacement solution (u_r, u_θ) in $\Omega_{\mathrm{p}1}$. It is possible to show that $\theta_1 + \theta_2 = \beta$ so that the displacement field in $\Omega_{\mathrm{p}2}$

can be deduced from that in Ω_{p1}, by merely replacing θ by $\theta - \beta$ and considering other coefficients which are determined through the displacement continuity at $\theta = \theta_2$. When the wedge has become completely plastic, then

$$\theta = \theta_1 = \theta_2 = \beta/2 \tag{A3.28}$$

and the results (A3.25)–(A3.26) are then reduced to those of the rigid–perfectly-plastic solution given in Prager and Hodge (1951, pp. 156–8) in the limit of an infinite strain.

A3.2 Elastoplastic torsion of a circular shaft

We consider the pure torsion problem of an elastoplastic circular shaft satisfying the isotropic Hooke law in the elastic regime and Mises' criterion in the plastic one. The notation is the one given in Fig. A3.3. The elastic solution can be found in Germain's book (1962, pp. 103–8)

Let $\mathscr{M}$ be the torque with axis parallel to the generators of the cylinder; α denotes the torsion angle and μ is the shear coefficient.

A3.2.1 Elastic solution

In this solution only the stress components σ_{23} and σ_{13} are nonzero. They satisfy the following *elastic torsion problem.*

$$\left.\begin{aligned} \sigma_{13} &= \partial\bar{\theta}/\partial x_2, & \sigma_{23} &= -\partial\bar{\theta}/\partial x_1, \\ \Delta\bar{\theta} + 2\alpha\mu &= 0, & \text{in } \Omega &= \{0 < r < a\}, \\ \bar{\theta}(x_1, x_2) &= 0, & \text{on } \partial\Omega &= \{r = a\}, \end{aligned}\right\} \tag{A3.29}$$

Fig. A3.3. Elastoplastic torsion of a circular shaft

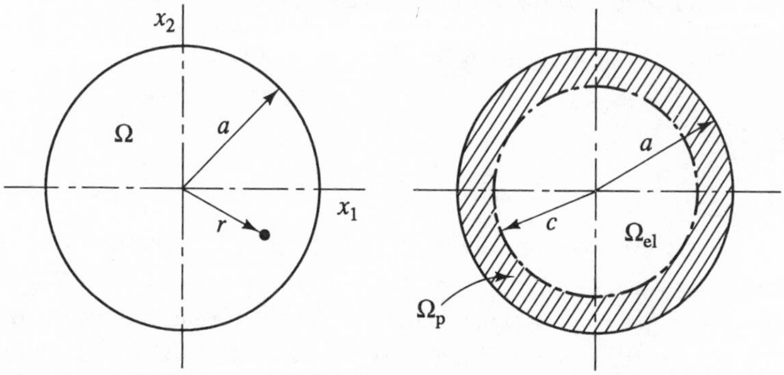

with

$$\mathscr{M} = 2\int_{\Omega} \bar{\theta}\, dx_1\, dx_2. \tag{A3.30}$$

The expressions $(A3.29)_1$ follow from the equilibrium condition $\sigma_{13,1} + \sigma_{23,2} = 0$. The function $\bar{\theta}$ is called the *stress function* (or potential). For a circular cross-section the solution of the problem (A3.29) is classically obtained as

$$\left.\begin{aligned} &2\bar{\theta} = \mu(a^2 - r^2) = \mu[a^2 - (x_1^2 + x_2^2)],\\ &\sigma_{13} = -\mu\alpha x_2, \qquad \sigma_{23} = \mu\alpha x_1,\\ &\mathscr{M} = \frac{\pi}{2}\mu\alpha a^4. \end{aligned}\right\} \tag{A3.31}$$

In elastoplasticity with Mises' criterion, this solution is valid only if

$$(\sigma_{13})^2 + (\sigma_{23})^2 \leqslant K_0^2, \tag{A3.32}$$

i.e., if

$$\alpha\mu a \leqslant K_0, \tag{A3.33}$$

since the yield limit is first reached on the exterior circle $r = a$. At the limit the torque takes the value

$$\mathscr{M}^* = \frac{\pi}{2} a^3 K_0. \tag{A3.34}$$

A3.2.2 Elastoplastic solution

It follows from (A3.33) that the plastic region starts to spread from the exterior and, by symmetry, develops in the annulus

$$\Omega_p = \{c < r < a\}, \tag{A3.35}$$

where c is an unknown. We shall proceed as if c were prescribed. In the plastic region where Mises' criterion is reached we have

$$\left.\begin{aligned} (\sigma_{13})^2 + (\sigma_{23})^2 = K_0^2, \qquad \mathbf{x} \in \Omega_p,\\ \sigma_{13,1} + \sigma_{23,2} = 0 \Rightarrow \sigma_{13} = \partial\bar{\theta}/\partial x_2, \qquad \sigma_{23} = -\partial\bar{\theta}/\partial x_1. \end{aligned}\right\} \tag{A3.36}$$

Therefore

$$|\mathbf{grad}\,\bar{\theta}| = K_0 \qquad \text{in } \Omega_p. \tag{A3.37}$$

This formula hints at the following *analogy* due to Prandtl (1903). The vector of components $(\partial\bar{\theta}/\partial x_1, \partial\bar{\theta}/\partial x_2, -1)$ represent the direction cosines of the surface $\bar{\theta}(x_1, x_2) = \text{const}$. Eqn (A3.37) tells us that this surface has a constant slope (i.e., the angle made by the normal to the surface and the vertical axis $0x_3$ is constant), and thus

$$\bar{\theta} = K_0(a - r) \qquad \text{in } \Omega_p. \tag{A3.38}$$

The strain function $\bar{\theta}$ is represented in Fig. A3.4 with the analogy of the 'sand hill' of Prandtl. Indeed, in the elastic region we have

$$2\bar{\theta} = \mu\alpha(c^2 - r^2) + C, \tag{A3.39}$$

where C is an integration constant determined by the continuity of $\bar{\theta}$ at $r = c$. That is, on account of (A3.38),

$$\bar{\theta} = \tfrac{1}{2}\mu\alpha(c^2 - r^2) + K_0(a - c). \tag{A3.40}$$

In the representation in Fig. A3.4 the conical part corresponds to (A3.38), $c < r < a$, while the parabolic cap corresponds to (A3.40), $0 < r < c$. The two regions are taken into account in the computation of $\mathcal{M}$. One obtains thus

$$\mathcal{M} = \frac{2}{3}\pi a^3 K_0 \left[1 - \frac{1}{4a^3}\left(\frac{K_0}{\mu\alpha}\right)^3\right]. \tag{A3.41$_a$}$$

The evolution of $\mathcal{M}$ with α is plotted in Fig. A3.5 taking account of the 'elastic limit' (A3.33) given by $\alpha^* = K_0/\mu a$. The final limit $\mathcal{M}^{**} = 2\pi a^3 K_0/3 = 4\mathcal{M}^*/3$ is obtained by using the above-mentioned analogy and computing the total volume of the 'sand hill'. It is easily checked that the displacement field has the *same* expression as in the purely elastic case, that is,

$$u_1 = -\alpha x_2 x_3, \qquad u_2 = \alpha x_1 x_3, \qquad u_3 = 0. \tag{A3.41$_b$}$$

Fig. A3.4. The stress function of the torsion problem. 'Sand hill'.

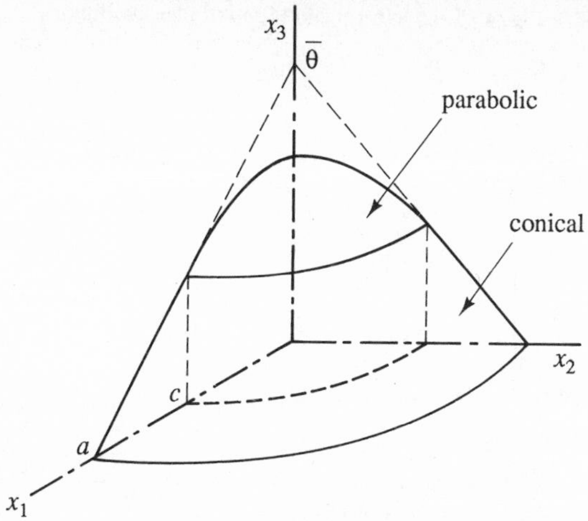

A3.3 Tube subjected to combined torsion and simple traction

This is the problem pictured in Fig. 1.12 in Chapter 1. It illustrates a case when different stress states are obtained depending on the path of loading, i.e., either torsion first and then traction, or traction first and then torsion. This difference is a typical feature of elastoplasticity. As in the previous two examples the material considered is Hookean and isotropic in the elastic regime and obeys Mises' criterion in the plastic regime. Moreover, *incompressibility* is assumed in *both* regimes.

In *simple traction* (of axis Ox_3) the only nonvanishing stress component is $\sigma_{33} = \sigma(r)$ while in *simple torsion* it is $\sigma_{23} = \tau(r)$ – with an appropriate choice of axes. The stress deviator components are then

$$\left.\begin{aligned} s_{11} &= s_{22} = -\frac{\sigma}{3}, \\ s_{33} &= \frac{2\sigma}{3}, \\ s_{23} &= \tau. \end{aligned}\right\} \tag{A3.42}$$

In the elastic case we have

$$s_{ij} = 2\mu\varepsilon_{ij} = 2\mu\varepsilon_{ij}^{\mathrm{d}}, \tag{A3.43}$$

while in plasticity

$$\dot{\varepsilon}_{ij} = \dot{\lambda} s_{ij} \quad (\text{with } \dot{\lambda} > 0). \tag{A3.44}$$

Fig. A3.5. Total torque in the elastoplastic solution

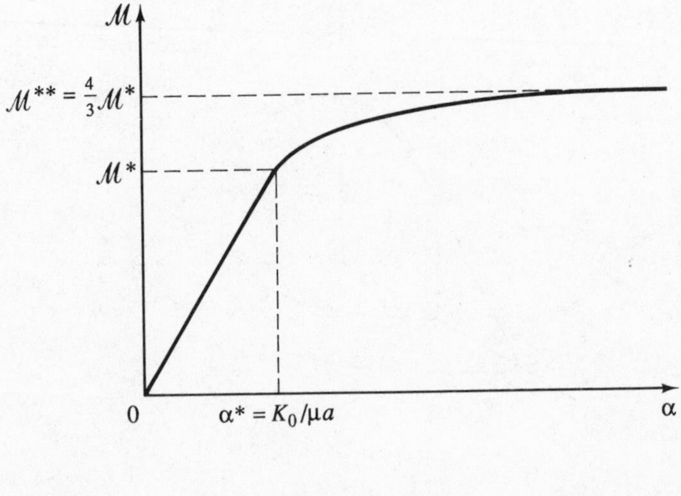

We set $\varepsilon_{33} = \varepsilon(r)$. Incompressibility requires that

$$\varepsilon_{22} = \varepsilon_{11} = -\frac{\varepsilon(r)}{2}, \tag{A3.45}$$

Furthermore, in torsion,

$$\varepsilon_{23} = \frac{r\alpha(r)}{2}. \tag{A3.46}$$

Transcribing the whole problem in velocities, we can state that

in elasticity

$$\left.\begin{aligned} \dot{\sigma} &= 3\mu\dot{\varepsilon}, \\ \dot{\tau} &= \mu r\dot{\alpha}, \end{aligned}\right\} \tag{A3.47}$$

in plasticity

$$\left.\begin{aligned} \dot{\sigma} &= 3\mu\dot{\varepsilon} - \dot{\lambda}\sigma, \\ \dot{\tau} &= \mu r\dot{\alpha} - \dot{\lambda}\tau, \end{aligned}\right\} \tag{A3.48}$$

with Mises' criterion reading

$$\left(\frac{\sigma}{\sqrt{3}}\right)^2 + \tau^2 = K_0^2. \tag{A3.49}$$

By time differentiation this yields

$$\frac{\sigma\dot{\sigma}}{\sqrt{3}} + \tau\dot{\tau} = 0, \tag{A3.50}$$

so that $\dot{\lambda}$ is given by

$$\dot{\lambda}K_0^2 = \mu(\sigma\dot{\varepsilon} + r\tau\dot{\alpha}). \tag{A3.51}$$

First path of loading Traction is first applied until the elasticity limit is reached (with $\alpha = 0$, $\tau = 0$), i.e., until

$$\sigma = K_0\sqrt{3}, \qquad \varepsilon = \frac{K_0}{\mu\sqrt{3}}. \tag{A3.52}$$

We set

$$\left.\begin{aligned} p &= \frac{\mu}{K_0}\sqrt{3}\cdot\varepsilon, \qquad & P &= \frac{\sigma}{K_0\sqrt{3}}, \\ q &= \frac{\mu a\alpha}{K_0}, & Q &= \frac{\tau}{K_0}, \\ \rho &= r/a, \end{aligned}\right\} \tag{A3.53}$$

so that at the end of the pure traction phase we have reached point A in Fig. A3.6, i.e.,

$$p = 1, \qquad P = 1, \qquad q = 0, \qquad Q = 0. \tag{A3.54}$$

At this point the whole body is in the plastic domain. Thereafter $\dot{\varepsilon} = 0$. We can write eqns (A3.48) and the condition (A3.49) as

$$\left.\begin{aligned} \dot{P} &= -PQ\rho\dot{q}, \\ \dot{Q} &= (1 - Q^2)\rho\dot{q}, \end{aligned}\right\} \tag{A3.55}$$

and

$$P^2 + Q^2 = 1. \tag{A3.56}$$

The solution of (A3.55) is obtained in the form

$$Q = \tanh \rho q, \qquad P = \operatorname{sech} \rho q = \frac{1}{\cosh \rho q}. \tag{A3.57}$$

We have thus the stress diagram in Fig. A3.7 (full lines) corresponding to point B in Fig. A3.6. We let the reader establish the equations corresponding to (A3.57) for the second stress-loading path which yields the dotted lines in Fig. A3.7 after final loading at point B in Fig. A3.6, but via point A'. The results differ essentially for $\rho \approx 1$, but the difference is already clearly visible for $\rho \approx 1/3$.

A3.4 Cyclic torsion of a composite with unidirectional fibres

The *anisotropic* medium considered is a composite material with unidirectional fibres which is described by a *unique* anisotropic elastoplastic con-

Fig. A3.6. Complex loading

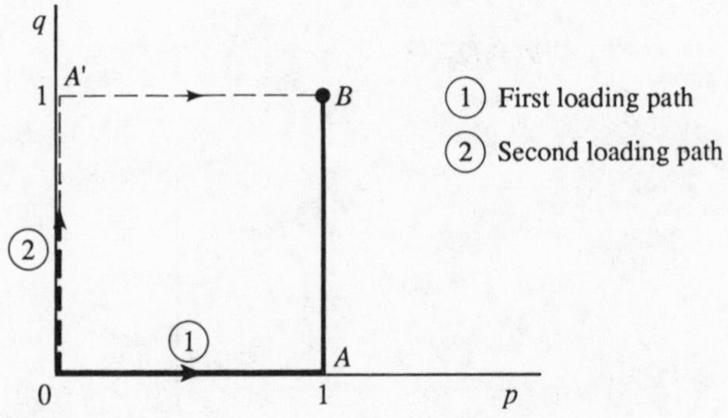

tinuum. That is, a homogenization of some kind has been performed and the result of this process is essentially a *transversely isotropic* macroscopic behaviour with axis directed along the unit vector **a** aligned with the parallel fibres.

A3.4.1 Basic equations

The expressions of the energy density $W(\boldsymbol{\varepsilon})$ and of a flow criterion generalizing those of Tresca and Mises have been given by Spencer (1972) and Rogers (1988). Let

$$\mathbf{A} = \mathbf{a} \otimes \mathbf{a}, \quad \text{i.e.,} \quad A_{ij} = a_i a_j. \tag{A3.58}$$

The (quadratic) elastic energy of such a composite is written as (tr = trace)

$$\begin{aligned} W(\boldsymbol{\varepsilon}; \mathbf{a}) &= \tfrac{1}{2}\lambda(\operatorname{tr}\boldsymbol{\varepsilon})^2 + \mu_{\mathrm{T}} \operatorname{tr}\boldsymbol{\varepsilon}^2 + \alpha \operatorname{tr}(\mathbf{A}\boldsymbol{\varepsilon})(\operatorname{tr}\boldsymbol{\varepsilon}) \\ &\quad + 2(\mu_{\mathrm{L}} - \mu_{\mathrm{T}})\operatorname{tr}(\mathbf{A}\boldsymbol{\varepsilon}^2) + \tfrac{1}{2}\beta(\operatorname{tr}\mathbf{A}\boldsymbol{\varepsilon})^2, \end{aligned} \tag{A3.59}$$

where λ and μ_{T} are the usual Lamé moduli (when $\mathbf{a} \neq \mathbf{0}$) and α, β and μ_{L} are three additional elasticity coefficients. The stress tensor is then given by

$$\begin{aligned} \boldsymbol{\sigma}(\boldsymbol{\varepsilon}; \mathbf{a}) = \frac{\partial W(\boldsymbol{\varepsilon}; \mathbf{a})}{\partial \boldsymbol{\varepsilon}} &= \lambda(\operatorname{tr}\boldsymbol{\varepsilon})\mathbf{1} + 2\mu_{\mathrm{T}}\boldsymbol{\varepsilon} + 2(\mu_{\mathrm{L}} - \mu_{\mathrm{T}})(\mathbf{A}\boldsymbol{\varepsilon} + \boldsymbol{\varepsilon}\mathbf{A}) \\ &\quad + \alpha[\mathbf{A}(\operatorname{tr}\boldsymbol{\varepsilon}) + (\operatorname{tr}\mathbf{A}\boldsymbol{\varepsilon})\mathbf{1}] + \beta\mathbf{A}\operatorname{tr}(\mathbf{A}\boldsymbol{\varepsilon}). \end{aligned} \tag{A3.60}$$

Let $\boldsymbol{\sigma}^{\mathrm{d}}$ be the deviator of $\boldsymbol{\sigma}$. Then we define **s** by

$$\mathbf{s} = \boldsymbol{\sigma} - \tfrac{1}{2}[\operatorname{tr}\boldsymbol{\sigma} - \operatorname{tr}(\mathbf{A}\boldsymbol{\sigma})]\mathbf{1} + \tfrac{1}{2}[\operatorname{tr}\boldsymbol{\sigma} - 3\operatorname{tr}(\mathbf{A}\boldsymbol{\sigma})]\mathbf{A}, \tag{A3.61}$$

Fig. A3.7. Stress state after loading at point B of Fig. A3.6. Solid line: first loading path (OAB). Broken line: second loading path ($OA'B$).

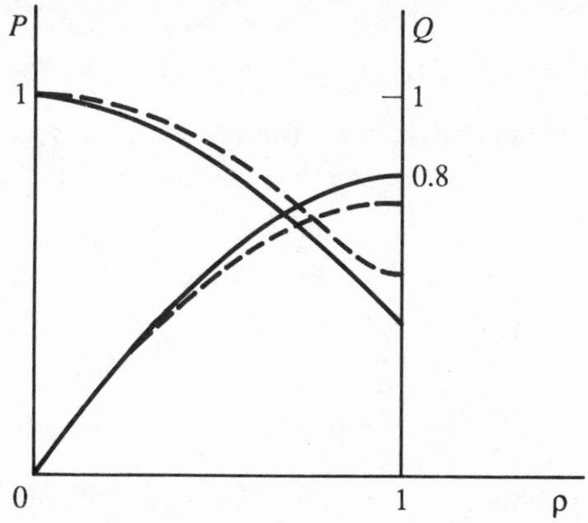

and the invariants I_1 to I_5 by (note that $\operatorname{tr}\mathbf{s} = 0$ and $\operatorname{tr}(\mathbf{As}) = 0$)

$$\left.\begin{aligned} I_1 &= \tfrac{1}{2}\operatorname{tr}\mathbf{s}^2 - \operatorname{tr}(\mathbf{As}^2), \quad I_2 = \operatorname{tr}\mathbf{As}^2, \quad I_3 = \operatorname{tr}\mathbf{s}^3, \\ I_4 &= \tfrac{3}{2}\operatorname{tr}(\mathbf{A}\boldsymbol{\sigma}^{\mathrm{d}}), \quad I_5 = \operatorname{tr}\boldsymbol{\sigma} - \tfrac{3}{2}\operatorname{tr}(\mathbf{A}\boldsymbol{\sigma}^{\mathrm{d}}). \end{aligned}\right\} \tag{A3.62}$$

Then the general plasticity criterion for such materials reads (compare to the expression (1.15) of the isotropic case)

$$f(I_1, I_2, I_3, I_4, I_5) \leqslant 0. \tag{A3.63}$$

For instance, we have a possible anisotropic generalization of Mises' criterion by

$$f_{\mathscr{M}}(\boldsymbol{\sigma}; \mathbf{a}) = \frac{I_1}{K_{\mathrm{T}}^2} + \frac{I_2}{K_{\mathrm{L}}^2} + \frac{I_4^2}{Y^2} - 1, \tag{A3.64}$$

or a generalization of Tresca's criterion by

$$f_{\mathrm{T}}(\boldsymbol{\sigma}; \mathbf{a}) = \left\{\begin{aligned} &(I_1/K_{\mathrm{T}}^2) - 1, \quad |I_2| \leqslant K_{\mathrm{L}}^2, \quad |I_4| \leqslant Y, \\ &(I_2/K_{\mathrm{L}}^2) - 1, \quad |I_1| \leqslant K_{\mathrm{T}}^2, \quad |I_4| \leqslant Y, \\ &(I_4^2/Y^2) - 1, \quad |I_1| \leqslant K_{\mathrm{T}}^2, \quad |I_2| \leqslant K_{\mathrm{L}}^2, \end{aligned}\right\} \tag{A3.65}$$

a form proposed by Lance and Robinson (1971).

A3.4.2 Torsion of a slice

Following Rogers (1988, 1990) we consider a thin plate sandwiched between two plate holders and subjected to a torsion with axis perpendicular to its faces, i.e., a torsion with axis Ox_3 while Ox_1 is the privileged axis along **a**. The sample is first loaded until the torsion angle α (per unit thickness) reaches its maximum α_{m} and then the load is reversed until α reaches $-\alpha_{\mathrm{m}}$, and so on. The displacement field is a pure torsion one such as in eqns $(\text{A3.41})_{\mathrm{b}}$, i.e.

$$u_1 = -\alpha x_2 x_3, \qquad u_2 = \alpha x_1 x_3, \qquad u_3 = 0. \tag{A3.66}$$

As the time parameter is arbitrary (there is no time scale!) we choose as 'plastic time' the torsion angle α, so that the 'velocity' deduced from (A3.66) is given by

$$v_1 = -x_2 x_3, \qquad v_2 = x_1 x_3, \qquad v_3 = 0, \tag{A3.67}$$

and the strain rate reads

$$\dot{\varepsilon}_{13} = -\frac{x_2}{2}, \qquad \dot{\varepsilon}_{23} = \frac{x_1}{2}, \qquad \dot{\varepsilon}_{11} = \dot{\varepsilon}_{22} = \dot{\varepsilon}_{33} = \dot{\varepsilon}_{12} = 0. \tag{A3.68}$$

The only nonvanishing components of $\boldsymbol{\sigma}$ are σ_{13} and σ_{23} which must satisfy the equilibrium equation

$$\sigma_{13,1} + \sigma_{23,2} = 0. \tag{A3.69}$$

The total torque about the Ox_3 axis is calculated by

$$\mathcal{M} = \int_\Omega (x_1\sigma_{23} - x_2\sigma_{13})\,\mathrm{d}s, \tag{A3.70}$$

where Ω is the area of the plate.

For *small* torques the whole sample deforms elastically, hence through (A3.60) the stress field is given by

$$\sigma_{13} = -\alpha\mu_\mathrm{L}x_2, \qquad \sigma_{23} = \alpha\mu_\mathrm{T}x_1, \qquad \text{other } \sigma_{ij}\text{'s} = 0, \tag{A3.71}$$

so that the computation of the expression (A3.70) yields

$$(\mathcal{M}/\alpha) = \mu_\mathrm{L}\mathcal{I}_1 + \mu_\mathrm{T}\mathcal{I}_2, \qquad \mathcal{I}_1 = \int_\Omega x_2^2\,\mathrm{d}s, \qquad \mathcal{I}_2 = \int_\Omega x_1^2\,\mathrm{d}s. \tag{A3.72}$$

For a *rectangular* sample, $\Omega = \{-a \leqslant x_1 \leqslant a, -b \leqslant x_2 \leqslant b\}$ *and eqn (A3.72) provides the value*

$$\mathcal{M} = \frac{4}{3}\alpha ab^3\left(\mu_\mathrm{L} + \mu_\mathrm{T}\frac{a^2}{b^2}\right). \tag{A3.73$_\mathrm{a}$}$$

Both $\mathcal{M}$ and shear stresses increase with α. Plastic flow will appear where the flow function reaches its critical value, which is zero in equation (A3.64) or (A3.65), i.e., when $\alpha = \alpha_\mathrm{c}$ such that

$$\frac{x_1^2}{a_\mathrm{c}^2} + \frac{x_2^2}{b_\mathrm{c}^2} = 1, \qquad a_\mathrm{c} = \left(\frac{K_\mathrm{T}}{\mu_\mathrm{T}}\right)\frac{1}{\alpha}, \qquad b_\mathrm{c} = \left(\frac{K_\mathrm{L}}{\mu_\mathrm{L}}\right)\frac{1}{\alpha}, \tag{A3.73$_\mathrm{b}$}$$

where both a_c and b_c decrease with increasing α. All material elements situated inside the ellipse Γ_c defined by eqn (A3.73) are elastically deformed with a stress field given by eqns (A3.71). Those elements that are situated outside Γ_c deform plastically when α increases, with a stress field satisfying the criterion

$$\frac{(\sigma_{13})^2}{K_\mathrm{L}^2} + \frac{(\sigma_{23})^2}{K_\mathrm{T}^2} - 1 = 0, \tag{A3.74}$$

while the flow law $\dot{\boldsymbol{\varepsilon}}^\mathrm{p} = \dot{\lambda}\partial f/\partial\boldsymbol{\sigma}$ reads

$$\frac{4\dot{\lambda}}{K_\mathrm{L}^2}\sigma_{13} + \frac{1}{\mu_\mathrm{L}}\dot{\sigma}_{13} = -x_2, \qquad \frac{4\dot{\lambda}}{K_\mathrm{T}^2}\sigma_{23} + \frac{1}{\mu_\mathrm{T}}\dot{\sigma}_{23} = x_1. \tag{A3.75}$$

For a material without hardening, these equations integrate to

$$\sigma_{13} = -\frac{K_L^2 x_2}{\sqrt{(x_1^2 K_T^2 + x_2^2 K_L^2)}}, \qquad \sigma_{23} = \frac{K_T^2 x_1}{\sqrt{(x_1^2 K_T^2 + x_2^2 K_L^2)}}, \tag{A3.76}$$

and the torque (A3.70) then is given by

$$\mathcal{M} = \alpha \int_{\Omega_{el}} (x_1^2 \mu_T + x_2^2 \mu_L)\, ds + \int_{\Omega_p} (x_1^2 K_T^2 + x_2^2 K_L^2)^{1/2}\, ds, \tag{A3.77}$$

where $\Omega_d = \Omega \cap \Gamma_c$ and $\Omega_p = \Omega - \Omega_{el}$.

A3.4.3 Material obeying a generalized Tresca criterion

For a material obeying the criterion (A3.65), Γ_c is a *rectangle* such that

$$|x_1| \leqslant a_c, \qquad |x_2| \leqslant b_c. \tag{A3.78}$$

The region Ω_p grows as α increases and we can follow this evolution in a very simple manner. In particular, when the elastic region Ω_{el} given by (A3.78) is entirely included within the boundary $\partial\Omega$, it is easily figured out that Ω_p is made of eight distinct regions (see Fig. A3.8).

For a simple rectangle for which $(b\mu_L/K_L) > (a\mu_T/K_T)$ we can evaluate $\mathcal{M}$ thus:

$$\mathcal{M} = \begin{cases} \dfrac{4}{3}ab^3\left(\mu_L + \mu_T\dfrac{a^2}{b^2}\right)\alpha, & 0 \leqslant \alpha \leqslant \dfrac{K_L}{b\mu_L} \quad (\text{eqn (A3.73)}), \\ 2ab^2K_L + \dfrac{4}{3}a^3 b\mu_L\alpha - \dfrac{2}{3}a\left(\dfrac{K_L}{\mu_L}\right)^3\dfrac{1}{\alpha^2}, & \dfrac{K_L}{b\mu_L} \leqslant \alpha \leqslant \dfrac{K_T}{a\mu_T}, \\ 2ab(aK_T + bK_L) - \dfrac{2}{3}\left[b\left(\dfrac{K_T}{\mu_T}\right)^3 + a\left(\dfrac{K_L}{\mu_L}\right)^3\right]\dfrac{1}{\alpha^2}, & \dfrac{K_T}{a\mu_T} \leqslant \alpha \leqslant \alpha_m. \end{cases} \tag{A3.79}$$

Fig. A3.8. Plasticization of an anisotropic plate loaded in torsion

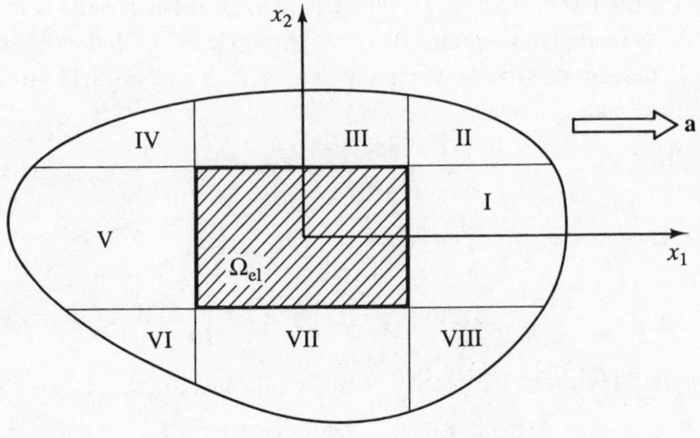

With increasing α, $\mathscr{M}$ goes asymptotically to

$$\mathscr{M}^{**} = 2ab^2\left(K_L + \frac{a}{b}K_T\right). \tag{A3.80}$$

This allows one to determine K_L and K_T from a series of experiments in torsion involving a variable ratio a/b.

In the case of a *cyclic test*, the sample first unloads elastically and may thereafter present a reverse plastic flow if the stress reaches a new threshold. For an *elastic unloading*, the stress and strain fields are obtained by superimposing an appropriate elastic solution onto an elastoplastic solution, itself obtained at $\alpha = \alpha_m$. Then the displacement field is still given by (A3.66) but the shearing stresses are more involved than before and given by

$$\left.\begin{aligned}\sigma_{13} &= \sigma_{13}(\alpha_m) - \mu_L(\alpha - \alpha_m)x_2,\\ \sigma_{23} &= \sigma_{23}(\alpha_m) + \mu_T(\alpha - \alpha_m)x_1.\end{aligned}\right\} \tag{A3.81}$$

In the domain Ω_{el}^m, as before, the stress is given by eqn (A3.71). But the situation is altogether different in Ω_p where only a careful examination permits us to see whether a reverse plastic flow has taken place. In the case of the generalized Tresca criterion (A3.65), it can be shown (Kaprielan and Rogers, 1987) that such a reverse plastic flow indeed takes place in regions numbered I to VIII in Fig. A3.8, as α reaches α_m. Then the new elastoplastic border has reached the same position as during the loading phase ($\alpha = \alpha_m$). The analysis becomes more and more cumbersome with additional cycles with an exception for Tresca's criterion and a rectangular sample, for then a stress cycle repeats itself.

A3.5 Problems with hardening

There is no doubt that adding hardening to the elastoplastic behaviour somewhat complicates the solution of problems. However, sometimes, solutions can be obtained in a closed form. This is true essentially when the loads are rather simple and the hardening law itself is very simple, such as the one represented in Fig. A3.9, where hardening beyond the elastic limit $\sigma_0 = E\varepsilon_0$ corresponds to a straight line of slope mE, $m < 1$. Problems both of the elastoplastic circular shaft (Section A3.2) and of the spherical envelope subjected to internal pressure (Subsection 6.1.3) can be re-examined for this more involved behaviour. This is left to the reader by way of exercise.

Here we simply re-examine the *torsion problem* of Section A3.2 in the presence of hardening. For *nonlinear* hardening the elastoplastic solution can be approached by means of the method of successive elastic solutions mentioned in Chapter 11. For *linear* hardening as sketched in Fig. A3.9 (with Mises' criterion) the solution is amenable to direct calculation. A closed-form solution can be obtained as follows. We shall work in polar coordinates (θ, r, z). The hardening curve in Fig. A3.9 is in fact given for Mises *equivalent* stress $\sigma_Y = (\frac{3}{2}\sigma^d_{ij}\sigma^d_{ij})^{1/2}$ versus an *equivalent* strain $\epsilon = \epsilon_t = \epsilon_e + \epsilon_p$ with $\epsilon \equiv (\frac{2}{3}\varepsilon_{ij}\varepsilon_{ij})^{1/2}$.

For the torsion problem the θz-component of the equation $\boldsymbol{\varepsilon} = \mathbf{S}:\boldsymbol{\sigma} + \boldsymbol{\varepsilon}^p$ reads (with $u_\theta = \alpha r z$)

$$\varepsilon_{\theta z} = \frac{1}{2}\alpha r = \frac{1}{2\mu}\sigma_{\theta z} + \varepsilon^p_{\theta z} \tag{A3.82}$$

or, using σ_Y and ϵ_p in lieu of $\sigma_{\theta z}$ and $\varepsilon^p_{\theta z}$,

$$\sigma_Y = \sqrt{3}\cdot\mu\alpha r - 3\mu\epsilon_p \tag{A3.83}$$

as $\sigma_Y = \sqrt{3}\cdot\sigma_{\theta z}$ and $\epsilon_p = (2/\sqrt{3})\varepsilon^p_{\theta z}$.

The problem is made nondimensional by introducing S, ρ, β, ϵ^*_p and $\epsilon^*_{p\theta}$ by

$$S = \sigma_Y/2\mu\varepsilon_0, \qquad \rho = \frac{r}{\theta}, \qquad \beta = \frac{\alpha a}{\varepsilon_0}, \qquad \epsilon^*_p = \frac{\epsilon_p}{\epsilon_0}, \qquad \epsilon^*_\theta = \frac{\varepsilon^p_{\theta z}}{\varepsilon_0}, \tag{A3.84}$$

so that eqn (A3.83) also reads

$$S = \frac{\sqrt{3}}{2}\beta\rho - \frac{3}{2}\epsilon^*_p. \tag{A3.85}$$

In the same representation the constitutive equation sketched out in Fig. A3.9 reads (note that $\mu = E/(1+\nu)$)

Fig. A3.9. Simple hardening law

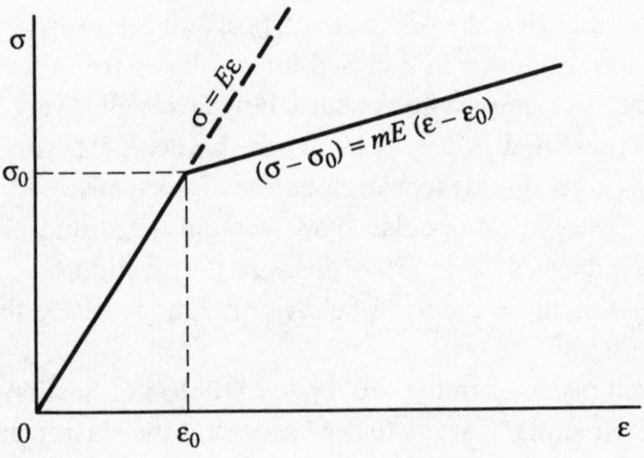

$$S = (1 + \nu)\left(1 + \frac{m}{1 - m}\epsilon_{\mathrm{p}}^{*}\right). \tag{A3.86}$$

Eqns (A3.85) and (A3.86) are the fundamental equations. Combining these two equations, we obtain

$$\epsilon_{\mathrm{p}}^{*} = \frac{\sqrt{3}\cdot\beta\rho - 2(1 + \nu)}{3 + 2(1 + \nu)m/(1 - m)}. \tag{A3.87}$$

But this holds good only in the plastic region Ω_{p}, i.e., if and only if $1 > \rho > \rho_{\mathrm{c}}$, where $\rho_{\mathrm{c}} = c/a$ (see Fig. A3.3). The plastic border $\rho = \rho_{\mathrm{c}}$ is determined by making $\epsilon_{\mathrm{p}}^{*} = 0$, i.e., for $S = 1 + \nu$, and thus eqn (A3.85) yields

$$\rho_{\mathrm{c}} = \frac{2(1 + \nu)}{\sqrt{3}\cdot\beta} \tag{A3.88}$$

a quantity that is *independent* of the stress–strain curve! Then β, corresponding to a torsion angle α, is found from eqn (A3.88) by setting ρ_{c} equal to 1 (plastic region starts to develop at the outer radius), i.e.,

$$\beta^{*} = \frac{2(1 + \nu)}{\sqrt{3}}, \qquad \text{or} \qquad \alpha^{*} = \frac{1}{\mu a}\frac{\sigma_0}{\sqrt{3}} \tag{A3.89}$$

which is the same as (A3.33) if we note that $K_0 = \sigma_0/\sqrt{3}$.

Once the equivalent plastic strain is known from eqn (A3.87), then the true plastic shear strain and the stress are computed from eqns $(\text{A3.83})_2$ and (A3.82). The total torque $\mathcal{M}$ is then evaluated by

$$\begin{aligned}\mathcal{M} &= \int_0^{2\pi} \mathrm{d}\theta \int_0^a \sigma_{\theta z} r(r\,\mathrm{d}r)\\ &= 4\pi\mu\varepsilon_0 a^3 \int_0^1 \sigma_\theta \rho^2\,\mathrm{d}\rho,\end{aligned} \tag{A3.90}$$

where

$$\sigma_\theta = \frac{\sigma_{\theta z}}{2\mu\varepsilon_0} = \begin{cases}\tfrac{1}{2}\beta\rho, & \rho \leqslant \rho_{\mathrm{c}}(\text{in } \Omega_{\mathrm{e}}),\\ \tfrac{1}{2}\beta\rho - \epsilon_{\mathrm{p}\theta}^{*}, & \rho \geqslant \rho_{\mathrm{c}}(\text{in } \Omega_{\mathrm{p}}).\end{cases} \tag{A3.91}$$

One obtains thus

$$\mathcal{M} = (2\mu\varepsilon_0 a^3) \times \left\{\frac{\pi\beta}{4} - \sqrt{3}\cdot\pi\left[\frac{1}{4}A(1 - \rho_{\mathrm{c}}^4) + \frac{1}{3}B(1 - \rho_{\mathrm{c}}^3)\right]\right\} \tag{A3.92}$$

with

$$A = \frac{\sqrt{3}\cdot\beta}{3 + 2(1 + \nu)m/(1 - m)}, \qquad B = -\frac{2(1 + \nu)}{3 + 2(1 + \nu)m/(1 - m)}.$$

With $K_0 = \sigma_0/\sqrt{3}$, it is readily checked that the result (A3.92) coalesces to $(A3.41)_a$ for $m = 0$. We leave as an exercise to the reader the plot of $\mathcal{M}(\beta)$ for various values of m and its comparison with Fig. A3.5.

Practical examples of elastoplasticity problems can be found in Calladine (1985), Hill (1950), Horne (1971), Johnson and Mellor (1962, 1973), Kachanov (1974), Mendelson (1968), Nádai (1950), Prager (1959), Sedov (1975), Zyczkowski (1981) for metallic structures, Baqué *et al.* (1973) in metal-forming processes, and Chen and Baladi (1985) and Halphen and Salençon (1987) in soil mechanics.

Appendix 4

Analytic computation of stress-intensity factors

The object of the appendix Here, a simple example of computation of stress-intensity factors by analytical means is given for crack modes I, II and III in plane problems. Further indications are also given concerning nonanalytic (computational, experimental) methods for evaluating these quantities.

A4.1 Plane problems in isotropic linear elasticity

To simplify the matter we consider only the so-called *plane* problems, defined by either one of the following conditions ($x_1 = x$, $x_2 = y$; $x_3 = z$ orthogonal to the 'plane'):

(a) *for the stress tensor* σ_{ij}

$$\left.\begin{array}{c} \sigma_{11} = \tilde{\sigma}_{11}(x, y), \qquad \sigma_{22} = \tilde{\sigma}_{22}(x, y), \qquad \sigma_{12} = \tilde{\sigma}_{12}(x, y), \\ \sigma_{33} = \tilde{\sigma}_{33}(x, y), \qquad \sigma_{13} = \sigma_{23} = 0, \end{array}\right\} \tag{A4.1}$$

(b) *for the small-strain tensor* ε_{ij}

$$\left.\begin{array}{c} \varepsilon_{11} = \tilde{\varepsilon}_{11}(x, y), \qquad \varepsilon_{22} = \tilde{\varepsilon}_{22}(x, y), \qquad \varepsilon_{13} = \tilde{\varepsilon}_{13}(x, y), \\ \varepsilon_{33} = \tilde{\varepsilon}_{33}(x, y), \qquad \varepsilon_{13} = \varepsilon_{23} = 0. \end{array}\right\} \tag{A4.2}$$

By using the compatibility condition (2.6), i.e.,

$$\varepsilon_{ij,kl} + \varepsilon_{kl,ij} = \varepsilon_{ik,jl} + \varepsilon_{jl,kl}, \tag{A4.3}$$

it is shown (exercise) that the displacement field is necessarily of the following type for a *plane problem* (this holds true completely independently of any constitutive equation):

$$\left.\begin{array}{l} u_1 = -A\dfrac{z^2}{2} + \omega_1(x, y), \\[2ex] u_2 = -B\dfrac{z^2}{2} + \omega_2(x, y), \\[2ex] u_3 = (Ax + By + C)z, \end{array}\right\} \tag{A4.4}$$

where A, B and C are constants of integration.

We now suppose that Hooke's law

$$\sigma_{ij} = \lambda\varepsilon_{kk}\delta_{ij} + 2\mu\varepsilon_{ij}, \qquad \varepsilon_{ij} = \tfrac{1}{2}(u_{i,j} + u_{j,i}) \tag{A4.5}$$

applies. Inverting the latter and using the compatibility condition (A4.3) we are led to the following condition *on stresses*:

$$\frac{\partial^2\sigma_{11}}{\partial y^2} + \frac{\partial^2\sigma_{22}}{\partial x^2} = 2\frac{\partial^2\sigma_{12}}{\partial x\partial y} + \nu\Delta\sigma, \qquad \sigma \equiv \sigma_{11} + \sigma_{22}, \tag{A4.6}$$

where ν is Poisson's ratio. Eqn (A4.6) plays the part of Beltrami's equations.

Plane-strain problems By *definition* we set

$$\varepsilon_{33} = 0, \qquad A = B = C = 0, \tag{A4.7}$$

from which there follow the fields

$$\left.\begin{aligned} &\sigma_{11} = \lambda(\varepsilon_{11} + \varepsilon_{22}) + 2\mu\varepsilon_{11}, \\ &\sigma_{22} = \lambda(\varepsilon_{11} + \varepsilon_{22}) + 2\mu\varepsilon_{22}, \\ &\sigma_{12} = 2\mu\varepsilon_{12}, \\ &\sigma_{33} = \lambda(\varepsilon_{11} + \varepsilon_{22}), \qquad \sigma_{23} = \sigma_{13} = 0, \\ &u_1 = \omega_1(x, y), \qquad u_2 = \omega_2(x, y), \qquad u_3 = 0. \end{aligned}\right\} \tag{A4.8}$$

Plane-stress problems By definition we set

$$\sigma_{33} = 0, \tag{A4.9}$$

and thus eqn (A4.5) applied to the case $i = 3, j = 3$ yields

$$\varepsilon_{33} = -\frac{\lambda}{\lambda + 2\mu}(\varepsilon_{11} + \varepsilon_{22}). \tag{A4.10}$$

The following fields are then deduced at once:

$$\left.\begin{aligned} &\sigma_{11} = \lambda^*(\varepsilon_{11} + \varepsilon_{22}) + 2\mu\varepsilon_{11}, \\ &\sigma_{22} = \lambda^*(\varepsilon_{11} + \varepsilon_{22}) + 2\mu\varepsilon_{22}, \\ &\sigma_{12} = 2\mu\varepsilon_{12}, \qquad \sigma_{13} = \sigma_{23} = 0, \end{aligned}\right\} \tag{A4.11}$$

where

$$\lambda^* = \frac{2\lambda\mu}{\lambda + 2\mu}.$$

In this case the problem, in general, remains three-dimensional in so far as displacements are concerned.

Airy function In the absence of inertial terms (quasi-statics) and body forces, the equilibrium equation div $\boldsymbol{\sigma} = \mathbf{0}$ or $\sigma_{ij,j} = 0$ is identically fulfilled with

$$\sigma_{11} = \frac{\partial^2 U}{\partial y^2}, \qquad \sigma_{22} = \frac{\partial^2 U}{\partial x^2}, \qquad \sigma_{12} = -\frac{\partial^2 U}{\partial x \partial y}, \tag{A4.12}$$

and U satisfying the equation

$$\Delta\Delta U = 0, \tag{A4.13}$$

where Δ is the Laplacian operator in the (x, y) plane. According to (A4.13), the function U, called the Airy function, is *biharmonic*.

In 1898, the mathematician E. Goursat showed that any biharmonic function could be expressed in terms of *analytic functions* (see, e.g., Cartan, 1961) of the complex variable $z = x + iy$. Indeed, letting $\bar{z} = x - iy$ denote the conjugate of z, we can rewrite eqn (A4.13) in the form

$$16\frac{\partial^4 U}{\partial z^2 \partial \bar{z}^2} = 0, \tag{A4.13)'$$

whence a possible expression for $U(z, \bar{z})$ is

$$U(z, \bar{z}) = \bar{z}\varphi(z) + z\bar{\varphi}(z) + \chi(z) + \bar{\chi}(z), \tag{A4.14}$$

where both φ and χ are analytic. This, in plane elasticity, leads to the complex representation introduced by G.V. Kolossov in 1909 and developed to a large extent by N.I. Muskhelishvili (1953). In this representation stresses and displacements are represented by

$$\left.\begin{aligned} \sigma_{11} + \sigma_{22} &= 2[\Phi(z) + \overline{\Phi(z)}] = 4\,\mathrm{Re}(\Phi(z)), \\ \sigma_{11} - \sigma_{22} + 2i\sigma_{12} &= 2[\bar{z}\Phi'(z) + \Psi(z)], \\ 2\mu(u_x + iu_y) &= \kappa\varphi(z) - \overline{z\varphi'(z)} - \overline{\psi(z)}, \end{aligned}\right\} \tag{A4.15}$$

with the conditions

$$\psi(z) = \chi'(z), \qquad \Phi(z) = \varphi'(z), \qquad \Psi(z) = \psi'(z) \tag{A4.16}$$

and the definitions

$$\kappa = \left\{\begin{aligned} &\frac{\lambda + 3\mu}{\lambda + \mu} = 3 - 4\nu \quad \text{(plane strains)}, \\ &\frac{\lambda^* + 3\mu}{\lambda^* + \mu} = \frac{3 - \nu}{1 + \nu} \quad \text{(plane stresses)}. \end{aligned}\right\} \tag{A4.17}$$

Boundary conditions on the boundary in $\mathbb{R}^2$ of the elastic region must also be prescribed (see Sedov, 1975, pp. 489–503 for greater detail).

A4.2 Stress-intensity factor at the crack tip

In a celebrated work Westergaard (1939) has applied the above-given representation to crack problems. In particular, he considered the following special case for the functions Φ and Ψ that appear in (A4.15):

$$\Phi(z) = \tfrac{1}{2}Z_1(z), \qquad \Psi(z) = -\tfrac{1}{2}zZ_1'(z), \tag{A4.18}$$

where $Z_1(z)$ is now called *Westergaard's function* (for mode I). There follows from $(\text{A4.15})_{1-2}$ that

$$\left.\begin{aligned} \sigma_{11} + \sigma_{22} &= 2\,\mathscr{R}e(Z_1), \\ \sigma_{22} - \sigma_{11} &= 2y\,\mathscr{I}m(Z_1'), \\ \sigma_{12} &= -y\,\mathscr{R}e(Z_1'), \end{aligned}\right\} \tag{A4.19}$$

where $\mathscr{R}e$ and $\mathscr{I}m$ denote the real and imaginary parts. As a consequence the solutions enjoy the following property at $y = 0$:

$$\sigma_{12} = 0, \qquad \sigma_{11} = \sigma_{22}. \tag{A4.20}$$

For plane strains we obtain the following displacement representation:

$$\left.\begin{aligned} 2\mu u_x &= (1 - 2\nu)\,\mathscr{R}e(Z_1^0) - y\,\mathscr{I}m(Z_1), \\ 2\mu u_y &= 2(1 - \nu)\,\mathscr{I}m(Z_1^0) - y\,\mathscr{R}e(Z_1), \end{aligned}\right\} \tag{A4.21}$$

when Z^0, is a function defined by the condition $Z_1 = \mathrm{d}Z_1^0/\mathrm{d}z$. The problem now is reduced to finding a representation of $Z_1(z)$.

We consider the case of a slit (crack) of length $2a$ in an infinite (x, y) plane (Fig. A4.1). This crack is subjected to the action of a normal *symmetric*

Fig. A4.1. Symmetrically charged crack

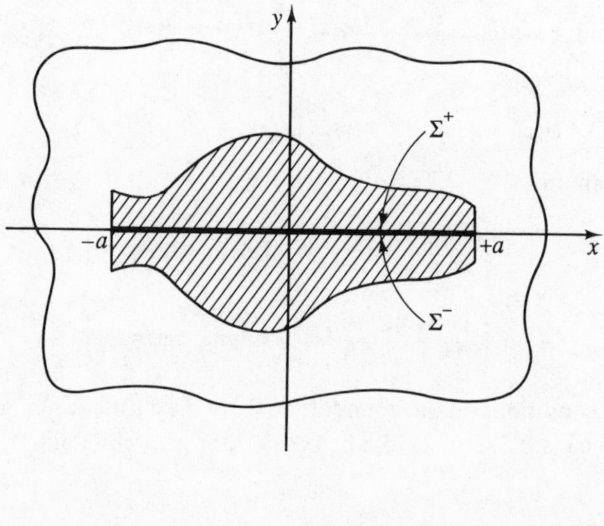

load (so that (A4.10) is satisfied). There remain the following boundary conditions on the lips of the crack:

$$\left.\begin{aligned}\sigma_{22}^{+} &= \sigma_{22}^{-} = -g(x),\\ \sigma_{12}^{+} &= \sigma_{12}^{-} = 0,\end{aligned}\right\} |x| \leqslant a,\ y = 0. \tag{A4.22}$$

Taking account of (A4.19) we find that (A4.22) yields

$$\mathscr{R}e(Z_1) = -g(x), \qquad |x| \leqslant a, \qquad y = 0. \tag{A4.23}$$

It is enough to find a function $Z_1(z)$ that (i) is sufficiently regular outside the segment $\{|x| \leqslant a, y = 0\}$, (ii) vanishes at infinity, and (iii) has a real part that satisfies (A4.23). Let us assume that the displacement defined by (A4.21) vanishes at infinity. Then Z_1 must decrease at least like $1/z^2$ at infinity. It is thus shown (Sedov, 1934) that

$$Z_1(z) = \frac{1}{\pi\sqrt{(z^2 - a^2)}}\int_{-a}^{+a} \frac{g(\xi)\sqrt{(a^2 - \xi^2)}}{z - \xi}\,d\xi, \tag{A4.24}$$

an expression that is also found in two-dimensional problems of hydrodynamics. It remains to consider special cases of loading $g(x)$.

In the neighbourhood of the right end of the crack ($y = 0, x = +a$) we set $z - a = re^{i\theta}$. For small $r = |z - a|$ we deduce from (A4.24) the following asymptotic expression:

$$Z_1 = \frac{K_I}{\sqrt{[2\pi(z - a)]}}, \qquad K_I = \frac{1}{\sqrt{(\pi a)}}\int_{-a}^{+a} g(\xi)\sqrt{\frac{a + \xi}{a - \xi}}\,d\xi. \tag{A4.25}$$

In polar coordinates the last expression yields directly Westergaard's solution for *mode I cracks*:

$$\left.\begin{aligned}
\sigma_{11} &= \frac{K_I}{\sqrt{(2\pi r)}}\cos\frac{\theta}{2}\left(1 - \sin\frac{\theta}{2}\sin\frac{3\theta}{2}\right) + 0\left(\sqrt{\frac{r}{a}}\right),\\
\sigma_{22} &= \frac{K_I}{\sqrt{(2\pi r)}}\cos\frac{\theta}{2}\left(1 + \sin\frac{\theta}{2}\sin\frac{3\theta}{2}\right) + 0\left(\sqrt{\frac{r}{a}}\right),\\
\sigma_{33} &= \nu(\sigma_{11} + \sigma_{22}), \qquad \sigma_{23} = \sigma_{13} = 0,\\
\sigma_{12} &= \frac{K_I}{\sqrt{(2\pi r)}}\sin\frac{\theta}{2}\cos\frac{\theta}{2}\cos\frac{3\theta}{2} + 0\left(\sqrt{\frac{r}{a}}\right),\\
u_x &= \frac{K_I}{\mu}\sqrt{\frac{r}{2\pi}}\cdot\cos\frac{\theta}{2}\left(1 - 2\nu + \sin^2\frac{\theta}{2}\right),\\
u_y &= \frac{K_I}{\mu}\sqrt{\frac{r}{2\pi}}\cdot\sin\frac{\theta}{2}\left[2(1 - \nu) - \cos^2\frac{\theta}{2}\right].
\end{aligned}\right\} \tag{A4.26}$$

For an infinite (x, y) plane, weakened by a rectilinear slit $\{|x| \leqslant a, y = 0\}$ and submitted to a traction $p_0 = \text{const.}$ at infinity, we have $\sigma_{22} = -p_0$ so that, using the technique of residues, from (A4.24) we find that

$$Z_1 = \frac{p_0}{\pi\sqrt{(z^2 - a^2)}} \int_{-a}^{+a} \frac{\sqrt{(a^2 - \xi^2)}}{z - \xi} \mathrm{d}\xi = \frac{p_0 z}{\sqrt{(z^2 - a^2)}} - p_0, \quad \text{(A4.27)}$$

which is the sum of two solutions of which only the first one corresponds to mode I crack, i.e.,

$$Z_1(z) = \frac{p_0 z}{\sqrt{(z^2 - a^2)}}, \quad \text{(A4.28)}$$

the coefficient K_{I} corresponding to (A4.25) can be computed. One obtains

$$K_{\mathrm{I}} = \frac{p_0}{\sqrt{(\pi a)}} \int_{-a}^{+a} \sqrt{\frac{a + \xi}{a - \xi}} \mathrm{d}\xi = \frac{p_0}{\sqrt{(\pi a)}} \int_{-a}^{+a} \frac{a + \xi}{\sqrt{(a^2 - \xi^2)}} \mathrm{d}\xi = p_0\sqrt{(\pi a)}. \quad \text{(A4.29)}$$

For *mode II cracks*, Westergaard introduces

$$\Phi(z) = -\tfrac{1}{2}\mathrm{i}Z_2(z), \qquad \Psi(z) = \tfrac{1}{2}\mathrm{i}zZ_2'(z) + \mathrm{i}Z_2(z), \quad \text{(A4.30)}$$

where $Z_2(z)$ is the unknown function which will be regular outside the crack and vanishes at infinity. One has then

$$\left.\begin{aligned} \sigma_{11} &= 2\,\mathcal{Im}(Z_2) + y\,\mathcal{Re}(Z_2'), \\ \sigma_{22} &= -y\,\mathcal{Re}(Z_2'), \\ \sigma_{12} &= \mathcal{Re}(Z_2) - y\,\mathcal{Im}(Z_2'), \end{aligned}\right\} \quad \text{(A4.31)}$$

and

$$\left.\begin{aligned} u_x &= \frac{2(1 - \nu)}{\mu}\mathcal{Im}(Z_2^0) + \frac{y}{2\mu}\mathcal{Re}(Z_2), \\ u_y &= -\frac{(1 - \nu)}{2\mu}\mathcal{Re}(Z_2^0) - \frac{y}{2\mu}\mathcal{Im}(Z_2), \end{aligned}\right\} \quad \text{(A4.32)}$$

where Z_2^0 is defined by $Z_2 = \mathrm{d}Z_2^0/\mathrm{d}z$.

For a crack submitted to an *antisymmetric tangential* load (mode II), the boundary conditions on the crack are of the type

$$\sigma_{11} = 0, \quad \sigma_{12} = -h(x), \quad |x| \leqslant a, \quad y = 0, \quad \text{(A4.33)}$$

that is

$$\mathcal{Re}(Z_2) = -h(x), \qquad |x| \leqslant a, \qquad y = 0. \quad \text{(A4.34)}$$

As $Z_2(z)$ is of order $1/z^2$ at infinity, $Z_2(z)$ is given by the same formula as (A4.24), but with $h(x)$ replacing $g(x)$.

In the neighbourhood of the right-end crack tip we shall get the following behaviour:

$$\left.\begin{aligned}
\sigma_{11} &= -\frac{K_{\mathrm{II}}}{\sqrt{(2\pi r)}}\sin\frac{\theta}{2}\left(2+\cos\frac{\theta}{2}\cos\frac{3\theta}{2}\right),\\
\sigma_{22} &= \frac{K_{\mathrm{II}}}{\sqrt{(2\pi r)}}\cos\frac{\theta}{2}\sin\frac{\theta}{2}\cos\frac{3\theta}{2},\\
\sigma_{12} &= \frac{K_{\mathrm{II}}}{\sqrt{(2\pi r)}}\cos\frac{\theta}{2}\left(1-\sin\frac{\theta}{2}\sin\frac{3\theta}{2}\right),\\
u_x &= \frac{K_{\mathrm{II}}}{\mu}\sqrt{\frac{r}{2\pi}}\cdot\sin\frac{\theta}{2}\left[2(1-\nu)+\cos^2\frac{\theta}{2}\right],\\
u_y &= \frac{K_{\mathrm{II}}}{\mu}\sqrt{\frac{r}{2\pi}}\cdot\cos\frac{\theta}{2}\left(1-2\nu+\sin^2\frac{\theta}{2}\right),
\end{aligned}\right\} \tag{A4.35}$$

with

$$K_{\mathrm{II}} = \frac{1}{\sqrt{(\pi a)}}\int_{-a}^{+a} h(\xi)\sqrt{\frac{a+\xi}{a-\xi}}\,\mathrm{d}\xi \tag{A4.36}$$

and

$$Z_2(z) = \frac{K_{\mathrm{II}}}{\sqrt{[2\pi(z-a)]}}. \tag{A4.37}$$

In the case of *mode III* we must consider $u_3(x, y) = w$, and this yields the condition $\Delta w = 0$ which can be satisfied by a displacement w of the form $w \approx \operatorname{Im} Z_3$. It is not difficult to show (see Chapter 7) that

$$\left.\begin{aligned}
&\sigma_{13} = -\frac{K_{\mathrm{III}}}{\sqrt{(2\pi r)}}\sin\frac{\theta}{2}, \qquad \sigma_{23} = \frac{K_{\mathrm{III}}}{\sqrt{(2\pi r)}}\cos\frac{\theta}{2},\\
&\sigma_{11} = \sigma_{22} = \sigma_{33} = \sigma_{12} = 0,\\
&w = u_3 = \frac{K_{\mathrm{III}}}{G}\sqrt{\frac{2r}{\pi}}\cdot\sin\frac{\theta}{2}, \qquad u_1 = u_2 = 0,\\
&K_{\mathrm{III}} = \lim_{|\xi|\to\infty}\left(\sqrt{(2\pi\xi)}\cdot Z_3\right),
\end{aligned}\right\} \tag{A4.38}$$

where G is the shear modulus.

We have then the general expressions for the three elementary modes of cracking and we must evaluate K_{I}, K_{II}, and K_{III} depending on the mode of loading and the geometry of the elastic body. As a simple example we may consider the uniaxial traction p_0 of a plane containing a rectilinear crack, the traction being active at an angle θ_0 from the x-axis (Fig. A4.2). By combining the stress fields acting on the crack, one obtains

$$\left.\begin{aligned} K_\mathrm{I} &= p_0\sqrt{(\pi a)}\cdot\sin^2\theta_0, \\ K_\mathrm{II} &= p_0\sqrt{(\pi a)}\cdot\sin\theta_0\cos\theta_0. \end{aligned}\right\} \tag{A4.39}$$

The reader will find in specialized monographs (e.g., Parton and Morozov, 1978, Labbens, 1980, Bui, 1977) the elaboration of some other examples, such as the three-dimensional case of a circular cylinder with a circumferential slit (Fig. A4.3) for which it is found that (see Parton and Morozov, 1978, pp. 121–36)

$$K_\mathrm{III} \approx \frac{3\mathscr{M}}{4\sqrt{\pi}}a^{-5/2}, \tag{A4.40}$$

where $\mathscr{M}$ is the applied torque in torsion and a is the internal radius of the slit, while

Fig. A4.2. Oblique uniaxial load

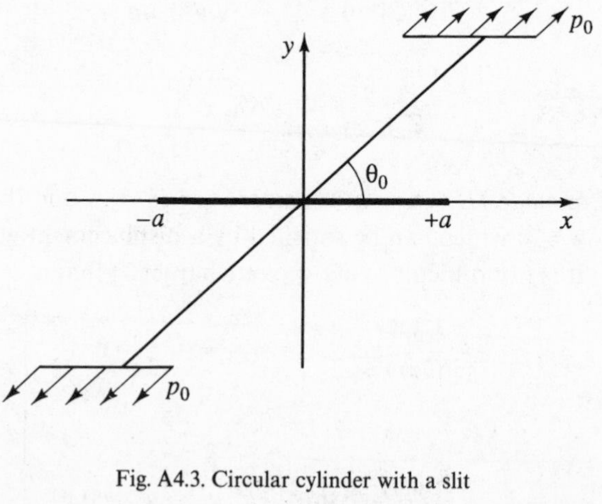

Fig. A4.3. Circular cylinder with a slit

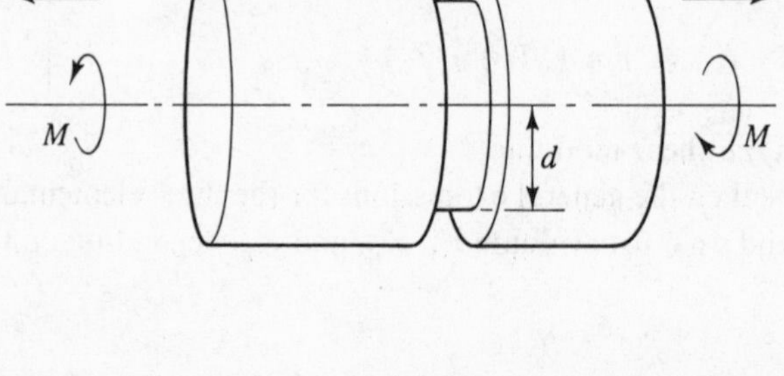

$$K_{\mathrm{I}} \approx \frac{q\sqrt{(\pi R)}}{2\alpha^{3/2}}, \tag{A4.41}$$

for a traction $P = q\pi R^2$ applied axially to the cylinder, α being a distance measured from the slit.

In practice, it is very seldom that stress-intensity factors can be computed analytically. Most often one needs the implementation of a numerical calculation.

A4.3 Remark on numerical computations of stress-intensity factors

The notion of stress-intensity factor is strongly related to that of singularity of fields. In using a numerical approach such as the finite-element method, one must therefore consider a rather thin grid wherever the singularity is expected (see, e.g., Fig. A4.4(a)). The reader may consult the book by Parton and Morozov (1978), where computation routines are provided and that of Bui (1977) who gives FEM programs for various test samples. At this point, we should emphasize that the traditional FEM is, in general, badly adapted to the analysis of singularities. In order to reach a better numerical result, it may be quite preferable to have recourse to special elements that are suited to the description of singular fields which, in effect, are asymptotic expressions of stresses that depend, for instance, on two parameters K_{I} and K_{II}, which are to be determined. Thus a classical grid is devised outside a circular nucleus of small extent (hatched in Fig. A4.4(b)). Then K_{I} and K_{II} are two additional parameters which complete the set of usual unknowns, at the N knots of the grid. The potential energy to be minimized, as in all elasticity problems, is made of two parts; the first obtained by

Fig. A4.4. Examples of FEM grids. (a) Classical grid (after Bui, 1977). (b) Grid with a nucleus.

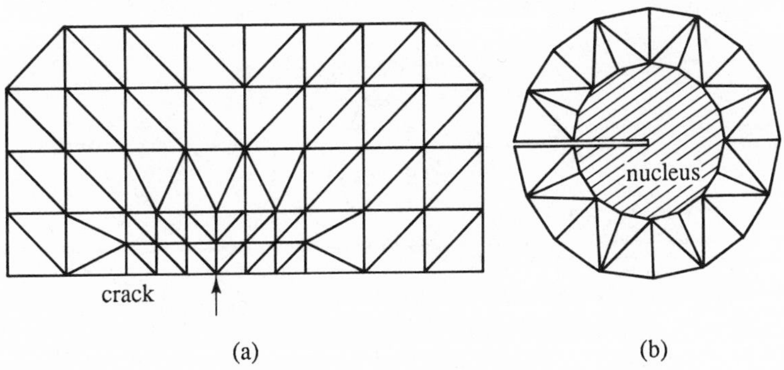

considering the solid body with a circular cavity on which surface tractions are prescribed, the expressions for which were given in Section A4.2 in asymptotic form and depend on K_{I} and K_{II}, e.g.,

$$\sigma_{\theta r}(K_{\mathrm{I}}, K_{\mathrm{II}}), \qquad \sigma_{rr}(K_{\mathrm{I}}, K_{\mathrm{II}}); \tag{A4.42}$$

and a second contribution that corresponds to the circular nucleus and is exactly computed. In addition one has minimization conditions with respect to the 'variables' K_{I} and K_{II}.

To conclude, it is most important to take notice of the recent approaches (related to the behaviour of composite materials) of Bazant and Estenssoro (1979), Babuska and Miller (1984), Ting (1986) and the monograph of Leguillon and Sanchez-Palencia (1987). The method developed by the last two authors is of particular interest. The main ideas of this method can be illustrated on a typical (elliptic) problem:

$$\left.\begin{aligned} -\Delta U &= f && \text{in } \Omega, \\ \frac{\partial U}{\partial n} &= 0 && \text{on } \partial\Omega_T, \\ U &= 0 && \text{on } \partial\Omega_u, \end{aligned}\right\} \tag{A4.43}$$

when Ω represents an angular sector ω which may be either acute or obtuse (a crack corresponds to $\omega = 2\pi$) – see Fig. A4.5. The variational formulation of (A4.43) is naturally given by (e.g., Duvaut and Lions, 1972):

$$\left.\begin{aligned} &\int_\Omega \nabla U \cdot \nabla V \,\mathrm{d}\Omega = \int_\Omega fV \,\mathrm{d}\Omega, \quad U \in \mathscr{V}, \\ &\forall V \in \mathscr{V}, \quad \mathscr{V} = \{V \in H^1(\Omega) \mid V|_{\partial\Omega_u} = 0\}. \end{aligned}\right\} \tag{A4.44}$$

Fig. A4.5. Elliptic problem concerning a region presenting an angular sector. (a) Acute angle; $\partial\Omega_u = \Sigma_1 \cup \Sigma_2$. (b) Crack.

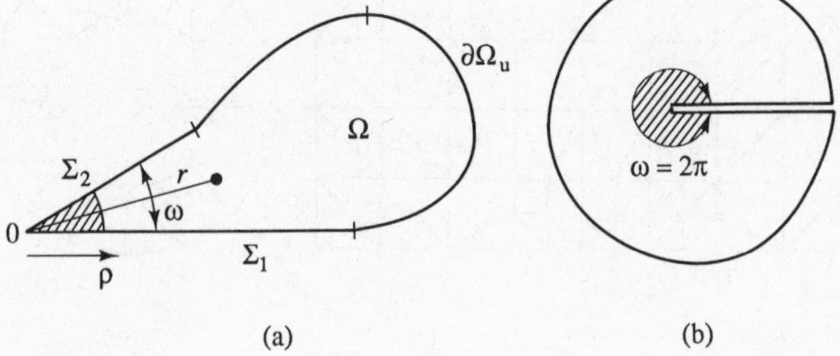

In the neighbourhood of the origin O where the boundary Ω represents two straight lines and Neumann's condition applies, the field U is such that

$$U = br^{\alpha}w(\theta) + U^{\text{regular}}, \tag{A4.45}$$

where $w(\theta) = (2/\omega)^{1/2} \cos \alpha\theta$, $\alpha = \pi/\omega$ is one of the normalized eigenfunctions of the problem (A4.43). We consider a function $V^{-\alpha}(x, y)$ which equals $r^{-\alpha}w(\theta)$ for small rs and such that $V^{-\alpha} \in H^1(\Omega)$, except for small rs; the latter satisfies the problem

$$\left.\begin{aligned} -\Delta V^{-\alpha} &= F \quad \text{in}\,\Omega, \qquad F = 0 \quad \text{for small}\, rs, \\ \frac{\partial V^{-\alpha}}{\partial n} &= 0 \qquad \text{on}\,\partial\Omega. \end{aligned}\right\} \tag{A4.46}$$

Applying Green's formula to the pair $(U, V^{-\alpha})$ for the region $\Omega_{\text{p}} = \{(r, \theta) \in \Omega, r > \rho\}$ and taking the limit as ρ goes to zero, we obtain

$$b = \frac{1}{2\alpha} \int_{\Omega} (FU - fV^{-\alpha})\, \mathrm{d}\Omega, \tag{A4.47}$$

which is none other than the *stress-intensity factor*. For examples of numerical computations in the elastic range, see Leguillon and Sanchez-Palencia (1987, pp. 121–8). Still other methods can be used to evaluate the stress-intensity factor. One uses the *Rice–Eshelby integral* (see eqns (7.41)) as intermediate. Experimental methods can also be implemented, such as *photoelasticity*, a method using the *analogies with electric conduction properties* (see, e.g., Hui and Ruina, 1985; Mukherjee *et al.*, 1982), or else the method of infrared thermography examined in Chapter 12 (see end of Section 12.3, and Sih, 1981).

Further reading

A Continuum mechanics and thermodynamics

Chadwick P., (1976), *Continuum Mechanics*, George Allen and Unwin, London.

Eringen A.C., (1967), *Mechanics of Continua*, John Wiley and Sons, New York (Second enlarged edition, Krieger, New York, 1980).

Germain P., (1973), *Cours de mécanique des milieux continus*, Masson, Paris.

Maugin G.A., (1988), *Continuum Mechanics of Electromagnetic Solids*, North-Holland, Amsterdam (Chapter 2 is a necessary and sufficient introduction to the thermomechanics of deformable solids).

Sedov L.I., (1975), *Mechanics of Continua*, Vol. 2, Mir Publishers, Moscow.

Truesdell C.A., (1969), *Rational Thermodynamics*, McGraw-Hill, New York (Second enlarged edition, Springer-Verlag, Berlin, 1986).

B The *classics* of plasticity theory

Hill R., (1950), *The Mathematical Theory of Plasticity*, Clarendon Press, Oxford.

Ilyushin A.A., (1948), *Plasticity* (in Russian), (Gostekhizdat, Moscow; French Translation, *Plasticité*, Eyrolles, Paris, 1956.

Kachanov L.M., (1974), *Fundamentals of the Theory of Plasticity*, Mir Publishers, Moscow.

Prager W., (1959), *An Introduction to Plasticity*, Addison-Wesley, Reading, Mass.

C Modern treatment of plasticity

Lubliner J., (1990), *Plasticity Theory* (Macmillan, New York).

Mandel J., (1971), *Plasticité classique et viscoplasticité*, (Lecture Notes, CISM, Udine, Italy), Springer-Verlag, Vienna.

Nguyen Quoc Son, (1973), Doctoral Thesis, Université Pierre-et-Marie Curie, Paris.

D Various mechanical behaviours

Lemaître J., and Chaboche J.L., (1990), *Mechanics of Solid Materials* (translation from the French), Cambridge University Press, Cambridge, UK.

Mandel J., (1978), *Propriétés mécaniques des matériaux*, Editions Eyrolles, Paris.

Reissner M., (1960), *Lectures on Theoretical Rheology*, North-Holland, Amsterdam.

Vyalov S.S., (1986), *Rheological Fundamentals of Soil Mechanics* (translation from the Russian), Mir Publishers, Moscow.

E Fracture

Atkins A.G., and May Y.N., (1985), *Elastic and Plastic Fracture*, Ellis Horwood, Chichester, UK.

Bui H.B., (1977), *Mécanique de la rupture fragile*, Masson, Paris.

Cherepanov G.P., (1979), *Mechanics of Brittle Fracture* (translation from the Russian), McGraw-Hill, New York.

Freund L.B., (1990), *Dynamic Fracture Mechanics*, Cambridge University Press, Cambridge, UK.

F Mathematical properties

Brezis H., (1973), *Opérateurs maximaux monotones et semi-groupes non linéaires dans les espaces de Hilbert*, North-Holland, Amsterdam.

Duvaut G., and Lions J.L., (1979), *Inequations in Physics and Mechanics* (translation from the French), Springer-Verlag, Berlin, Heidelberg, New York.

Ekeland I., and Temam R., (1976), *Convex Analysis and Variational Problems* (translation from the French), North-Holland, Amsterdam.

Sewell M.J., (1987), *Maximum and Minimum Principles*, Cambridge University Press, Cambridge, UK.

G Historical perspective

Timoshenko S.P., (1953), *History of the Strength of Materials* (with a brief account of the history of theory of elasticity and theory of structures), McGraw-Hill, New York (Reprint by Dover, New York, 1983).

Bibliography

Abe H. (1975), A systematic formulation of nonlinear thin elastic–plastic shell theories, *Int. J. Engng. Sci.*, **13**, 1003–14.

Aker B. and Heyman J. (1969), *Plastic Design of Frames, I, Fundamentals*, Cambridge University Press, Cambridge, UK.

Argon A.S. (Editor, 1975), *Constitutive Equations in Plasticity*, M.I.T. Press, Cambridge, Mass.

Asaro R.J. (1983), Crystal plasticity, *ASME Trans. J. Appl. Mech.*, **50**, 921–34.

Atkins A.G. and May Y.M. (1985), *Elastic and Plastic Fracture*, Ellis Horwood, Chichester, UK.

Babuska I. and Miller A. (1984), The post processing approach to the finite element method, Part 2: The calculation of stress intensity factors, *Int. J. Num. Math. Engng.*, **20**, 1111–29

Ball J.M. (1977), Convexity conditions and existence theorems in nonlinear elasticity, *Arch. Rat. Mech. Anal.*, **63**, 337–403

Bamberger Y. (1981), *Mécanique de l'ingénieur*, Vol. 2, Hermann, Paris.

Baqué P., Felder E., Hyafil J. and D'Escatha Y. (1973), *Mise en forme des métaux: Calculs par la plasticité.* Dunod, Paris.

Barenblatt G.I. (1972), The mathematical theory of equilibrium cracks in brittle fracture, in *Advances in Applied Mechanics*, Vol. 7, pp. 55–129, ed. G.S. Sih, Academic Press, New York.

Barré de Saint Venant J. (1871), Mémoire sur l'établissement des équations différentielles des mouvements intérieurs opérés dans les corps solides ductiles ..., *J. Math. Pures et Appl.*, **16**, 308–16.

Barthélémy B. (1980), *Notions pratiques de la mécanique de la rupture*, Eyrolles, Paris.

Bauschinger J. (1886), *Mitt. Mech. Lab. Munich* (Yearly report) (cited by Timoshenko, 1953, p. 280).

Bazant A.P. and Estenssoro L.F. (1979), Surface singularity and crack propagation, *Int. J. Solids Structures*, **15**, 405–26.

Bell J.F. (1968), *The Physics of Large Deformation of Crystalline Solids*, Springer Tracts in Natural Philosophy, Vol. 14, Berlin, Heidelberg, New York.

Bell J.F. (1973), The experimental foundations of solid mechanics, in *Handbuch der Physik*, Bd. VIa/1, ed. C.A. Truesdell, pp. 1–799, Springer-Verlag, Berlin, Heidelberg, New York.

Berveiller M. and Zaoui A. (1978), An extension of the selfconsistent scheme to plastically-flowing polycrystals, *J. Mech. Phys. Solids*, **26**, 325–44.

Bingham E.C. (1922), *Fluidity and Plasticity*, McGraw-Hill, New York.

Boehler J.P. (Editor, 1985), *Plastic Behaviour of Anisotropic Solids*, Editions du CNRS, Paris.

Borkowski A. (1988), *Analysis of Skeletal Structural Systems in the Elastic and Elastic–Plastic Range*, (Translated from the Polish), P.W.N., Warsaw, and Elsevier, Amsterdam.

Bouc R. and Nayroles B. (1985), Méthodes et résultats en thermographie infrarouge, *J. Méch. Théor. Appl.*, **4**, 27–58.

Bretheau Th., Hervé E. and Zaoui A. (1991), On the influence of the phase connectivity on the yield point and the plastic flow of two-phase materials, in *Continuum Models and Discrete Systems*, Vol. 2, ed. G.A. Maugin, pp. 84–92, Longman, London.

Brézis H. (1973), *Opérateurs maximaux monotones et semi-groupes de contraction dans les espaces de Hilbert*, North-Holland, Amsterdam.

Bridgman P.W. (1943), *The Nature of Thermodynamics*, Harvard University Press, Cambridge, Mass.

Bridgman P.W. (1952), *Studies in Large Plastic Flow and Fracture*, Harvard University Press, Cambridge, Mass.

Buchdahl H.A. (1960), *The Concept of Classical Thermodynamics*, Cambridge University Press, Cambridge, UK.

Budiansky B. and Pearson C.E. (1956/57), The variational principle and Galerkin's procedure for nonlinear elasticity, *Quart. Appl. Math.*, **14**, 328–31.

Bufler H. (1988), Nonlinear elasticity and complementary variational principles, General Lecture at *27th Polish Solid Mechanics Conference*, Rytro, Poland, Sept. 1988.

Buhan P. de, and Salençon J. (1983), Determination of a macroscopic yield criterion for a multilayered material.

Bui H.D. (1970), Evolution de la frontière du domaine élastique des métaux avec écrouissage plastique et viscoplastique des monocristaux et comportement élastoplastique d'un agrégat de cristaux cubiques, *Mém. Artillerie Franç., Sci. Tech. Armement*, **1**, 141–65.

Bui H.D. (1977), *Mécanique de la rupture fragile*, Masson, Paris.

Bui H.D. (1981), Discussion of Session 6, in *Three-dimensional Constitutive Relations and Ductile Fracture*, ed. S. Nemat-Nasser, pp 357–62, North-Holland, Amsterdam.

Bui H.D. (1982), Rupture et endommagement: Applications aux matériaux fragiles, in *Mécanique de la rupture*, Séminaire C.I.S.C., Saint Rémy les Chevreuses, pp. 35–48, CILF, Paris.

Bui H.D., Ehrlacher A. and Nguyen Quoc Son (1979), Propagation de fissure en thermoélasticité dynamique, *C.R. Acad. Sci. Paris*, **289B**, 211–14.

Bui H.D., Ehrlacher A. and Nguyen Quoc Son (1981), Etude expérimentale de la dissipation dans la propagation des fissures par thermographie infrarouge, *C.R. Acad. Sci. Paris*, **293**, 1015–18.

Bui H.D., Zaoui A. and Zarka J. (1973), Sur le comportement élastoplastique et viscoplastique des monocristaux et polycristaux métalliques de structure cubique à faces centrées, in *Foundations of Plasticity*, ed. A. Sawczuk., pp. 51–75, Noordhoff, Leyden.

Cadek U. (1988), *Creep in Metallic Materials*, Academia, Prague.

Calcotte L.R. (1968), *Introduction to Continuum Mechanics*, van Nostrand, Princeton, NJ.

Calladine C.R. (1985), *Plasticity for Engineers*, Ellis Horwood, Chichester, UK.

Carathéodory C. (1909) Untersuchungen über die Grundlagen der Thermodynamik, *Math. Ann.*, **67**, 355–86.

Carathéodory C. (1925), *Berl. Ber.*, pp. 39–47.

Cartan H. (1961), *Théorie élémentaire des fonctions analytiques d'une ou plusieurs variables complexes*, Hermann, Paris.

Casal P. (1978), Interpretation of the Rice integral in continuum mechanics, *Lett. Appl. Engng. Sci.*, **16**, 335–47.

Chaboche J.L. (1974), Une loi différentielle d'endommagement de fatigue avec accumulation non linéaire, *Rev. Française Méc.*, 50–1.

Chaboche J.-L. (1989a), Continuum damage mechanics–I: General Concepts, *Trans. A.S.M.E. J. Appl. Mech.*, **55**, 59–64.

Chaboche J.-L. (1989b), Continuum damage mechanics–II: Damage growth, crack initiation, and crack growth, *Trans. A.S.M.E. J. Appl. Mech.*, **55**, 65–72.

Chakrabarty J. (1987), *Theory of Plasticity*, McGraw-Hill, New York.

Chen W.R. and Baladi G.Y. (1985), *Soil Plasticity, Theory and Implementation*, Elsevier, Amsterdam.

Cherepanov G.P. (1967), Crack propagation in continuus media, *Prikl. Mat. Mekh.*, (Appl. Math. Mech., USSR), **31**, No. 3, 467–88.

Cherepanov, G.P. (1968), Cracks in solids, *Int. J. Solids Structures*, **4**, 811–31.

Choquet-Bruhat Y. (1968), *Géométrie différentielle et systèmes extérieurs*, Dunod, Paris.

Chrysochoos A. (1985), Bilan énergétique en élastoplasticité en grandes déformations, *J. Méc. Théor. Appl.*, **4**, 489–514.

Ciarlet P.G. (1988), *Mathematical Elasticity*, Vol. I: *Three-Dimensional Elasticity*, North-Holland, Amsterdam.

Cleja-Tigoiu S. and Soos E. (1990), Elastoviscoplastic models with relaxed configurations and internal state variables, *Appl. Mech. Rev.*, **43**, 131–51.

Clifton R.J. (1972), On the equivalence of $\mathbf{F}^e\,\mathbf{F}^p$ and $\mathbf{F}^p\,\mathbf{F}^e$, *ASME Trans. J. Appl. Mech.*, **39**, 287–9.

Cocks A.C.F. and Leckie F.A. (1987), Creep constitutive equations for damaged materials, in *Advances in Applied Mechanics*, ed. T.Y. Wu and J.W. Hutchinson, Vol. 25, pp. 183–238, Academic Press, New York.

Cohn M.Z. and Maier G. (Editors, 1979), *Engineering Plasticity by Mathematical Programming*, Pergamon Press, Oxford.

Cordebois J.P. (1983), Critères d'instabilité plastique et endommagement ductile en grandes déformations, *Thèse de Doctorat d'Etat*, Université Pierre-et-Marie Curie, Paris.

Cordebois J.R. and Sidoroff F. (1982), Damage induced elastic anisotropy, in *Mechanical Behaviour of Anisotropic Solids*, ed. J.P. Boehler, (Euromech Coll, 115, 1979), pp. 761–74, Editions du CNRS, Paris.

Cottrell A.H. (1953), *Dislocations and Plastic Flow of Crystals*, Clarendon Press, Oxford.

Critescu N. and Suliciu I. (1976), *Viscoplasticitate* (in Rumanian), Editura technica, Bucharest. [English translation, Viscoplasticity, Norton and Hoff. 1980]

Day Y. (1972), *The Thermodynamics of Simple Materials with Memory*, Springer-Verlag, Berlin, Heidelberg, New York.

Desbordes D. (1977), Contribution à la théorie et au calcul de l'élastoplasticité asymptotique, Thèse de Doctorat d'Etat ès Sci. Math., Université de Marseille, Marseille.

Drucker D.C. (1951), A more fundamental approach to plastic stress–strain relations, in *Proc. First US National Congress of Applied Mechanics*, pp. 487–91, ASME, New York.

Drucker D.C. (1988), Conventional and unconventional plastic response and representation, *Appl. Mech. Rev.*, **41**, 151–67.

Drucker D.C., Prager W. and Greenberg H.J. (1952), Extended limit design theorems for continuous media, *Quart. Appl. Math.*, **9**, 301–9.

Duhem, P. (1911), *Traité d'énergétique ou de thermodynamique générale*, Gauthier-Villars, Paris.

Duvaut G. and Lions J.L. (1972), *Les Inéquations en mécanique et en physique*, Dunod, Paris (English Translation: *Inequations in Physics and Mechanics*, Springer-Verlag, Berlin, 1979).

Dvorak G. and Rao M. (1976), Axisymmetric plasticity theory of fibrous composites, *Int. J. Engng. Sci.*, **14**, 361–73.

Edelen D.G.B. (1973), On the existence of symmetry relations and dissipation potentials, *Arch. Rat. Mech. Anal.*, **51**, 218–227.

Ekeland I. and Temam R. (1974), *Analyse convexe et problèmes variationnels*, Dunod, Paris (English Translation: *Convex Analysis and Variational Problems*, North-Holland, Amsterdam, 1976).

Epstein M. and Maugin G.A. (1990a), Sur le tenseur de moment matériel d'Eshelby en élasticité non linéaire, *C.R. Acad. Sci. Paris*, **II-310**, 675–8.

Epstein M. and Maugin G.A. (1990b), The energy–momentum tensor and material uniformity in finite elasticity, *Acta Mechanica*, **83**, 127–33.

Eringen A.C. (1967) *Mechanics of Continua*, John Wiley and Sons, New York (Second enlarged Edition: Krieger, New York, 1980).

Eringen A.C. and Maugin G.A. (1989), *Electrodynamics of Continua*, Vol. I, Springer-Verlag, New York.

Eshelby J.D. (1951), The force on an elastic singularity, *Phil. Trans. Roy. Soc. London*, **A244**, 87–112.

Francfort G., Leguillon D. and Suquet P. (1983), Homogénéisation des milieux viscoélastiques linéaires de Kelvin–Voigt, *C.R. Acad. Sci. Paris*, **I-206**, 287–90.

Freund L.B. (1990), *Dynamic Fracture Mechanics*, Cambridge University Press, Cambridge, UK.

Germain P. (1962), *Mécanique des milieux continus*, Masson, Paris.

Germain P. (1973), *Cours de mécanique des milieux continus*, Vol. 1, Masson, Paris.

Germain P. (1982), Sur certaines définitions liées à l'énergie en mécanique des solides, *Int. J. Engng. Sci.*, **20**, 245–60.

Germain P., Nguyen Quoc Son and Suquet P. (1983), Continuum thermodynamics, *ASME Trans. J. Appl. Mech.*, **105**, 1010–20.

Glowinski R. and Lanchon H., (1973), Torsion élastoplastique d'une barre cylindrique de section multiconnexe, *J. Mécanique*, **12**, 151–72.

Golebiewska-Herrmann A. (1981), On conservation laws of continuum mechanics, *Int. J. Solid Structures*, **17**, 1–9.

Grabacki J. (1989), On continuous descriptions of the damage, *Mech. Teor. Stosow.* (Poland), **27**, 271–91.

Green A.E. and Naghdi P.M. (1965), A general theory of an elastic–plastic continuum, *Arch. Rat. Mech. Anal.*, **18**, 251–81.

Greenberg H.J. (1949), On the variational principles of plasticity, *Grad. Div. Appl. Math. Brown Univ. Report*, A-11-54, March 1949.

Griffith A.A. (1920), The phenomena of rupture and flow in solids, *Phil. Trans. Roy. Soc. London*, **221**, 16–98.

Groot S.R. de, and Mazur P. (1962), *Non Equilibrium Thermodynamics*, North-Holland, Amsterdam.

Gurson A.L., (1977), Continuum theory of ductile rupture by void nucleation and growth, Part I: Yield criteria and flow rules for porous ductile media, *J. Eng. Mater. Technol.*, **99**, 2–15.

Gvozdev A.A (1938, Russian original): English Translation by R.M. Haythornthwaite: The determination of the value of the collapse load of statically indeterminate systems undergoing plastic deformation, *Intern. J. Mech. Sci.*, **1**, (1960), 322–55.

Haigh B.P. (1920), *Engineering*, **109**, 158 (not seen by Author).

Halphen B. (1975), Sur le champ des vitesses en thermoélasticité finie, *Int. J. Solids Structures*, **11**, 947–60.

Halphen B. and Nguyen Quoc Son (1975), Sur les matériaux standards généralisés, *J. Mécanique*, **14**, 39–63.

Halphen B. and Salençon J. (1987), *Elasto-plasticité*, Presses de l'ENPC, Paris.

Hashin Z. (1983), Analysis of composite materials: A survey, *A.S.M.E. Trans. J. Appl. Mech.*, **50**, 481–505.

Havner K.S. (1987), On the continuum mechanics of crystal slip, in *Continuum Models of Discrete Systems*, ed. A.J.M. Spencer, pp. 47–59, A.A. Balkema, Rotterdam.

Hellinger E. (1914), Die allgemein Ansätze der Mechanik der Kontinua, *Eng. Math. Wis.*, **4**, 602–94.

Hencky H. (1923), Über die einige statisch bestimmte Fälle des Gleichgewichts in plastischen Körpern, *Zeit. Angew. Math. Mech.*, **3**, 241.

Hencky H. (1924), Zur Theorie plastischer Deformationen und der hierdurch im Material herfogerufenen Nachspannungen, *Zeit. angew. Math und Mech.*, **4**, 323–34.

Heyman J. (1971), *Plastic Design of Frames, 2, Applications*, Cambridge University Press, New York.

Hill R. (1948), A variational principle of maximum plastic work in classical plasticity, *Q.J. Mech. Appl. Math.*, **1**, 18–28.

Hill R. (1950), *The Mathematical Theory of Plasticity*, Clarendon Press, Oxford.

Hill R. (1951), On the state of stress on a plastic-rigid body at the yield point, *Phil. Mag.*, **42**, 868–75.

Hill R. (1954), On the limit set by plastic yielding, *J. Mech. Sol*, **2**, 278–85.

Hill R. (1963), Elastic properties of reinforced solids: Some theoretical principles, *J. Mech. Phys. Solids*, **11**, 357–72.

Hill R. (1965a), A self-consistent mechanics of composite materials, *J. Mech Phys. Solids*, **12**, 213–22.

Hill R. (1965b), Micro-mechanics of elastoplastic materials, *J. Mech. Phys. Solids*, **13**, 89–101.

Hill R. (1967), The essential structure of constitutive laws for metal composites and polycrystals, *J. Mech. Phys. Solids*, **15**, 79–95.

Hodge P. and Prager W. (1949), A variational principle for plastic materials with strain hardening, *J. Math. and Phys.*, **27**, 1.

Hodge P.G., Jr. (1959), *Plastic Analysis of Structures*, McGraw-Hill, New York.

Hodge P.G., Jr. (1963), *Limit Analysis of Rotationally Symmetric Plates and Shells*, Prentice-Hall, Englewood Cliffs, NJ.

Honeycombe R.W.K. (1968), *The Plastic Deformation of Metals*, Edward Arnold, London.

Horne M.R. (1971), *Plastic Theory of Structures*, Nelson, London.

Houlsby, G.T. (1981), A study of plasticity theories and their applicability to soils, *Ph. D. Thesis*, University of Cambridge, U.K.

Huber M.T. (1904), *Czasopismo tech.*, Vol. *15*, 1904, Lwów; Poland (cited by Timoshenko, 1953, p. 369).

Hui C.Y. and Ruina A. (1985), Eddy current flow near cracks in thin plates, *ASME Trans. J., Appl. Mech.*, **52**, 841–6.

Ilyushin A.A. (1943), Some problems in the theory of plastic deformation (in Russian), *Prikl. Mat. Mekh.*, **7**, 245–72.

Ilyushin A.A. (1948), *Plasticity: Elastic–plastic Deformations*, (in Russian), Gostekhizdat, Moscow, Leningrad.

Ince E.E. (1944), *Ordinary Differential Equations*, Dover (Reprint), New York.

Irwin G.R. and Kries J.A. (1951), *Weld. J. Research Suppl.*, **33**, 1935.

Johnson C. (1976), Existence theorems for plasticity problems, *J. Math. Pures et Appl.*, **55**, 431–44.

Johnson W. and Mellor P.B. (1962), *Plasticity for Mechanical Engineers*, van Nostrand, London.

Johnson, W. and Mellor P.B. (1973), *Engineering Plasticity*, van Nostrand Reinhold Co., London.

Kachanov L.M. (1958), Time of rupture process under Deep conditions, *Izv. Akad. Nauk SSSR*, **8**, 26.

Kachanov L.M. (1974), *Fundamentals of Plasticity Theory* (in English), Mir Publishers, Moscow.

Kachanov L.M. (1986), *Introduction to the Theory of Damage*, Martinus Nijhoff, The Hague. [New revised printing, 1990]

Kaprielan P.V. and Rogers T.C. (1987), Cyclic elastic–plastic torsion of fibre reinforced materials, *Internal Report* JRS 1/87, Dept of Theoretical Mechanics, Univ. of Nottingham, U.K.

Kazinczy G. (1914), Experiments with clamped girders, *Betonszmele*, **2**, Nos 4, 5, and 6.

Kestin J. (1966), *A Course in Thermodynamics*, Blaisdell, Waltham, Mass.

Kestin J. and Bataille J. (1975), L'interprétation physique de la thermodynamique rationnelle, *J. Mécanique*, **14**, 365–86.

Kestin J. and Rice J.M. (1970), Paradoxes in the applications of thermodynamics to strained solids, in *A Critical Review of Thermodynamics*, ed. E.B. Stuart, B. Gal'Or and A.J. Brainard, pp. 275–98, Mono Book Corp., Baltimore, Md.

Kleiber M. (1989), *Incremental Finite Element Modelling in Non-linear Solid Mechanics*, (Translation from the Polish), PWN, Warsaw, and Ellis Horwood Ltd, Chichester, UK.

Knowles J.K. and Sternberg E. (1972), On a class of conservation laws in linearized and finite elastostatics, *Arch. Rat. Mech. Anal.*, **44**, 185–211.

Koiter W.T. (1960), General theorems for elastic–plastic solids, in *Progress in Solid Mechanics*, ed. I.N. Sneddon and R. Hill, North-Holland, Amsterdam, pp. 167–221.

König J.A. (1987), *Shakedown of Elastic–plastic Structures*, (Translation from the Polish), PWN, Warsaw, and Elsevier, Amsterdam.

Krajcinovic D. (1989), Damage mechanics, *Mechanics of Materials*, **8**, 117–97.

Krajcinovic D. and Lemaître J. (1987, Editors), *Continuum Damage Mechanics: Theory and Applications*, (CISM Lectures, Udine), Springer-Verlag, Vienna.

Kröner E. (1958), *Kontinuumstheorie der Versetzungen und Eigenspannungen*, Springer-Verlag, Berlin, Göttingen, Heidelberg.

Kröner E. (1963), Dislocation: a new concept in the continuum theory of plasticity, *J. Math. and Phys.*, **42**, 27–37.

Kröner E. (1972), *Statistical Continuum Mechanics*, Springer-Verlag, Vienna (CISM, Lecture Notes, Udine, Italy).

Kwiecinski M. (1989), *Collapse Load Design of Slab-Beam Systems*, (Translation from the Polish) PWN, Warsaw, and Ellis Horwood, Chichester, UK.

Labbens R. (1980), *Introduction à la Mécanique de la rupture*, Editions Pluralis, Paris.

Lance R.H. and Robinson D.N. (1971), A maximum shear-stress theory of plastic failure of fiber reinforced materials, *J. Mech. Phys. Solids*, **19**, 49–50.

Lee E.H. (1952), On the significance of the limit load theorem for an elastic–plastic body, *Phil. Mag.*, **43**, 549–60.

Lee E.H. (1969), Elastic plastic deformation at finite strain, *ASME Trans. J. Appl. Mech.*, **36**, 1–6.

Lee E.H. and Agah-Tehrani A. (1987), The fusion of physical and continuum-mechanical concepts in the formulation of constitutive relations for elastic–plastic materials, in *Non-classical Continuum Mechanics*, eds. R.J. Knops and A.A. Lacey, pp. 245–59, Cambridge University Press, Cambridge, UK.

Leguillon D. and Sanchez-Palencia E. (1983), On the behaviour of a cracked elastic body with (or without) friction. *J. Méca. Théor. Appl.*, **1**, 195–209.

Leguillon D. and Sanchez-Palencia E. (1987), *Computation of Singular Solutions in Elliptic Problems in Elasticity*, Masson, Paris, and J. Wiley, New York.

Lemaître J. (1985), Coupled elasto-plasticity and damage constitutive equations, *Comp. Math. Appl. Mech. and Engng.*, **51**, 31–49 (Proc. Fenomech 84, Stuttgart, 1984).

Lemaître J. and Chaboche J.L. (1985), *Mécanique des matériaux solides*, Dunod,

Paris (English translation: *Mechanics of Solid Materials*, Cambridge University Press, Cambridge, UK, 1990).

Le Nizhery D. (1980), Calcul à la rupture des matériaux composites, in *Problèmes non linéaires de la Mécanique*, ed. W.K. Nowacki, pp. 359–70, PWN, Warsaw.

Lescouarch Y. (1983), *Calcul en plasticité des structures*, Editions COTEGO, Paris.

Lévy M. (1871), Extrait du mémoire sur les équations générales des mouvements intérieurs des corps solids ductiles au delà des limites où l'élasticité pourrait les ramener à leur premier état, *J. Math. Pures Appl.*, **16**, 369–72.

Lippman H. and Mahrenholtz O. (1967), *Plastomechanik der Unformung metallischer Werkstoffe*, Springer-Verlag, Berlin.

Loret B. (1983), On the effects of plastic rotation in the finite deformation of anisotropic elastoplastic materials, *Mechanics of Materials*, **2**, 287–304.

Maier G. and Munro J. (1982), Mathematical programming applications to engineering plastic analysis, *Appl. Mech. Rev.*, **35**, 1631–43.

Mandel J. (1965), Energie élastique et travail dissipé dans les modèles, *Cahiers Groupe Français de Rhéologie*, **1**, 9–13.

Mandel J. (1966), *Cours de mécanique des milieux continus*, Gauthier-Villars, Paris.

Mandel J. (1971), *Plasticité classique et viscoplasticité*, Springer-Verlag, Vienna (CISM Lecture Notes, Udine, Italy).

Mandel J. (1973a), Equations constitutives et directeurs dans les milieux plastiques et viscoplastiques, *Int. J. Solids Structures*, **9**, 725–40.

Mandel J. (1973b), Thermodynamics and plasticity, in *Foundations of Thermodynamics*, ed. J.J.D. Domingos, M.N.R. Ninas, and J.H. Whitelaw, pp. 283–304, John Wiley and Sons, New York.

Mandel J. (1978), *Propriétés mécaniques des matériaux*, Editions Eyrolles, Paris.

Mandel J. (1980), Variables cachées, puissance dissipée, dissipativité normale, in: *Thermodynamique des comportements rhéologiques, Sci. et Techn. Armement*, Special Issue, pp. 37–49, January 1980.

Manjoine M.J. (1944), Influence of rate of strain and temperature on yield stresses of mild steel, *J. Appl. Mech. Trans. ASME*, **11**, 1–211.

Marigo J.J. (1981), Formulation d'une loi d'endommagement d'un matériau élastique, *C.R. Acad. Sci. Paris*, **II-292**, 1309–12.

Marigo J.J. (1985), Modeling of brittle and fatigue damage for elastic materials by growth of microvoids, *Eng. Fracture Mech.*, **21**, 861–74.

Marigo J.J., Mialon P., Michel J.C. and Suquet P. (1987), Plasticité et homogénéisation: un exemple de prévision des charges limites d'une structure hétérogène périodique, *J. Méca. Théor. Appl*, **6**, 47–75.

Martin J.B. (1975), *Plasticity: Fundamentals and General Results*, MIT Press, Cambridge, Mass.

Massonet C. and Save M. (1963), *Calcul plastique des constructions*, Vol. 1 and 2, Centre Belgo-luxembourgeois d'Information de l'Acier, Brussels. (English translation: *Plastic Analysis and Design*, Blaisdell, Waltham, Mass.).

Maugin G.A. (1980), The principle of virtual power in continuum mechanics – Application to coupled fields, *Acta Mechanica*, **35**, 1–70.

Maugin G.A. (1987), Thermodynamique à variables internes et applications, *Sém.*

Thermodynamique des Processus Irréversibles, Institut Français du Pétrole, Rueil-Malmaison.

Maugin G.A. (1988), *Continuum Mechanics of Electromagnetic Solids*, North-Holland, Amsterdam.

Maugin G.A. (1990), Internal variables and dissipative structures, *J. Non-Equilibrium Thermodynamics*, **15**, 173–92.

Maugin G.A. and Cadet S. (1991), Existence of solitary waves in martensitic alloys, *Int. J. Engng. Sci.*, **29**, 243–55.

Maugin G.A. and Drouot R. (1983), Internal variables and the thermodynamics of macromolecule solutions, *Int. J. Engng. Sci.*, **21**, 705–24.

Maugin G.A. and Drouot R. (1988), Thermodynamic modelling of polymers in solution, in *Constitutive Laws and Microstructures*, ed. D.R. Axelrad and W. Muschik, pp. 137–61, Springer-Verlag, Berlin.

Maugin G.A. and Epstein M. (1991), The electro-elastic energy–momentum tensor, *Proc. Roy. Soc. Lond.*, **A433**, 299–312.

McLaughlin P.V. (1970), Limit behavior of fibrous materials, *Int. J. Solids Structures*, **6**, 1357–1376.

Meeking R.M. and Rice J.M. (1975), Finite formulation of large elasto-plastic deformation, *Int. J. Solids Structures*, **11**, 601–16.

Melan E. (1938), Zur Plastizität des räumlichen Kontinuums, *Ing. Arch.*, **9**, 116–25.

Mendelson A. (1968), *Plasticity Theory and Applications*, MacMillan, London.

Mendelson A. and Spero S.W. (1962), A general solution for the elastoplastic thermal stresses in a strain-hardening plate with arbitrary material properties, *J. Appl. Mech. Trans. ASME.*, **29**, 151–8.

Mercier B. (1977), Sur la théorie et l'analyse numérique de problèmes de plasticité, *Thèse de Doctorat d'Etat ès Sci. Mathématiques*, Université Pierre-et-Marie Curie, Paris.

Michel J.C. (1984), Homogénéisation de matériaux élastoplastiques, *Thèse*, Université Pierre-et-Marie Curie, Paris.

Mises R. von (1913), Mechanik der festen Körper im plastisch deformablen Zustand, *Gött. Nach. Math. -Phys. Kl.*, 582–92.

Miyoshi T. (1985), *Foundations of the Numerical Analysis of Plasticity*, North-Holland, Amsterdam, and Kinokuniya, Tokyo.

Mohr O. (1900), *Z. Ver. deut. Ing.*, p. 1524 (cited by Timoshenko, 1953, p. 286).

Moreau J.J. (1966), Fonctionelles convexes, *Polycopié Collège de France*, Paris.

Moreau J.J. (1971), On unilateral constraints, friction and plasticity, in *Lecture Notes CIME*, Bressanone, Edizioni Cremonese, Rome.

Moreau J.J. (1977), Evolution problems associated with a moving convex in a Hilbert space, *J. Diff. Equa.*, **26**, 347–74.

Mroz Z. (1963), Non-associated flow laws in plasticity, *J. Mécanique*, **2**, 21–42.

Mroz Z. (1967), On the description of anisotropic work hardening, *J. Mech. Phys. Solids*, **15**, 163–75.

Mukherjee S., Morjaria M. and Moon F.C. (1982), Eddy current flows round cracks in thin plates for nondestructive testing, *ASME Trans. J. Appl. Mech.*, **49**, 389–95.

Müller I. (1985), *Thermodynamics*, Pitman/Longman, London.

Mura T. (1982), *Micromechanics of Defects in Solids*, Martinus Nijhoff, The Hague, Boston, Mass.

Murakami Y. (Editor, 1987), *Stress-Intensity Factors Handbook*, Two Volumes, Pergamon Press, Oxford.

Murakami S. and Ohno N. (1981), A continuum theory of creep and creep damage, in *Creep in Structures*, eds, A.R.S. Ponter and D.R. Hayhurst, pp. 433–43 (IUTAM Symposium, Leicester, UK, 1980), Springer-Verlag, Berlin.

Murch S.A. and Naghdi P.M. (1958), On the infinite elastic perfectly-plastic wedge under uniform surface tractions, in *Proc. 3rd U.S. Nat. Congress of Mechanics*, ASME, pp. 611–24.

Mushkelishvili N.I. (1953), Some basic problems in the mathematical theory of elasticity, Noordhoff, Groningen.

Nádai A. (1950), *Theory of Flow and Fracture of Solids*, McGraw-Hill, New York.

Naghdi P.M. (1957), Stresses and displacements in an elastic – plastic wedge, *ASME Trans. J. Appl. Mech.*, **24**, 98–104.

Naghdi P.M. (1990), A critical review of the state of finite plasticity, *Zeit. angew. Math. Phys.*, **41**, 315–94.

Nayroles B., Bouc R., Caumon H., Chezeaux J.C. and Giacometti E. (1981), Télé-thermographie infrarouge et mécanique des structures, *Int. J. Engng. Sci.*, **19**, 929–47.

Neal B.G. (1977), *The Plastic Methods of Structural Analysis*, Chapman and Hall, London.

Neal B.G. and Symonds P.S. (1952), The rapid calculation of the plastic collapse load of a framed structure, *Proc. Inst. Civil Engrs* (*London*), **1**, 58–71.

Neale K.W. (1981), Phenomenological constitutive laws in finite plasticity, *S.M. Archives*, **6**, 79–128.

Necas J. and Hlavacek I.L. (1981), *Mathematical Theory of Elastic and Elastico-Plastic bodies: An Introduction*, Elsevier, Amsterdam.

Nemat-Nasser S. (1972), General variational principles in nonlinear and linear elasticity with applications, in *Mechanics Today*, ed., S. Nemat-Nasser, Vol. 1, pp. 214–61, Pergamon Press, New York.

Nemat-Nasser S. (1979), Decomposition of strain measures and their rates in finite deformation elastoplasticity, *Int. J. Solids Structures*, **15**, 155–6.

Nguyen Quoc Son (1973), Contribution à la théorie de l'élasto-plasticité avec écrouissage, *Thèse de Doctorat d'Etat*, Université Pierre-et-Marie Curie, Paris.

Nguyen Quoc Son (1976), Loi de comportement élasto-plastique des plaques et des coques minces, *Internal Report No. 1-LMS*, Ecole Polytechnique, Palaiseau, France, October, 1976.

Nguyen Quoc Son (1977), On the elastic–plastic initial–boundary value problem and its numerical integration. *Int. J. Num. Meth. Engng.*, **11**, 817–32.

Nguyen Quoc Son (1980), Méthodes énergétiques en mécanique de la rupture, *J. Mécanique*, **19**, 363–86.

Nguyen Quoc Son (1981a), Problèmes de plasticité et de rupture, *Cours polycopié de DEA* (Analyse Numérique, Université de Paris-Sud, Orsay).

Nguyen Quoc Son (1981b), A thermodynamic description of the running crack problem, in *Three-dimensional Constitutive Relations and Ductile Fracture*, ed. S. Nemat-Nasser, pp. 315–30, North-Holland, Amsterdam.

Nguyen Quoc Son (1982), Effets thermiques en fond de fissures, in *Mécanique de la rupture* (Séminaire Saint-Rémy les Chevreuses, June 1982), pp. 85–106, CILF, Paris.

Noether E. (1918), Invariante Variationsprobleme, *Gött. Nach. Math.-Phys. Kl.*, **2**, 235.

Norton F.H. (1929), *Creep of Steel at High Temperature*, McGraw-Hill, New York.

Nowacki W.K. (1978), *Stress Waves in Non-elastic Solids*, (Translation from the Polish), Pergamon Press, Oxford.

Nowacki W. (1986), *Thermoelasticity* (2nd revised edition, translated from the Polish), Pergamon Press, Oxford.

Oden J.T. (1972), *Finite Elements of Nonlinear Continua*, McGraw-Hill, New York.

Odqvist F.K.G. (1933), Die Verfestigung von flusseisenahnlichen Körpern. Ein Beitrag zur Plastizitätstheorie, *Zeit. angew. Math. und Mech.*, **13**, 360–3.

Odqvist F.K.G. (1966), *Mathematical Theory of Creep and Creep Rupture*, Oxford University Press, London.

Ogden R.W. (1984), *Nonlinear Elastic Deformations*, Ellis Horwood, Chichester, UK.

Olszak W. and Sawczuk A. *Inelastic Behaviour in Shells*, Noordhoff, Groningen, The Netherlands (1967).

Orowan E. (1934), *Zeit. Phys.*, **89**, 605.

Owen D.R.J. and Hinton E. (1980), *Finite Elements in Plasticity: Theory and Practice*, Pineridge Press Ltd, Swansea, UK.

Paris P.C. and Sih G.C. (1965), *Finite Elements in Plasticity: Theory and Practice*, Pineridge Press Ltd, Swansea, UK.

Parton V.Z. and Morozov E.M. (1978), *Elastic–Plastic Fracture Mechanics* (in English), Mir Publishers, Moscow.

Perzyna P. (1966), Fundamental problems in viscoplasticity, in *Advances in Applied Mechanics*, ed. C.S. Yih, vol. 9, pp. 243–377, Academic Press, New York.

Phillips A. (1968), Yield surfaces of pure aluminium at elevated temperatures, in *Thermoinelasticity*, ed. B.A. Boley, pp. 241–58 (IUTAM Symposium, East Kilbride, 1968), Springer-Verlag, Vienna.

Phillips A. and Tang J.-L. (1972), The effect of loading path on the yield surface at elevated temperatures, *Int. J. Solids Structures*, **8**, 463–74.

Polukhin R., Gorelik S. and Vorontsov V. (1983), *Physical Principles of Plastic Deformation* (Translation from the Russian), Mir Publishers, Moscow.

Prager W. (1949), Recent developments in the mathematical theory of plasticity, *J. Appl. Phys.*, **20**, 235–41.

Prager W. (1955), The theory of plasticity–A survey of recent achievements, *Proc. Inst. Mech. Engrs* (London), **169**, 41–57.

Prager W. (1957), On ideal locking materials, *Trans. Soc. Rheology*, **1**, 169–75.

Prager W. (1959), *An Introduction to Plasticity*, Addison-Wesley, Reading, Mass.

Prager W. and Hodge P.G. (1951), *Theory of Perfectly Plastic Solids*, John Wiley and Sons, New York.

Prandtl L.T. (1903), Zur Torsion von prismatischen Staeben, *Phys. Zeit.*, **4**, 758–9.

Prandtl L.T. (1924), Spannungsverteilung in plastischen Körpern, in *Proc. 1st Intern. Congr. Mechanics*, Delft, pp. 43–54.

Rabier P.J. (1989), Some remarks on damage theory, *Int. J. Engng. Sci.*, **27**, 29–54.

Rabotnov Yu. N. (1963), On the equations of state for creep, in *Progress in Applied Mechanics* (Prager Anniversary Volume), MacMillan, New York, p. 307 (not seen by Author).

Rabotnov Yu. N. (1969), *Creep Problems in Structural Members* (Translation from the Russian), North-Holland, Amsterdam.

Rabotnov Yu. N. (1980), *Elements of Hereditary Solid Mechanics* (Translation from the Russian), Mir Publishers, Moscow.

Reissner E. (1953), On a variational theorem for finite elastic deformations, *J. Math. and Phys.*, **32**, 129–35.

Reuss A. (1939), Berücksichtigung der elastischen Formanderung in der Plastizitätstheorie, *Zeit. Angew. Math. und Mech.*, **10**, 26–274.

Rice J.M. (1968a), A path-independent integral and the approximate analysis of strain concentration by notches and cracks, *ASME Trans. J. Appl. Mech.*, **33**, 379–85.

Rice J.M. (1968b), Mathematical analysis in the mechanics of fracture, in *Fracture – A Treatise*, ed. H. Liebowitz, Vol., 8, pp. 191–311, Academic Press, New York.

Rice J.M. (1970), On the structure of stress–strain relations for time-dependent plastic deformations in metals, *ASME, Trans. J. Appl. Mech.*, **37**, 728–37.

Rice J.M. (1971), Inelastic constitutive equations for solids: An internal variable theory and its application to metal plasticity, *J. Mech. Phys. Solids*, **19**, 433–55.

Richmond O. and Spitzig W.A. (1980), Pressure dependence and dilatancy of plastic flow, in *Theoretical and Applied Mechanics* (Proceedings 15th IUTAM, Toronto), eds. F.P.J. Rimrott and B. Tabarrok, pp. 377–86, North-Holland, Amsterdam.

Rockafellar R.T. (1970), *Convex Analysis*, Princeton University Press, Princeton, NJ.

Rogers T.C. (1988), Yield criteria, flow rules and hardening in anisotropic plasticity, in *Proc. IUTAM-ICM Symposium on Yielding, Damage and Failure of Anisotropic Solids*, ed. J.P. Boehler, Mech. Eng. Publ., London (1989).

Rogers T.C. (1990), Plasticity of fibre-reinforced materials, in *Continuum Models and Discrete Systems*, Vol. 1, ed., G.A. Maugin, pp. 87–102, Longman, London.

Salençon J. (1974), *Théorie de la plasticité pour les applications à la mécanique des sols*, Editions Eyrolles, Paris.

Salençon J. (1979), *Le Comportement plastique, in Evolution et théories modernes en élasticité et plasticité*, pp. 119–21, Editions du Bâtiment et des Travaux Publics, Paris.

Salençon J. (1983)), *Calcul à la rupture et analyse limite*, Presses de l'ENPC, Paris.

Save M.A. and C.E. Massonet (1972), *Plastic Analysis Design of Plates, Shells and Disks*, North-Holland, Amsterdam.

Sawczuk A. (1989), *Mechanics and Plasticity of Structures*, (Translation from the Polish), PWN, Warsaw, and Ellis Horwood Ltd, Chichester, UK.

Schleicher F. von (1926), *Zeit. angew. Math. Mech.*, **6**, 199.

Schmid E. (1924), Neuere Untersuschungen an Metallkristallen, *Proc. First Intern. Cong. Appl. Mech.*, eds. C.B. Biezeno and J.M. Burgers, pp. 342–53, Technische Boekhandel en Drukkereij J. Waltman, Delft.

Sedov L.I. (1934), *Trudy TZAGI*, **187** (in Russian).

Sedov L.I. (1975), *Mécanique des milieux continus* (in French), Vol. 2., Editions de Moscou (Mir), Moscow.

Seeger A. (1958), Kristallplastizität, in *Handbuch der Physik*, ed. S. Flügge, Bd VII/2, Springer-Verlag, Berlin.

Sewell M.J. (1987), *Maximum and Minimum Principles*, Cambridge University Press, Cambridge, UK.

Shu L.S. and Rosen B.W. (1967), Strength of fiber reinforced composites – A limit analysis method, *J. Composite Materials*, **1**, 366–81.

Sidoroff F. (1973), The geometrical concept of intermediate configuration and elastic–plastic finite strain, *Arch. Mech.*, **25**, 299–309.

Sidoroff F. (1976), Variables internes en viscoélasticité et viscoplasticité, *Thèse de Doctorat d'Etat ès Sci. Mathématiques*, Université Pierre-et-Marie Curie, Paris.

Sidoroff F. (1982), Incremental constitutive equations for large strain elasto-plasticity, *Int. J. Engng. Sci.*, **20**, 19–28.

Sidoroff F. (1984), Ecrouissage cinématique et anisotropie induite en grandes déformations élasto-plastiques, *J. Méca. Théor. Appl.*, **3**, 117–33.

Sih G.S. (1973), *Handbook of Stress Intensity Factors*, Lehigh University, Bethlehem, Pa.

Sih G.S. (Editor, 1981), *Experimental Evaluation of Stress Concentration and Intensity Factors*, Vol., 7 of *Mechanics of Fracture*, Martinus Nijhoff, The Hague, Boston, Mass.

Sobotka Z. (1989), *Theory of Plasticity and Limit Design of Plates* (Translation from the Czech), Elsevier, Amsterdam.

Sokolnikoff I.S. (1956), *Mathematical Theory of Elasticity*, McGraw-Hill, New York.

Spencer A.J.M. (1972), *Deformation of Fibre-Reinforced Materials*, Clarendon Press, Oxford.

Stolz C. (1982), Contribution à l'étude de grandes transformations en élastoplasticité, *Thèse de Docteur-Ingénieur*, ENPC, Paris.

Stolz C. (1987), Anélasticité et stabilité, *Thèse de Doctorat d'Etat ès Sciences Mathématiques*, Université Pierre-et-Marie Curie, Paris.

Suquet P. (1979), Un espace fonctionnel pour les équations de la plasticité, *Ann. Fac. Sci. Toulouse*, **1**, 77–87.

Suquet P. (1981a), Sur les équations de la plasticité: existence et régularité des solutions, *J. Mécanique*, **20**, 3–40.

Suquet P. (1981b), Méthodes d'homogénéisation en mécanique des solides, in *Comportements Rhéologiques et Structures des Matériaux*, ed. C. Huet et A. Zaoui, pp. 87–128, Presses de l'ENPC, Paris.

Suquet P. (1982), Plasticité et homogénéisation, *Thèse de Doctorat d'Etat ès Sci. Mathématiques*, Université Pierre-et-Marie Curie, Paris.

Suquet P. (1983), Analyse limite et homogénéisation, *C.R. Acad. Sci. Paris*, **II-293**, 1355–8.

Suquet P. (1987), Elements of homogenization for inelastic solid mechanics, in *Homogenization techniques for Composite Materials*, ed. E. Sanchez-Palencia and A. Zaoui, pp. 193–278, Springer-Verlag, Berlin, Heidelberg, New York (CISM Lecture Notes, Udine, Italy).

Symonds P.S. (1951), Shakedown in continuous media, *J. Appl. Mech.*, **18**, 18–35.

Symonds P.S. and Neal B.G. (1951), Recent progress in the plastic method of structural analysis, *J. Franklin Inst.*, **252**, 383–407, 469–92.

Taylor G.I. (1947), A connexion between the criterion of yield and the strain ratio relationships in plastic solids, *Proc. Roy. Soc. Lond*, **A191**, 441–6.

Taylor, G.I. and Quinney H. (1931), The plastic distortion of metals, *Phil. Trans., Roy. Soc. Lond.*, **A230**, 323–62.

Temam R. (1988), *Mathematical Problems in Plasticity* (translation from the French), Gauthier-Villars, Paris.

Temam R. and Strang G. (1980), Functions of bounded variations, *Arch. Rat. Mech. Anal.*, **75**, 7–21.

Teodosiu C. (1975), A physical theory of the finite elastic–plastic behaviour of single crystals, *Engin. Transactions*, (Poland), **23**, 151–84.

Teodosiu C. (1982), *Elastic Models of Crystal Defects*, Springer-Verlag, Berlin.

Teodosiu C. and Sidoroff F. (1976), A theory of finite elastoplasticity in single crystals, *Int. J. Engng. Sci.*, **14**, 165–76.

Thomas T.Y. (1961), *Plastic Flow and Fracture in Solids*, Academic Press, New York.

Timoshenko S.P. (1953), *History of the Strength of Materials*, McGraw-Hill, New York (reprint by Dover, New York, 1983).

Ting T.C.T. (1986), Explicit solution and invariance of the singularities at an interface crack in anisotropic composites, *Int. J. Solids Structures*, **2**, 965–83.

Todhunter I. and Pearson K., (1893), *A History of the Theory of Elasticity and of the Strength of Materials*, Vol. II, Parts 1 and 2, Cambridge University Press, Cambridge, UK (Reprint by Dover, New York, 1960).

Tresca H.E. (1872), Mémoire sur l'écoulement des corps solides, in *Mémoires Présentés par Divers Savants, Acad. Sci. Paris*, **20**, 75–135 (cited by Timoshenko, 1953, p. 233).

Truesdell C.A. (1969), *Rational Thermodynamics*, McGraw-Hill, New York (New enlarged edition, Springer-Verlag, Berlin, 1984).

van der Giessen E. (1989a), Continuum models of large deformation plasticity, Part I: Large Deformation Plasticity and the Concept of a Natural Reference State, *Eur. J. Mech. A/Solids*, **8**, 15–34.

van der Giessen E. (1989b), Continuum models of large deformation plasticity, Part II: A kinematic model and the concept of a plastically induced orientational structure, *Eur. J. Mechanics A/Solids*, **8**, 89–108

Vinogradov G.V. and Malkin A. Ya. (1980), *Rheology of Polymers* (translation from the Russian), Mir Publishers, Moscow.

Vyalov S.S. (1986), *Rheological Fundamentals of Soil Mechanics* (translation from the Russian), Mir Publishers, Moscow.

Washizu K. (1982), *Variational Methods in Elasticity and Plasticity*, Third Edition, Pergamon Press, Oxford.

Westergaard H.M. (1920), On the resistance of ductile materials to combined stresses, *J. Franklin Inst.*, **189**, 627–40.
Westergaard H.M. (1939), Bearing pressures and cracks, *J. Appl. Mechanics*, **6A**, 49–53.
Wilmanski K. (1991), Remarks on rate-type monocrystal plasticity, *Cont. Mech. and Thermodynamics* (to be published).
Woehler A. (1858), *Z. Bauwesen*, **8**, 641–52; (1860), ibid., **10**, 583–616, (1866), ibid., **16**, 67–84; (1870), ibid., **20**, 73–106 (cited by Timoshenko, 1953, p. 168).
Wood R.H. (1961), *Plastic and Elastic Design of Slabs and Plates*, Thames and Hudson, London.
Woods L.C. (1975), *Thermodynamics of Fluid Systems*, Clarendon Press, Oxford.
Zaoui A. (1987). Approximate statistical modelling and applications, in *Homogenization Techniques for Composite Media*, ed. E. Sanchez-Palencia and A. Zaoui, pp. 337–95, Springer-Verlag, Berlin, Heidelberg, New York (CISM, Lecture Notes, Udine, Italy).
Zarka J. (1972), Généralisation de la théorie du potentiel plastique multiple en viscoplasticité, *J. Mech. Phys. Solids*, **20**, 179–95.
Zener C. (1948), *Elasticity and Anelasticity of Metals*, The University of Chicago Press, Chicago, Ill.
Ziegler H. (1963), Some extremum principles in irreversible thermodynamics with applications to continuum mechanics, in *Progress in Solid Mechanics*, ed. I.N. Sneddon and R. Hill, Vol. 4, pp. 91–193, North-Holland, Amsterdam.
Ziegler H. (1968), A possible generalization of Onsager's theory, in *Proc. IUTAM Symposium on Irreversible Aspects of Continuum Mechanics*, ed. H. Parkus and L.I. Sedov, pp. 411–24, Springer-Verlag, Berlin.
Ziegler H. (1986), *An Introduction to Thermomechanics*, North-Holland, Amsterdam.
Ziegler H. and Wehrli C. (1987), The derivation of constitutive relations from the free energy and the dissipation functions, in *Advances in Applied Mechanics*, Vol, 25, pp. 183–238, ed. T.Y. Wu and J.W. Hutchinson, Academic Press, New York.
Zienkiewicz O.C. (1971), *The Finite Element Method*, McGraw-Hill, New York.
Zyczkowski M. (1981), *Combined Loadings in the Theory of Plasticity*, PWN-Polish Scientific Publishers, Warsaw (contains one of the most extensive bibliographies on plasticity).

The following proceedings aimed at professionals in the research field of plasticity give a nice overview of developments in the 1970s and 1980s:

Sawczuk A. (Editor, 1973), *Foundations of Plasticity*, Noordhoff, Leyden.
Sawczuk A. (Editor, 1974), *Problems of Plasticity*, Noordhoff, Leyden.
Sawczuk A. and Bianchi G., (Editors, 1985), *Plasticity Today: Modelling, Methods and Applications*, Elsevier, London.

Index